Thermodynamics and
Pattern Formation
in Biology

Thermodynamics and Pattern Formation in Biology

Editors
I. Lamprecht A. I. Zotin

W DE G

Walter de Gruyter · Berlin · New York 1988

Editors

Prof. Dr. Ingolf Lamprecht
Institute of Biophysics
Free University Berlin
D-1000 Berlin 33

Professor Dr. A. I. Zotin
Institute of Developmental Biology
Academy of Science of the U.S.S.R.
Moscow

Library of Congress Cataloging in Publication Data

Thermodynamics and pattern formation in biology.

Translated from Russian
Bibliography: p.
1. Biophysics. 2. Thermodynamics. I. Lamprecht, Ingolf, 1933- . II. Zotin, A. I. (Aleksandr Il'ich), 1926-
III. Title: Pattern formation in biology.
QH505.I3934 1985 574.19'1 88-23550
ISBN 0-89954-07-1 (U.S.)

Deutsche Bibliothek Cataloging in Publication Data

Thermodynamics and pattern formation in biology / ed. I. Lamprecht ; A. I. Zotin. – Berlin ; New York : de Gruyter, 1988
ISBN 3-11-011368-6
NE: Lamprecht, Ingolf [Hrsg.]

 Printing: Gerike GmbH, Berlin. – Binding: Lüderitz & Bauer, Berlin. – Printed in Germany.

This monograph deals with various aspects of the formation of spatial and temporal structures in living and inanimate systems and its thermodynamic background. Prominent scientists were invited to contribute to these questions from the point of view of theoretical physics, mathematics, synergetics, chemistry, biochemistry, biophysics and developmental biology.

Starting with general physical and mathematical considerations on nonlinear systems this book turns in its second part to pattern formation in chemical and biochemical systems with special emphasis on oscillating (bio)chemical reactions and their transformation into spatial structures. The third part concentrates on morphogenesis, tissue geometry and a mechanochemical autowave model.

The present monograph will be helpful for all those biochemists, biophysicists, physiologists and developmental biologists dedicated to the appearance of pattern in living systems, but moreover to theoretical physicists and mathematicians too, interested in recent questions of biological thermodynamics and mathematics.

Preface

After publishing the three volumes 'Thermodynamics of Biological Processes' (1978), 'Thermodynamics and Kinetics of Biological Processes' (1982) and 'Thermodynamics and Regulation of Biological Processes' (1985) we felt that a fourth and last volume on pattern formation during biological processes would complete this series and throw a bridge to the first monograph with its special section on dissipative structures. Again a number of internationally recognised scientists from different fields of natural sciences were invited to contribute to this book looking to the formation of structure from the point of view of the more theoretically and mathematically oriented researcher as well as of the experimentalist.

The first monograph of this series 'Thermodynamics of Biological Processes' (1978) concentrated on thermodynamic aspects of developmental biology, dissipative structures, classification and evolution of organisms. It was especially devoted to constitutive processes connected with the development of organisms and thus to biological problems. The second volume 'Thermodynamics and Kinetics of Biological Processes ' (1982) was orientated towards the theoretical view how living systems react on external influences and adapt to changing environmental conditions. It described the present state of statistical and phenomenological thermodynamics of systems far from equilibrium. The third volume 'Thermodynamics and Regulation of Biological Systems' (1985) was again more concerned with biological questions such as control and regulation, evolution and ontogenesis, but also with synergetics (self-organisation) and information theory which might serve as a satisfactory future basis of thermodynamics of organised systems.

The present monograph of this series 'Thermodynamics and Pattern Formation of Biological Processes' deals with questions connected with the emergence of temporal and spatial structures in living units and traces such pattern formation back to inanimate systems. For long time structure creating abilities of organisms seemed to contradict the Second Law of Thermodynamics predicting the permanent increase of entropy and thus decrease of order. The implicit assumption of isolated thermodynamic systems was of course not valid in biology so that animals as open systems could 'feed on negative entropy' (Schrödinger, 1944). It was just the thermodynamics of irreversible processes far from equilibrium which explained the appearance of 'order by fluctuation' (Prigogine, 1973). Such 'dissipative structures' in contrast to the static equilibrium ones emerge as result of instabilities in systems which crossed a critical threshold or a critical distance from equilibrium.

In its first part this monograph deals with general problems connected with structure formation, expecially in nonlinear systems. The second part is dedicated to chemical and biochemical systems with emphasis on autocatalytic reactions. Special interest is paid in four chapters to the well known oscillations in glycolysis. The last section concentrates on morphogenesis, self-organisation and pattern generation in small units such as organelles as well as in whole organisms.

Emergence of order in living and inanimate systems is one of the most fascinating and thrilling topics in modern natural sciences. The similarity of structures found in hydrodynamics, meteorology, geology or metabolizing cytoplasmic extracts to those first described intensively by Bénard (1904) as well as the abundance of oscillating reactions point to a fundamental theory which might be given by thermodynamics of systems far from equilibrium. We do hope that this monograph will raise further discussion among the representatives of various disciplines and stimulate scientists from different fields to common investigations in this new branch of biology.

We thank all authors of this monograph for their readiness to contribute to the fourth volume, for their helpful discussions and their engagement towards the ideas of this book.

Again we are deeply indepted to Dr. A. E. Beezer/Egham for thoroughly checking the English translation.

Berlin, June 1988

I. Lamprecht
A. I. Zotin

Contents

General Problems

Some Basic Principles for the Formation of Structure in Nonlinear Systems of Thermodynamic Kind

H.-G. Busse

Systems in physical and biological sciences display some remarkable common properties, e. g.

1. they are composed of subsystems;
2. the subsystems may interact;
3. the elements of a system show a striking similarity and can be classified into a few types;
4. composite systems display non-additive properties (i. e. they are more than the sum of their components).

However, at least one property is not common to the two. Systems in life sciences develop into more structured states than those of physical systems, since the latter tend to assume homogeneous states (usually equilibria). This difference will be considered in more detail from a systematic approach to the theory of thermodynamics of irreversible processes.

Thermodynamics and Pattern Formation in Biology

Reversal of entropy production

Thermodynamic systems consist of interacting components. The laws, which describe the interaction, are mathematically formulated in a vector space. Primarily, the physical-chemical laws of the theory of thermodynamics of irreversible processes will be considered. The quantity of thermodynamics, which potentially is the most useful, is the rate of entropy production. This quantity is usually defined in the following way:

$$\sigma = \sum_i J_i \cdot X_i \,.$$

In electronics, the flux (current) is forced by the electric field. The entropy (heat) produced in this process is related to the product of the current and the strength of the electric field.

The Second Law of Thermodynamics states, that

$$\sigma \geq 0$$

at least in a thermodynamically closed system. Since, σ is a sum of products, one may decompose it into two major terms

$$\sigma = \sigma_{ENV} + \sigma_{SYS} \geq 0 \,.$$

The second law, $\sigma \gg 0$, remains valid if

$$\sigma_{SYS} < 0 \,, \quad \sigma_{ENV} \gg 0 \,,$$

where σ_{SYS} is produced by a subsystem and σ_{ENV} is the rate of entropy production of the remainder of the system, i. e. the environment of the subsystem.

This partitioning of the rate of entropy production has the following consequences:

1. The subsystem cannot exist without its environment. It has to interact with the environment to negotiate a compensation of its negative rate of entropy production by a positive contribution in the environment. Hence, the subsystem modifies its environment imposing its negative entropy upon it. Hence, the subsystem is an open system.

2. At thermodynamic equilibrium, $\sigma = 0$ not only for the total system but also for each of its subsystems. Hence, $\sigma_{SYS} = 0$ and $\sigma_{ENV} = 0$. Conditions, which allow $\sigma_{SYS} < 0$, are reached when a threshold has been exceeded on the path from the equilibrium where $\sigma_{SYS} < 0$.

3. Since the subsystem usually contains dissipative processes and deposits its entropy in the environment, the parameters of either the environment or the subsystems must, within definite time, change to values which are incompatible with the condition $\sigma_{SYS} < 0$. Hence, the life time of such states of subsystems is limited.

4. Usually, the second law, $\sigma \geqslant 0$, is considered as responsible for the decay of order in a system. Therefore, in the subsystem, in which $\sigma_{SYS} < 0$, structure can be generated.

5. At equilibrium, the processes may be considered as linear. Therefore, the principle of superposition is valid. (Here, superposition also applies to the dynamics of linear processes.) Since a subsystem may be closed, $\sigma \geqslant 0$ must be valid for each of them. Hence, nonlinearity is a prerequisite for the existence of subsystems with $\sigma_{SYS} < 0$.

6. Nonlinearity adds some further properties to systems, e. g.
   - multistationary states, limit cycles, deterministic chaos; and other strange phenomena may occur.
   - The system can exist in several steady states, which usually are insensitive to small perturbations. However, it may contain domains in which the subsystem can interact with its environment, i. e. where it can adapt

Table 1

Physical chemical process	Normal effect	Reverse effect	Phenomena observed
diffusion	homogeneous spatial distribution of particles	generation of a of a spatial distribution of particles	spatial structure in chemical and biological systems (1)
chemical reaction	approach to equilibrium	onset of processes directed away from equilibrium	chemical oscillations (subsystems return to their previous states (2))
friction	mechanical movement ceases	spontaneous arisal of movement	movement of particles by chemical forces (3)
mixing	homogeneous distribution of particles	separation of particles	demixing in chemical solutions (4). (Perhaps,involved in cell division)
Ohm's law	charge equilibrium	generation of a charge distribution	charge distribution in plasma (5)

(1) Nicolis, Prigogine, 1977 (2) Jessen, 1978 (3) Busse, 1976 (4) Preston et al., 1980
(5) Neff et al.,1980

Subsystems operating under the condition $\sigma_{sys} < 0$ may be identified by observation of phenomena listed in statement 4 because if the condition

$$\sigma_{sys} = J \cdot X < 0$$

only is applied to a single process, then the thermodynamic driving force, X, causes a thermodynamic flux in the opposite direction, e. g. the diffusion flux of particles is directed against their concentration gradients. Hence, the new rules are simply the reverse of the well known thermodynamic laws of physical chemistry. In addition to the direction of the flux, the change of the thermodynamic order is also reversed according to this law, e. g. normally the diffusion results in a homogeneous distribution in space, but the reverse of this statement requires the rise of a spatially structured distribution of components. Such phenomena have been observed in chemically reacting systems (Zhabotinskii, Zaikin, 1971). In Table 1, some other, similar laws of physical chemistry and their effects under various conditions are listed.

Note that $J \cdot X < 0$ in a subsystem implies $\sum_{i} J_j^{Sys} \cdot X_j^{Sys} < 0$, where J_j^{Sys} and X_j^{Sys} are the flux and force respectively which act on its j-th component. The apparent reversal of the flux with respect to the force X may arise from a series of processes that mimic the reverse.

Physical and biological examples of the reversal of entropy

Two phenomena in this domain of thermodynamics have been kinematically illustrated by film (Busse, 1976): reversed diffusion and reversed friction. In the case of diffusion in a homogeneous chemical medium the appearance of spots depicting concentration gradients were observed. These spots spread and finally produce a set of concentric rings representing high concentrations of a compound. The rings are separated by spaces in which the concentration of this substance is low. The concentric rings may be recognised as a structure generated by diffusion (Winfree, 1980) in a homogeneous medium. Here, the latter process formed, and not as usual eliminated, concentration gradients.

In the same chemical system under different conditions, several other structures can be demonstrated. Thus, instead of concentric rings, radial stripes (Fig. 1) originating from a center are observed. As soon as a certain distance is covered, the stripes branch off, thus maintaining an equal density of stripes over the new area which has become available (Busse, 1968; Zhabotinskii, 1971). Such radial structures may be important in biology, e. g. to establish a supply system of the blood vessels, since evidence has been presented which shows that chemical compounds are involved in the vascularization of tissues (Wissler et al., 1982; Meinhardt, 1982). Also here, the formation of structure may be considered as a result of a potential for the creation of order in a previously homogeneous system.

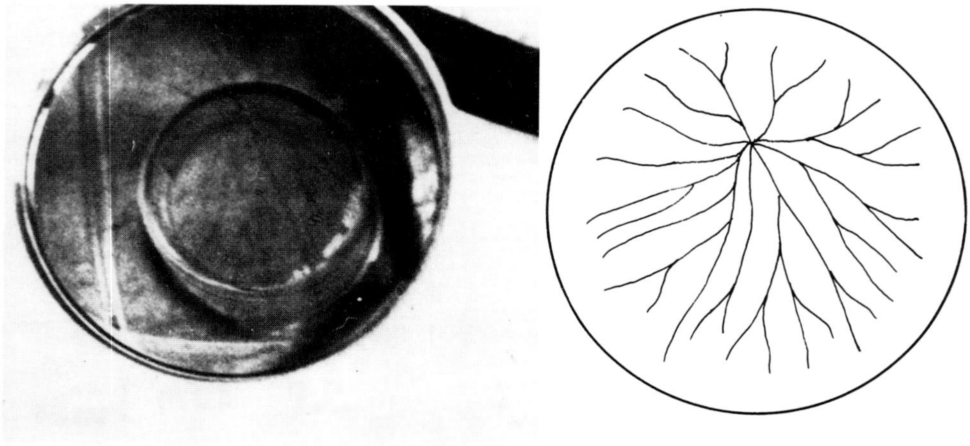

Fig. 1 Photo (left) and sketch (right) of a radially expanding spatial structure in the Belousov-Zhabotinskii medium with malonic acid as the organic compound. The stripes start radially from a centre in the upper third in the Petri dish. The Petri dish is supported by a flask to allow illumniation from below. They seem to cover the solution phase at a nearly equal density by emitting new stripes as soon as a certain distance between adjacent stripes is reached. The stripes are visualised by the indicator ferroin.

The same chemical medium may suddenly turn the radial structure into an equally spaced hexagonal structure covering the whole area. The liquid layer in the experiment was about 3 mm and the surface of the layer was in free contact with the air since the Petri dish was not covered. The hexagons had a diameter of about 5 mm (Busse, 1968). In this case, ordering of the area by diffusion manifested itself in the regularity of the hexagons. Instead of sets of concentric rings, the chemical medium may exhibit spirals. Most often, the spirals appeared in pairs in which one spiral turned clockwise whereas the other was rotating in the opposite direction. Such spirals were also observed during the aggregation phase of the slime mould *Dictyostelium discoideum* (Gerisch, 1971). The similarity between the pattern of the slime mould in the aggregation phase and those in the chemical medium was the basis for the study of the biochemical processes participating in the ordered aggregation of cells to an organism, in that case a slime mould.

Moreover, the effects of nonlinearity mentioned in 5, clearly gives the slime mould an evolutionary advantage. At the border of two domains of rings, nonlinearity forbids superposition. Hence, the domains are fitted together as far as possible. In the case of the slime mould, the border defines the area within which cells of an organism exist. Cells leave the border centripetally to form an assembly of cells at the centre. Hence, the border fringe is the area which at first is depleted of cells. The coordinated movement of the amoeba is another biological process in which order apparently is achieved by the principles mentioned. Here, not only the friction of cells against the supporting material is overcome but also the direction of movement prescribed for each cell by its enviromenment becomes apparent. The polarity of the cell which predisposes it to move in a certain direction, is modified by an external concentration gradient. Thus, the individual cell is directed toward the centre for the assembly of an integral organism. The environmental gradient field is believed to be chemically generated in a way similar to that found in chemical oscillating reaction systems. Another example which demonstrates some basic principles of the generation of structure in biological material are experiments on chicken limbs (Wolpert, 1971). Here also, biochemical compounds are applied to induce specific types of limbs (Tickle et al., 1982) and morphogenetic fields are supposed to be responsible for the appearance of structure. This example is especially impressive since it demonstrates how the anatomy of a limb can be artificially manipulated (e. g. furnished with a desired number of digits). Since the processes in the limbs of the chicken are believed to resemble those in human legs and arms, it may be feasible to construct human extremities in a similar fashion. Moreover, this example shows that the genetic information is only partly in control of the process of structure formation and that both the environment and internal interaction play a crucial role.

Selected Remarks to Nonlinear Systems

The analyst tries to elucidate the mechanistic details behind such phenomena. In general, this is a difficult task since the systems are complicated and since the phenomenon itself seldom significantly restricts the manifold of possible mechanisms. For this reason, several models often exist which explain the experimental data reasonably well. The models developed for slime moulds and chemical excitable media (Tyson, 1976) excel in clarity in this respect.

As yet, phenomena found in chemical systems have to be used to explain strikingly similar phenomena in cellular systems. However, there is an important difference between the chemical and cellular constituents of such systems. Chemicals are preserved whereas cell may perish and vanish from the system. This difference leads to a new principle for the formation of structure by elimination of undesirable individuals. Let us, for example, consider the construction of a system which is capable of recognising all its constituents. A generation of individuals is developed which form a subsystem. These individuals recognise their mates by markers. If all such individuals of the subsystem are removed (eliminated), then the subsystem may be used to identify foreign individuals. Such a selection principle spontaneously generates defense systems, e. g. an immunesystem. However, this principle might be of more general importance to the evolution of structures in systems where there is a sequence of events, such as growth of individuals, approval of the newly matured individual, initiation of the next growth phase, etc. The conservation individuals in the selection phase restricts the evolution of structure in the system, since misfits have to be removed.

Concluding Remarks

The sign of the rate of entropy production of the phenomena described lends support to the idea of classifying subsystems into two types: one in which this rate is positive and another in which it is negative. In both cases, the structure of the system may be recognised by its symmetry. In the close-to-equilibrium case, crystals are prominent examples and in the far- from-equilibrium case, e. g. spirals in chemically excitable medium and the hexagonal pattern of the Bénard (Chandrasekhar, 1961) experiment are typical. The justification for the application of symmetry to explain the properties inherent to nonlinear processes far from equilibrium is given in the literature (Steeb, 1977). Symmetry represents an invariance of quantities in these systems against operations applied to the system. Balance equations, e. g. of energy or chemical compounds, are of this kind. However, the border operators (Perelson, Oster, 1979) which limit a subsystem to a definite domain also seem to fall into this category. The boundary sets of deterministic chaos (Haken, 1981) also represent borders (Mandelbrot, 1983) - perhaps fractals. Nevertheless, analogies between the phenomena observed in biological systems and those in physical systems appear to be identified.

Finally, the difference between physical and biological systems should be emphasised. As long as the components of the system are molecules, the interaction between the components are of physico-chemical nature. In contrast, the interaction between cells (or individuals) seem only in primitive cases to be of a purely physico-chemical nature, but in general tends to be like a language. The interaction is therefore termed cell communication and the laws used to describe a language are different from those used to describe physico-chemical relationships. This is probably the reason why life systems seem to differ from physical systems. However, the laws of entropy should also apply to this kind of interaction. The interpretation of some of the phenomena dealt with in this paper seem to justify this view.

The extent to which results obtained in the field of biology are tansferable to societies of individuals is a question to be answered in the future. At least the statistical theory of thermodynamic systems which in general operates with models, here seems to offer a basis which might help to present a clearer view of processes which generate structure in societies under the force of a negative rate of entropy production.

Acknowledgement

Professor B. Havsteen should be gratefully acknowledged for his steady interest in and his valuable discussions about this topic.

Topologically Stable Patterns in Condensed Matter

V. P. Mineev, G. E. Volovik

Introduction

Nonlinear patterns in condensed matter physics, as well as in particle physics and cosmology may be roughly divided into three large groups. The first one contains topologically stable textures, defects of structural order in ordered media. The stability of these objects is guaranteed by the conservation law for some quantities of topological origin known as topological charges. As a result these textures are stable both in motion and in statics, for their existence no external energy source is required and dissipation does not destabilize them. These are dislocations in solid crystals, disclination lines and hedgehogs in liquid crystals, quantized vortices in superfluids, domain walls in ferromagnets, topological solitons, boojums and monopoles in different condensed matter, hypothetical magnetic monopoles in particle physics, etc.

The patterns of the second group are dynamical solitons, nonlinear solitary waves. They are always in motion and are stable only if dissipation may be neglected: they degrade by dissipation. Examples are Russell solitary waves in water, nonlinear waves in plasma physics, light pulses in nonlinear optics, solitonic solutions in general relativity, etc. The main tool in the investigation of this phenomenon in dissipationless nonlinear wave equations is the inverse scattering method.

Thermodynamics and Pattern Formation in Biology

The third group contains essentially nonequilibrium objects which appear in matter with dissipation by the action of steady external currents. These dissipative structures, like autooscillations, roll patterns in Rayleigh-Bénard convection, Taylor vortices in Couette flow in liquids, Belousov-Zhabotinskii reactions in chemical systems, pulses in neurons, all that represents the subject of synergetics, are quite familiar to biologists.

Of course there are no exact boundaries between the patterns of these three groups. For example, a dynamic soliton would not degrade by dissipation if an external energy source is introduced and in this way it transforms to the pattern of third group. Dissipative structures can exist in ordered media, e. g. in the AC Josephson effect in superconductors, the periodic motion of topological defects, Abrikosov vortices, appear under constant external electric current.

We concentrate here on the topologically stable patterns, which exist only in ordered media with spontaneously broken symmetry: in crystals with broken translational symmetry, liquid crystals with broken rotational symmetry, ferro- and antiferromagnets with magnetic symmetry broken and in superfluids and superconductors where gauge symmetry is broken. These systems are characterized by long range order, infinite in an ideal system. As a result the ordered medium is the simplest example of 'self-organisation'. A solid crystal, growing from a small crystallization nucleus, repeats the internal structure and crystallographic axes orientation over the entire space, thus producing crystalline long range order. However for real self-organisation purposes this pure crystal is too rigid, since it contains only one structural state and thus minimal information volume.

To increase the information space, i. e. the number of stationary states in a crystal, defects in crystal structure, dislocations and vacancies, should be introduced. These new stationary states are rather stable, e. g. the stability of a dislocation line is enormous: due to topological conservation laws the energetic barrier, which must be overcome in the process of elimination of the dislocation, is infinite. This property of topologically stable patterns is already used for construction of memory cells in ferromagnetic films, where the carrier of information is the so-called bubble, the topologically stable domain.

The next step towards the larger number of stationary states and lower barriers (large amount of information and more flexibility in learning and 'unlearning') is the transition to less ordered systems, such as quasi-crystals, then glasses and their analogue in magnetic systems, spin glasses. In these systems with strong short range order and weakened long range order there are enormous numbers of stationary states. Attempts to construct a model of flexible memory in such a system seems to be worthwhile.

As for the basis of life nature seems to have chosen another system with more subtle correlation between long and short range order, the linear chain of molecules. The order and self-organisation in this complicated system is under investigation, therefore we discuss here the more simple system, the ordered condensed matter, which possibly can help in the construction of analogies. From the variety of ordered systems with long range order we have chosen liquid crystals, since in their physical properties they are closer to living system materials than any other.

Disclinations in nematic liquid crystals

A nematic liquid crystal, or simply nematic, is a liquid with uniaxial anisotropy. Such a liquid consists of molecules, which usually have the form of a stick, however disk-like molecules can also produce nematic states. The interactions between the molecules tend to orient them parallel to each other, while thermal motion makes for chaotic orientation of molecular axes at high temperature where the molecules form an ordinary isotropic liquid (Fig. 1 a). Below some critical temperature T_c interaction prevails over thermal noise (typical values of T_c are in the region of room temperature and phase transition in the nematic phase occurs, where long range orientational order appears (Fig. 1 b). Though thermal fluctuations still tend to disorient the molecules, they cannot destroy the coherent common anisotropy axis which has the same direction over all space. Mathematically this common direction of molecular orientation is described by unit vector $\vec{n}$ called the director. This specific name reflects the fact, that the states of nematic with opposite vectors, $\vec{n}$ and $-\vec{n}$, are physically indistinguishable. Thus $\vec{n}$ and $-\vec{n}$ describe the same state. Translational order in nematics is absent as in an ordinary isotropic liquid: the positions of molecules are chaotic.

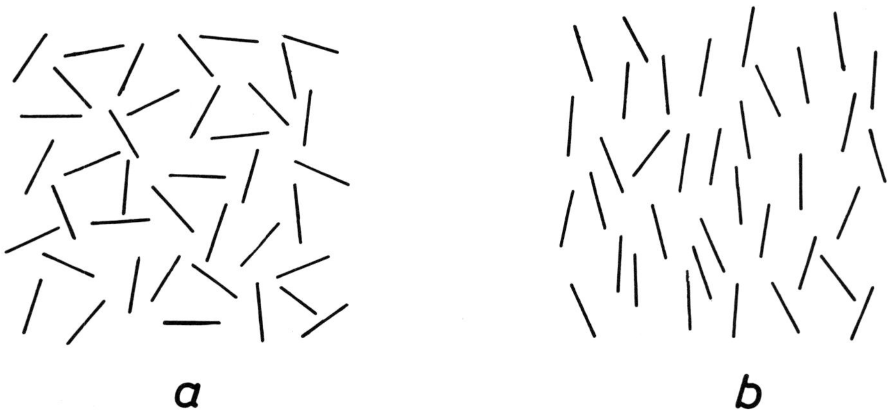

Fig. 1 a - above T_c (in the isotropic liquid phase): chaotic orientation of stick-like molecules; b - below T_c (in nematic phase): orientational long range order appears, acquires uniaxial anisotropy

In an ideal nematic the $\vec{n}$-field is constant over the volume, and the orientation of $\vec{n}$ is arbitrary if there are no external forces acting on the orientation of molecular axes. Thus in an ordered phase the system may be in a continuous set of equilibrium states, which differ from each other by orientation of $\vec{n}$. Under the influence of external fields (electric and magnetic fields, laser pulse, etc.) and of boundary conditions on the walls of vessel the $\vec{n}$-field becomes nonuniform in space, $\vec{n} = \vec{n}(\vec{r})$; these nonuniform distributions of director are called textures. The texture may be continuous and singular, in the latter case the $\vec{n}$-vector field has linear or point singularities, on which the orientation of $\vec{n}$ is not determined. Such singular lines in the director field are known as disclinations. They are visible with a microscope as long thin threads floating in a liquid, the thread (νημα in Greek) gave the name to this type of liquid crystal.

The simplest examples of $\vec{n}$-field distribution around the singular thread are shown in Fig. 2, these are planar textures: the $\vec{n}$-field orientation is parallel to some plane, the plane of the figure. The widest point is the intersection of the singular line with the plane of figure. The singularities of the flat vector field are characterized by the winding number N, which is the number of full rotations carried out by vector $\vec{n}$ in a positive direction when passing around the singular line along the closed contour γ . This index is N = 1 for the texture in Fig. 2 a, N = -1 for that in Fig. 2 b and N = 2 in Fig. 2 c. Since $\vec{n}$ and $-\vec{n}$ are equivalent, textures with half integer N are possible: N = 1/2 for the texture in Fig. 2 d and -1/2 for that in Fig. 2 e.

The nonuniform distribution of $\vec{n}$-vector is energetically unfavourable in nematics if external forces are absent. Therefore, for pure energetic reasons singular lines will be dispersed after some deformation of the $\vec{n}$-field. However such a continuous process of the transformation of the singular line into a uniform state may be prohibited for topological reasons. Let us first consider the possible deformation of the disclination on line with N = 1. Fig. 3 a shows this line not from the 'top' but from the 'side'. Continuous deformation (Fig. 3 b) transforms the singular line to the uniform $\vec{n}$-field without any singularity (Fig. 3 c). This deformation, appearing like an umbrella folding, was described as a 'flow into third dimension' since the $\vec{n}$-field escapes from the plane of its initial orientation. Thus the N = 1 disclination is unstable towards flow in the third dimension. Are the other disclinations stable and how can one distinguish stable and unstable singularities?

Here it should be noted that, in principle, it is possible to eliminate any given disclination line by creating the singular surface in the intermediate states. This process is shown in Fig. 4. But this 'melting' of nematic ordering in a half plane requires that an enormous energetic barrier be overcome and this is proportional to the surface area and thus is infinite in infinite liquid crystals. Thus only continuous deformations like in Fig. 3 should be considered and this is the subject of topology.

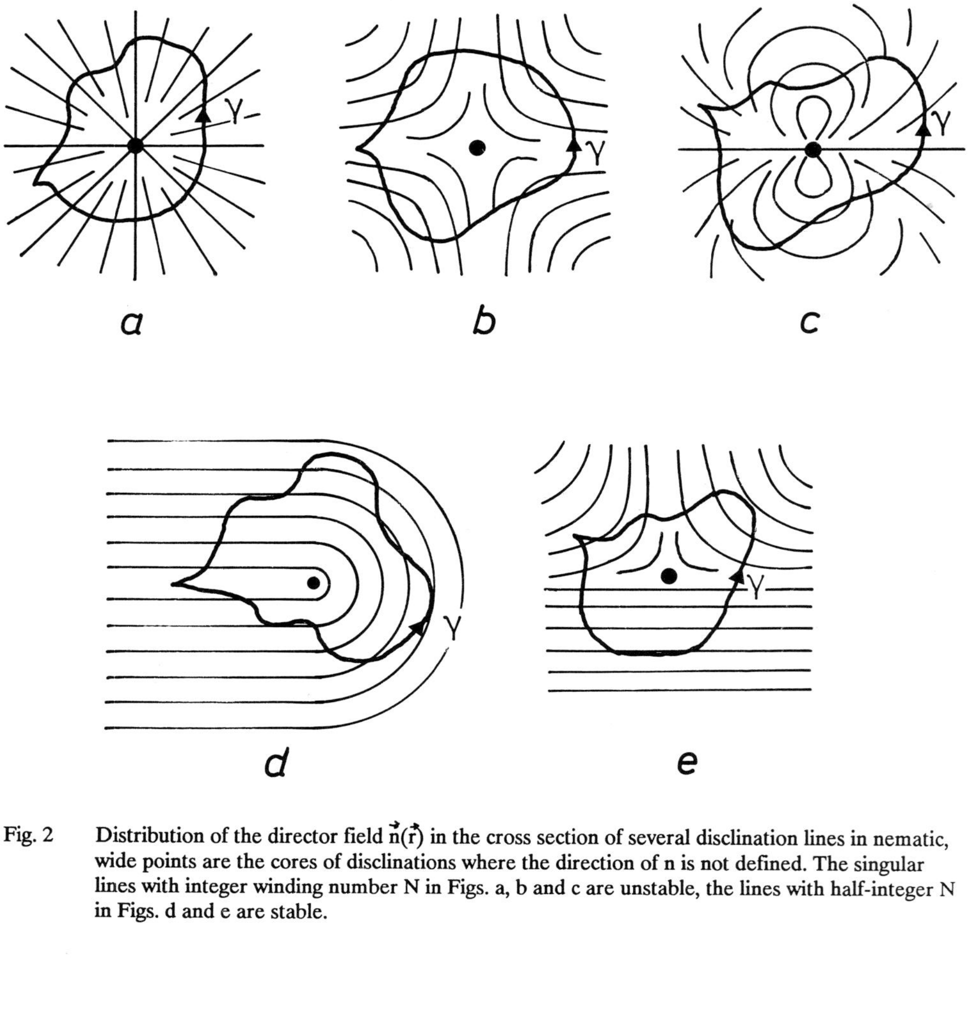

Fig. 2 Distribution of the director field $\vec{n}(\vec{r})$ in the cross section of several disclination lines in nematic, wide points are the cores of disclinations where the direction of n is not defined. The singular lines with integer winding number N in Figs. a, b and c are unstable, the lines with half-integer N in Figs. d and e are stable.

The key to the topological solution of the disclination stability problem is the mathematical concept of mapping. Here the mapping is given by the director field $\vec{n}(\vec{r})$ which maps the coordinate space $\vec{r}$ onto the space of equilibrium states, the spherical surface of unit radius where $\vec{n}$ takes all its possible values (Fig. 5). For a complete description of the space of equilibrium states one has to take into account the physical equivalence of diametrically opposite points on a sphere, $\vec{n}$ and $-\vec{n}$. Such a sphere, with diametrically opposite points connected, is topologically equivalent to the projective plane denoted as RP^2.

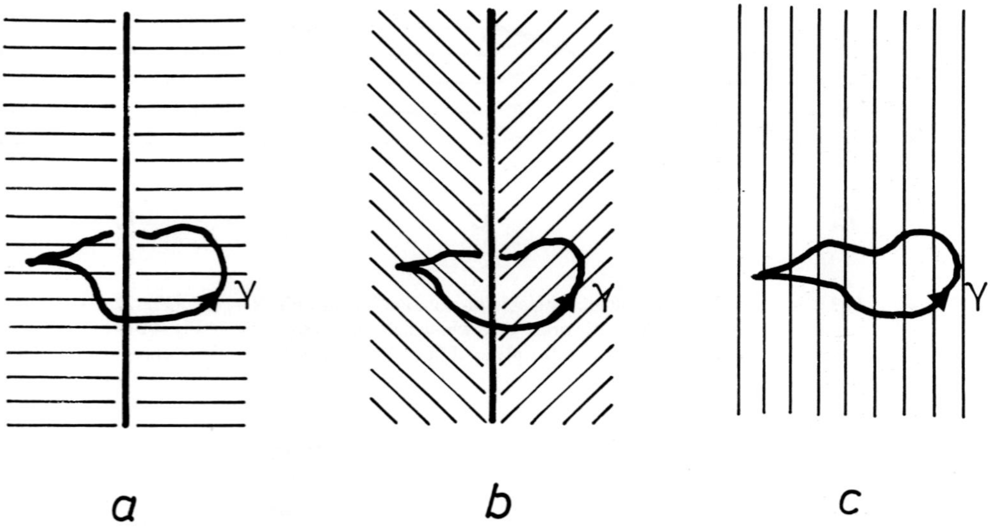

Fig. 3 The process of flow of singularity into 'third' dimension, i. e. a continuous transformation of the unstable disclination line with N = 1 (the view from above is in Fig. 2 a, side-view is in Fig. 3 a, solid line is the core of disclination) into the uniform state (Fig. 3 c). The tilting angle of the dirctor with respect to the singular line changes from $\pi/2$ to 0, this recalls the folding of an umbrella.

To check the topological stability of a given disclination line let us surround the line by the closed contour γ and consider the mapping of this contour onto the space RP^2.

Since to each point $\vec{r}$ on γ there corresponds the point $\vec{n}(\vec{r})$ on RP^2 one obtains the closed contour Γ on RP^2 (Fig. 6). Any deformation of the $\vec{n}$-field around the disclination line causes the deformation of the contour Γ onto RP^2. In particular the flow into the third dimension in Fig. 3 is accompanied by contraction of Γ into a point (Fig. 7).

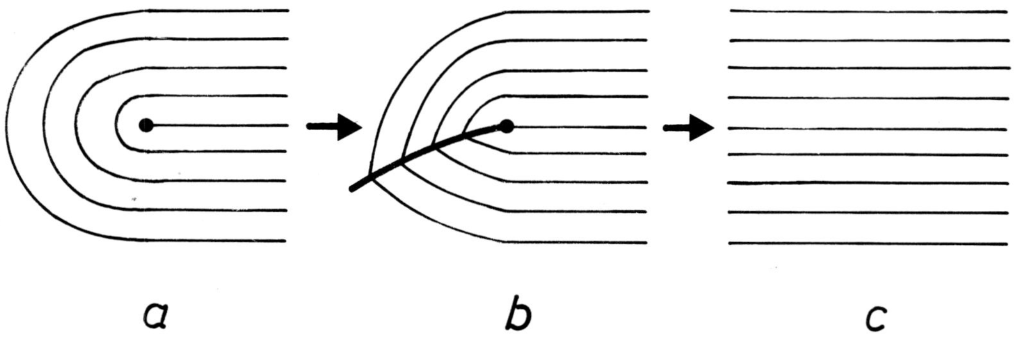

Fig. 4 The elimination of the topologically stable disclination line in with N = 1/2 (Figs. 2 d, 4 a) requires breaking of the nematic order on a whole half-plane (solid line in Fig. 4 b is the intersection of this half-plane with the plane of figure), which costs an infinite energy for infinite systems.

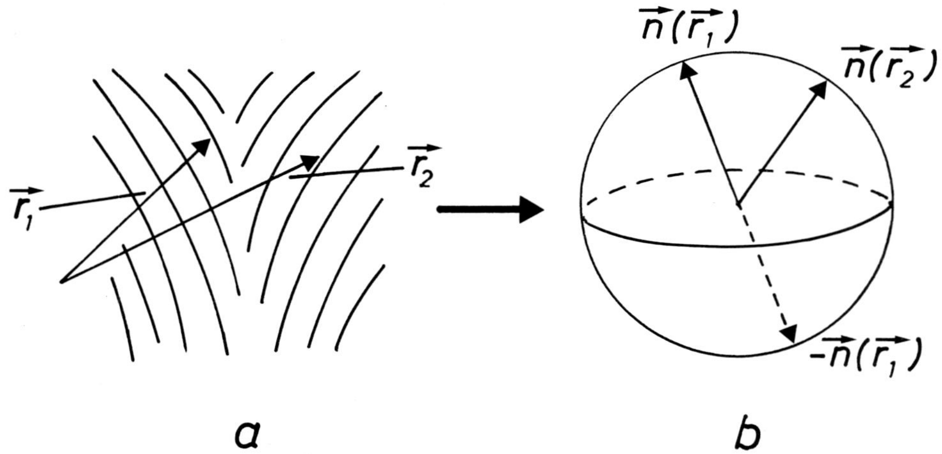

Fig. 5 The vector function $\vec{n}(\vec{r})$ in the nematic texture realises the mapping of the coordinate space to the space of equilibrium states of nematic, RP^2, which is the unit sphere of vector $\vec{n}$ with identified diametrically opposite points. Points $\vec{r}_1$ and $\vec{r}_2$ in texture are mapped to the points $\vec{n}(\vec{r}_1)$ and $\vec{n}(\vec{r}_2)$ on unit sphere; $\vec{n}(\vec{r}_1)$ and $-\vec{n}(\vec{r}_1)$ are equivalent.

Straightforward generalisation reveals that unstable singular lines are described by contractable contours Γ on RP^2. Let us call the whole class of contractable contours or, which is the same thing, the class of topologically unstable disclinations as Γ_0 class. It is easy to check that the disclinations on Figs. 2 b, 2 c as well as all the disclination with integer N belong to this class: they are described by the contours of the class Γ_0 and therefore may be continuously deformed into each other and into nonsingular state.

The disclinations on Figs. 2 c, 2 e belong to another class of contours, $\Gamma_{1/2}$, which connect diametrical opposite points on a sphere: since these points are equivalent the $\Gamma_{1/2}$ contours are also closed. As distinct from the contours of the Γ_0 class, the $\Gamma_{1/2}$ contours cannot be contracted into points. In this respect they are similar to the noncontractable contours on a torus surface. Thus the corresponding disclinations of the class $\Gamma_{1/2}$ are topologically stable: they can transform into each other but not to the nonsingular state Fig. 8).

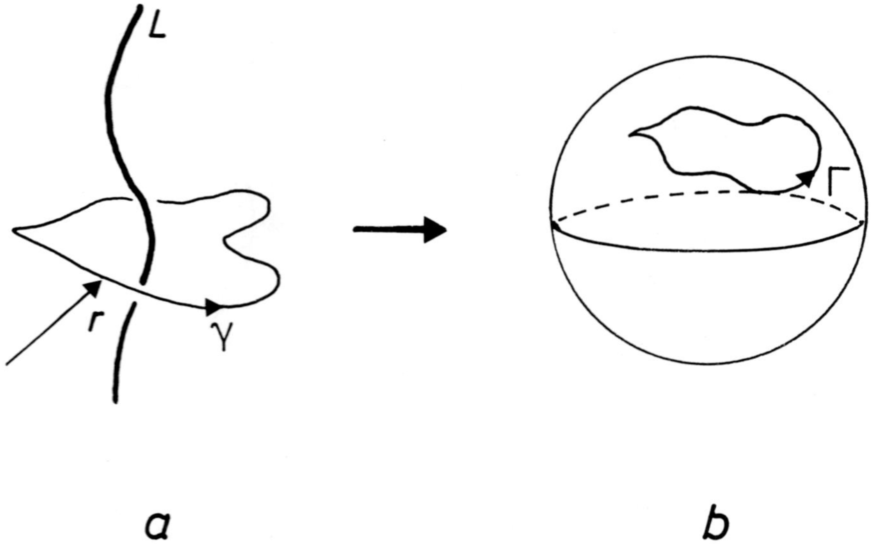

Fig. 6 The stability of the disclination line L is checked by mapping of contour γ embracing the disclination line (Fig. 6 a) onto the space of equilibrium states. If the image Γ of contour γ is such as shown on Fig. 6 b (i. e. it may be contracted to a point) the disclination is unstable.

Note that the existence of noncontractable closed contours in the space of equilibrium states of nematics and therefore of topologically stable linear defects in this substance is the direct consequence of the equivalence of states $\vec{n}$ and $-\vec{n}$. In the substances, described by unit vector without identification of the opposite directions, as in ferromagnets, there are no stable linear singularities, since all the contours on the spherical surface are contractable.

The space RP^2 contains only two classes of contours, $\Gamma_{1/2}$ and Γ_0 . Thus only two classes of topologically equivalent disclination lines exist in nematics, stable and unstable. The stable line of the class $\Gamma_{1/2}$ cannot have an end in the bulk liquid, i. e. the nematic thread cannot be torn to pieces. This may be checked by following the evolution of the contour γ along the disclination. Thus such a line forms a closed disclination loop; or may start and finish on the wall if the boundary conditions are favourable for that. On the contrary for unstable disclinations topology allows breaking of the thread.

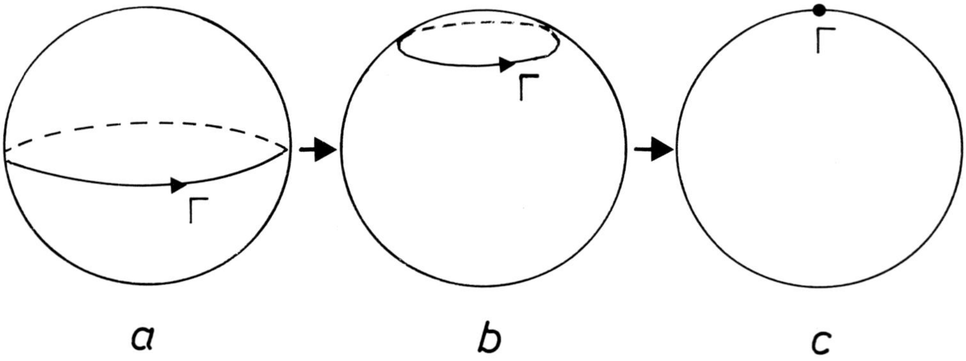

Fig. 7 The deformation of the contour Γ to a point in the process of the flow of unstable disclination with N = 1 into the third dimension. The sequence of states on Figs. 3 a, 3 b, 3 c in this process corresponds to the sequence of contours Γ on Figs. 7 a, 7b, 7c.

Topology also allows discussion of the process of coalescence of two singular lines. The resulting disclination is described by the contour γ which is the product of two contours corresponding to the initial disclinations. Let us consider the coalescence of two stable lines. The RP^2 image of the contour $\gamma = \gamma_2 \cdot \gamma_1$ surrounding both lines (the product means the contour which first moves along γ_1 and then along γ_2) is the contour $\Gamma_{1/2} \Gamma_{1/2}$. As is evident from Fig. 9 this product belongs to the Γ_0 class. Thus the coalescence of any two stable disclination lines results in their annihilation. Thus we have the following topological law for the fusion and fission of the disclination $\Gamma_{1/2} \cdot \Gamma_{1/2} = \Gamma_0$, $\Gamma_{1/2} \cdot \Gamma_0 = \Gamma_{1/2}$, which corresponds to a multiplication law for the mathematical group consisting of two elements.

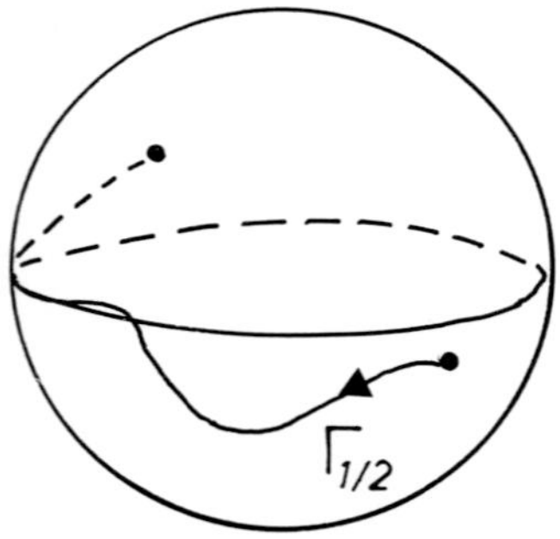

Fig. 8 The closed contour of class $\Gamma_{1/2}$ in RP^2, which cannot be contracted into a point. It connects two diametrically opposite points which are equivalent in RP^2, therefore this contour is closed. The disclination lines described by contours of this class are stable. The disclinations with half-integer N on Figs. 2 d and 2 e correspond to a contour of this class $\Gamma_{1/2}$, therefore they may be continuously transformed to each other but not to the uniform state.

This may be generalised for linear defects in any condensed matter with long range order. All the defects are divided into classes. Each class represents an element of some mathematical group, the fundamental group of the space of equilibrium states in the matter. The processes of fission and fusion of the defects are regulated by the multiplication law for this group, denoted as π_1. The group $\pi_1(RP^2)$ consists of only two elements, Γ_0 and $\Gamma_{1/2}$, which in an additive representation may be described by numbers, topological charges, 0 and 1/2 with the summation law $1/2 + 1/2 = 0$, $0 + 1/2 = 1/2$.

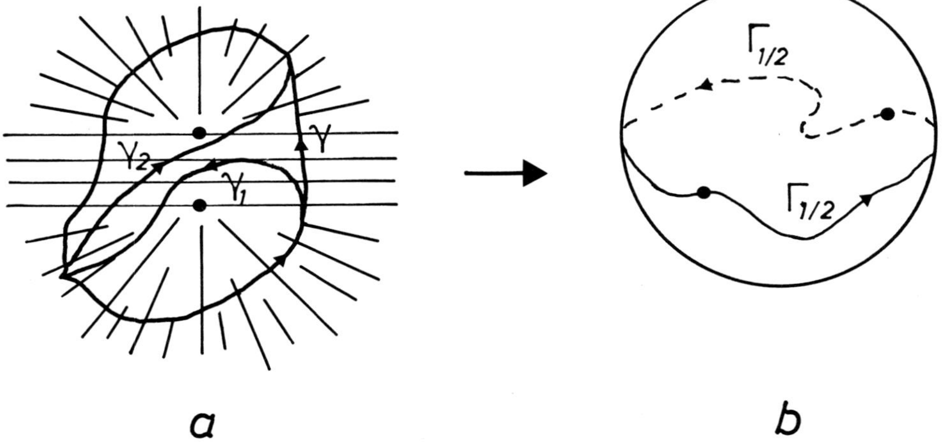

Fig. 9 The contour γ embracing two stable disclination lines in nematic is mapped to contour $\Gamma_{1/2} \cdot \Gamma_{1/2}$ which is contractable on RP^2. Therefore coalescence of any two stable disclinations results in their annihilation.

Hedgehogs in nematics

In addition to the singular lines in the n-field isolated singular points are also possible in nematic liquid crystals. The simplest example of such a point is the 'hedgehog' with spines radially directed: $\vec{n}(\vec{r}) = \vec{r}/r$ (Fig. 10 a). How can one check its topological stability towards deformations into nonsingular state? For this purpose one must surround the singular point by a spherical surface S and investigate its image in RP^2. For the simplest hedgehog the image of the S on RP^2 is all the RP^2 surface, with one to one correspondence between points on sphere S and the unit sphere of $\vec{n}$-vector. Thus this radial field achieves the first degree mapping of sphere to sphere. If one deforms the $\vec{n}$-field of the hedgehog, 'combing' the spines, the image of S which is the closed film covering RP^2 is also deformed and may even form folds on the unit sphere. However it is impossible to contract this covering film to one point on RP^2 and hence eliminate continuously the hedgehog: the hedgehog is topologically stable.

Unlike linear defects in nematics, there are many different classes of topologically stable point-like defects. They are described by the topological invariant, the integer N, which describes the degree of sphere to sphere mapping: how many times the image of S covers the unit sphere of the $\vec{n}$-vector. If $N = 0$ the corresponding hedgehog is unstable.

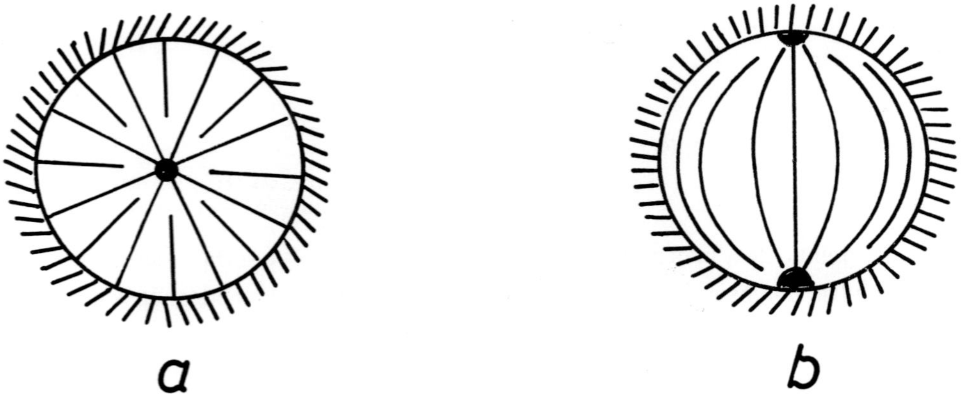

Fig. 10 Two patterns in nematic drop with different boundary conditions. In the case of normal boundary conditions for $\vec{n}$ on the wall (Fig. 10 a) there is a hedgehog with the topological invariant $N = 1$ in the centre of the drop. In the case of tangential boundary conditions (Fig. 10 b) there are two point defects on the surface, boojums.

What is the topological law for fission and fusion processes with the hedgehogs? The answer is very easy for ferromagnets, where the space of the equilibrium states is a unit sphere without identification of different points: the topological charge of the resulting defect is the sum $N_1 + N_2$ of the charges of the initial point singularities. The situation in the nematic is more complicated: because of identification of the diametrically opposite points. The hedgehog with its spine outwards and thus with N = 1 is indistinguishable from the hedgehog with its spines inwards, N = -1. Thus the singular points in nematics are described by the modulus of N. The problem then arises as to what is the result of fusion of two hedgehogs with topological charges

$$|N_1| \text{ and } |N_2| : |N_1 + N_2| \text{ or } |N_1 - N_2| ?$$

The answer is: both of them. The result depends on the prehistory of their meeting: motions before meeting are not topologically equivalent in the presence of the disclination line. Thus passing along the different sides of the disclination before meeting, two hedgehogs with $|N| = 1$ will either annihilate or form a hedgehog with double charge.

Patterns in nematic droplets

Another point defect in nematic liquid crystals, the boojum, is a surface defect (see Fig. 10 b with two boojums in the nematic drop). The boojum is not a hedgehog which came from the bulk liquid to the wall and stuck to it, though such a hedgehog on the surface can exist (Fig. 11 c). The boojum can live only on the surface of ordered matter and, in principle, due to topological constraints imposed by the boundary conditions, cannot sink into the bulk. Its existence and topological stability are the result of a subtle intercoupling between the topology of the space of the equilibrium states in the bulk nematic, RP^2, the topology of the space of the equilibrium states in the surface layers of nematic and the topology of the vessel itself, which is a spherical surface in the case of droplet. Boojums cannot exist if the boundary condition for $\vec{n}$-vector is strictly normal. Only if the director has a nonzero angle with the normal to the surface and thus has continuous degrees of freedom on the surface, do boojums become topologically stable.

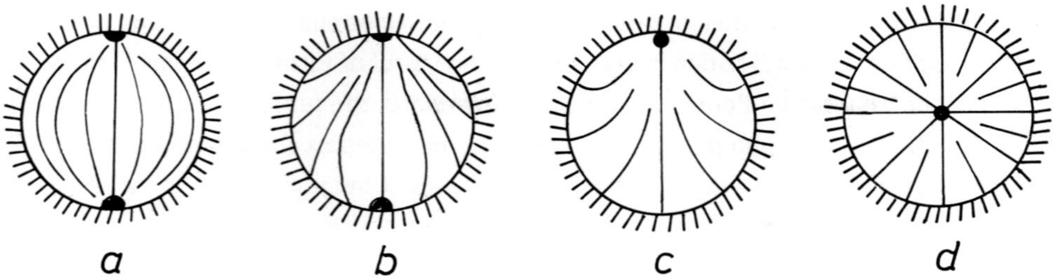

Fig. 11 The transformation of the texture on Fig. 10 b to the texture on Fig. 10 a under continuous change of boundary conditions. One of the boojums disappears while the other turns to the hedgehog and sinks into the bulk nematic.

The boundary conditions for the director on the phase boundary between nematic and isotropic liquids may be varied by changing the external parameters, e. g. the temperature. This gives the reversible mechanism of pattern transformation. Fig. 11 shows the process of the transformation of the bipolar texture of Fig. 10 b with two boojums to the hedgehog with $|N| = 1$ in Fig. 10 a by continuous change of the boundary conditions. In this process one of the boojums, on the south pole, gradually becomes thinner and completely disappears at the moment when the normal boundary conditions are achieved. On the contrary the boojum on the north pole becomes fatter and under normal boundary conditions suddenly turns into a hedgehog (Fig. 11 c), which then sinks into the bulk liquid to the centre of the droplet. This recalls the Baker's caution in the 'Hunting of the Snark' by Lewis Carroll: 'Beware of the day your Snark be a Boojum!'.

The number of topologically stable boojums on the vessel walls essentially depends on the vessel topology. For example at least two boojums with unit winding number should exist in a droplet, or one double boojum if they coalesce on the surface. A nematic texture without any surface defects can exist in the vessel of torus form. This is the result of the well known Poincare theorem which relates the sum of the surface singular point indices (winding numbers of boojums) with the Eulerian characteristics of the surface. Let us recall that the Eulerian characteristics of the polyhedron is $\chi = N_v - N_e + N_f$, where N_v, N_e and N_f are the number of vertices, edges and faces respectively. For all singly connected polyhedrons as well as for the spherical surface $\chi = 2$, which may be checked for a cube: $\chi = 8 - 12 + 6 = 2$, while for a torus $\chi = 0$ (no boojums, or the algebraic sum of the boojum's winding numbers is zero).

Fig. 11 gives a simplified version of the real process observed in the experiment, which also includes the surface disclination loop (Fig. 12), which like the boojum is a specifical surface object. The surface disclination loop exists only at intermediate boundary conditions: it disperses at tangential boundary conditions and turns into a bulk disclination loop at normal boundary conditions. Fig. 12 shows the processes of mutual transformation of boojums, hedgehogs and disclination loops, regulated by topological conservation laws.

Thus even in the simplest case of the nematic droplet one encounters four types of topologically stable objects: boojum, hedgehog, surface and bulk disclination loops, which can form a variety of stable patterns under different external influences. These patterns are affected by temperature, electric and magnetic fields, etc.

However, the most interesting effects occur when the nematic is decorated by chiral molecules (molecules which are not identical to their mirror image, i. e. the space parity is violated). This produces a new class of liquid crystals, cholesteric liquid crystals, which are spiralled nematics (Fig. 13). When the helical pitch becomes smaller than the droplet radius the cholesteric layers form a system of concentric spheres. Because of topological constraints this pattern has an unusual property, which again follows from the Poincare theorem: since the director field in the layer is tangential to the sphere it must have two singular lines with N = 1 or one singular line with N = 2 starting from the centre of the sphere (Fig. 14).

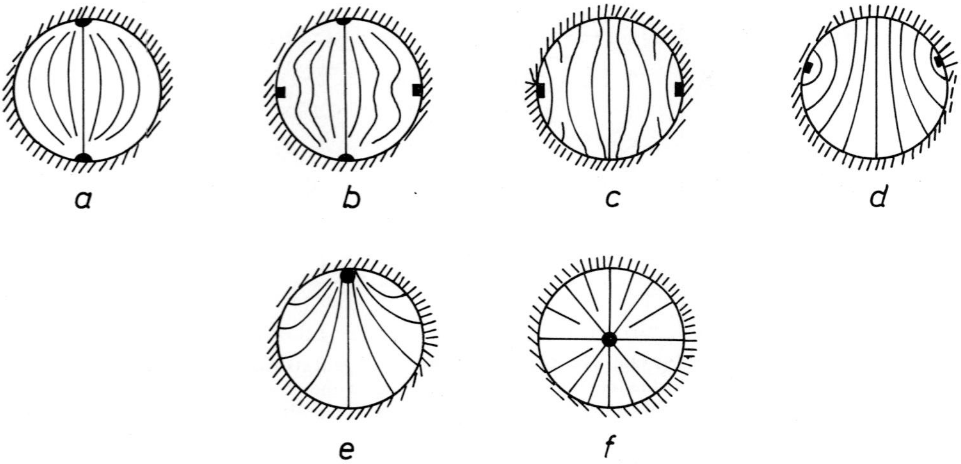

Fig. 12 The transformation of the textures with intermediate creation of the surface disclination loop on the equator of the drop. Both boojums gradually disappear, while the disclination loop moves to the north pole and transforms to the hedgehog.

Such singularities in the n-vector field in cholesterics are known as cholesteric disclinations of the χ type. In addition to χ -disclination cholesterics have two more types of singular lines, λ -lines and τ -lines: this is due to more complicated long range order, which also includes translational ordering. Thus depending on the experimental conditions two patterns exist in the cholesteric droplet: two χ -lines starting from the centre or one χ -line with double strength. The latter pattern has much in commmon with the hypothetical magnetic monopole, predicted by Dirac in

1931, which however has not so far been observed among elementary particles. Double χ -line recalls the Dirac string attached to a monopole.

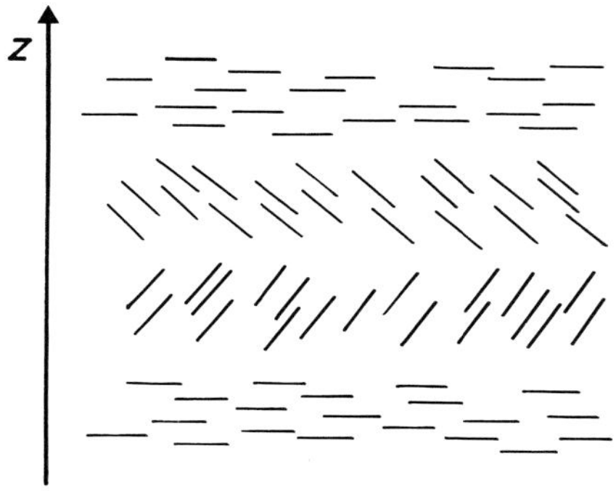

Fig. 13 The cholesteric liquid crystal. The axis of the cholesteric spiral is oriented along z while the director field is confined in the x, y plane. Moving along z the observer sees the gradual rotation of molecules. This helical structure is periodic: on the period of the spiral structure, the pitch, the molecules rotate by π around the axis of the spiral.

Topological interaction of defects

In addition to the singular lines and the monopole the cholesteric allows the existence of other topologically stable defects of structure: topological solitons, confocal textures, etc. The variety of patterns appears due to different combinations of these textures. One of the interesting results of topological intercoupling between different defects may be seen in the example of the biaxial liquid crystal which has the same three types of disclination lines as the cholesteric. This is a liquid with two anisotropy axes as distinct from the ordinary uniaxial nematic. Biaxial nematics have similar orientational order as cholesterics but without translational long range order and thus without the additional constraints imposed by translational order on the structure of the disclinations.

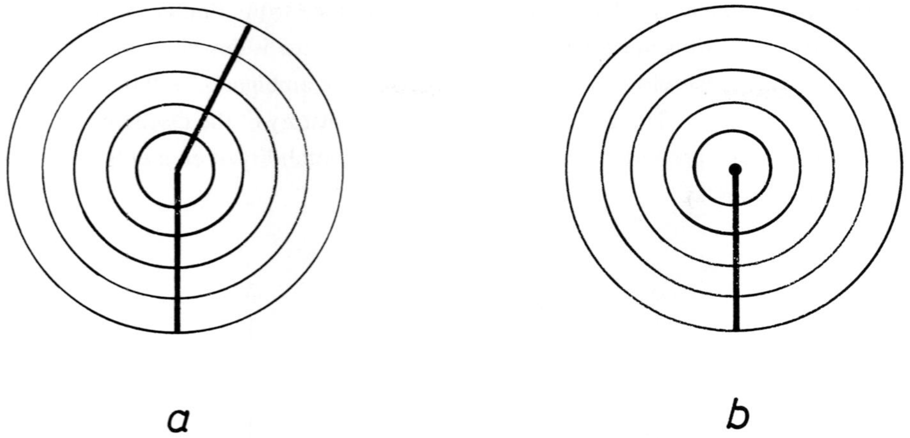

Fig. 14 **Two different patterns in a cholesteric droplet: a - monopole with two strings and b - monopole with one double string. The topological structure of the latter texture recalls that of Dirac magnetic monopole.**

In the case of biaxial nematics the χ , λ and τ disclinations form an unusual mathematical group which governs the processes of topological interaction and mutual transformation of these linear defects. Each type of line, like the hedgehog in uniaxial nematics is described by two elements: $\chi \to (i, -i)$, $\lambda \to (j, -j)$, $\tau \to (k, -k)$, where i, j, k are the quaternionic units with the following multiplication law:

$$i^2 = j^2 = k^2 = -1,$$
$$ij = -ji = k, \quad jk = -kj = i, \quad ki = -ik = j \cdot$$

According to the correspondence between the product of the elements of the fundamental group and the result of fission of the defects one may show from the last equation that the coalescence of χ and τ disclinations results in the formation of the λ -line. From the first equation it follows, simply that the fusion of two identical disclinations, e. g. χ and χ , gives χ disclination of double strength, which is topologically equivalent (may be continuously transformed to) to double λ -line and double τ-line. But this is not completely correct: the χ -line is described by two elements, i and -i, therefore the result of fusion may be ambiguous: either $i \cdot i = -1$ or annihilation $i \cdot (-i) = 1$. This again depends on the prehistory of the disclinations motion. This strange property results from the nonpermutability of elements in the group product: jk ≠ kj.

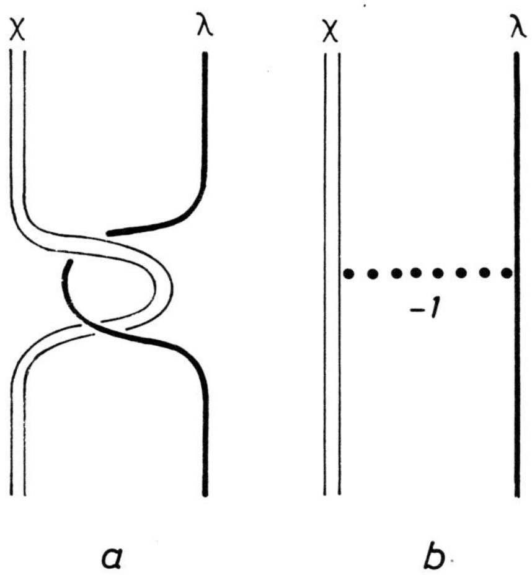

Fig. 15 The topological obstruction for crossing χ and λ lines. After crossing the singular line of the class -1 appears between these disclinations, which ties them together. This is one of the examples of topological confinement.

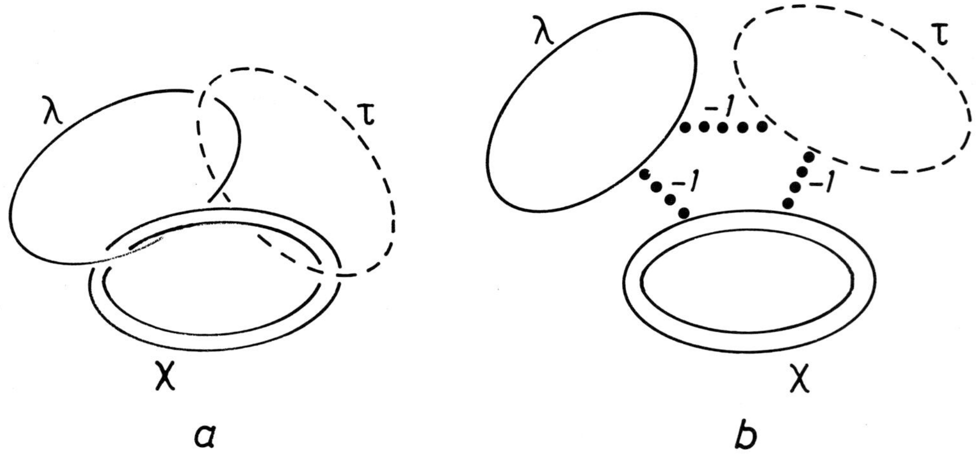

Fig. 16 Topological confinement of 3 'quarks': χ , λ and τ disclination loops cannot be liberated from their mutual prison, since they are tied by disclinations of class -1.

There exists another exotic consequence of nonpermutability which may have some analogy with confinement of quarks inside hadrons. Let two nonparallel rectilinear χ and λ disclinations move towards each other until they cross. In ordinary uniaxial nematics the disclination lines would move freely after interaction during crossing. But in substance with noncommutative groups the lines will be caught on each other: they will be tied by the line of class -1 which confines them together (Fig. 15). In the same manner the knot formed by three different disclination loops (Fig. 16) cannot be broken because of topological constraints - not even by the sword of Alexander the Great.

Conclusion

In ordered systems (with long range order) there exist nonuniform textures and defects of high stability due to specific conservation laws imposed by topology. The topological interaction of defects and textures produces a variety of stable patterns which reversibly change under the influence of external fields. The processes of fission, fusion and mutual transformation of defects are regulated by topological conservation laws.

New Types of Order and Stochastic Regimes in Nonlinear Media

T. S. Akhromeyeva, S. P. Kurdjumov, G. G. Malinetskii, A. B. Potapov, A. A. Samarskii

Hundreds of models have been proposed over the past few years to describe and explain some biological processes in various nonlinear media. But qualitatively behaviour (and this is of the greatest importance for numerous biological applications) proved to be very similar for most of them and independent of specific detail.

As a rule, one investigates diffusion instabilities that give rise to stationary dissipative structures (typical examples are numerous models of morphogenesis) and simple autowave processes (like 'switching' waves in the Kolmogorov-Petrovskii-Piskunov model (1937) and waves in excitable media). These properties of different models have been given a biological interpretation and these have been confirmed experimentally. The main goal of this paper is to attract attention of biologists to the new types of order and chaotic regimes that have been discovered during studies of mathematical models with biological applications.

A number of important general properties typical of a wide class of nonlinear systems have become clear during the studies of these and other models in recent years.

1. It was shown that in many cases simple dissipative structures can be combined to form a more complex structure. For a number of problems the principles of organisation of structures can be established for both one-dimensional and multi-dimensional geometries (they are substantially different).

Thermodynamics and Pattern Formation in Biology

2. Localised processes are of great importance for a wide class of nonlinear media. In this case many interesting features are due to the internal properties of the media rather than the boundary conditions.
3. There are methods of influencing a nonlinear system (sometimes called a 'resonant perturbation') and which give rise to a process qualitatively different from that observable after other types of influence. Correspondence to the internal properties of a system rather than to the intensity of this influence is significant (in nonlinear media it can be a certain configuration of initial data or the presence of spatial symmetry).
4. Usually processes in nonlinear dissipative media (having infinite number of degrees of freedom) can be effectively described by dynamic systems of small dimension for the order parameters. Their determination can be achieved not only because of the slaving of damped modes by nondamped modes, but also by the localisation of dissipative structures, to the existence of 'fast' processes that are more substantial than 'slow'.

It is possible that the regularities mentioned above as new types of order will prove to be useful in the mathematical simulation of biological processes and in the explanation of different phenomena. Below we shall illustrate this via several examples of nonlinear systems.

Elementary and complex thermal structures

Let us consider a nonlinear medium where diffusion processes and a nonlinear source exist:

$$T_t = (k(T)\,T_x)_x + Q(T), \quad k(T) = k_0 T^{\sigma} \;;\; Q(T) = q_0\, T^{\beta} \tag{1}$$

$$-\infty < x < \infty, \qquad k_0, q_0, \beta, \sigma > 0 \qquad \beta > \sigma + 1.$$

This model may have different interpretations, but many authors consider T(x, t) as a temperature of the medium, and its combustion being simulated by the source Q(T). For simplicity let Q(T) and k(T) be power functions of temperature.

Nonlinearity of k(T) provides a finite speed (in contrast to a linear equation) of propagation of perturbations. The intensity of the source Q(T) increases when moving away from the equilibrium state (T = 0). Such systems are of interest not only in plasma physics and the theory of combustion (Kurdjumov, 1982), but also in biology. For example, natural selection of features, advantageous for survival, results in '... the process running with an increasing speed, up to some limit - the system does not just depart from thermodynamic equilibrium, but departs further and further with increasing speed, because more perfect forms arise faster and earlier than others can conquer them' (Shnol, 1979).

The fact that a source is a power function of temperature may be considered as an approximation of the initial part of Q(T) in systems close to the Kolmogorov-Petrovskii-Piskunov model (Fig. 1 a).

Behaviour typical of solutions of equations (1) is depicted in Fig. 1 b. The heat proves to be localised within a domain G_L, outside of which the temperature is equal to zero. This results in splitting of the medium into limited combustion regions.

A wide class of solutions of equations (1), in particular the one shown in Fig. 1 b, exists only for a finite time t_f. During this time the function T becomes infinite at one or several points, or sometimes in the whole domain. Such regimes may serve as an idealisation of the situation where the most rapid processes developed within a finite time are of particular interest (for long times this model is not applicable and it is necessary to take into account features limiting temperature increase). Certainly, in real processes described by model (1) the function T(x, t) will be limited, but its growth by several orders of magnitude permits one to observe many interesting effects predicted by the theory (Kurdjumov, 1982; Samarskii, Mikhailov, 1976; Elenin et al., 1983).

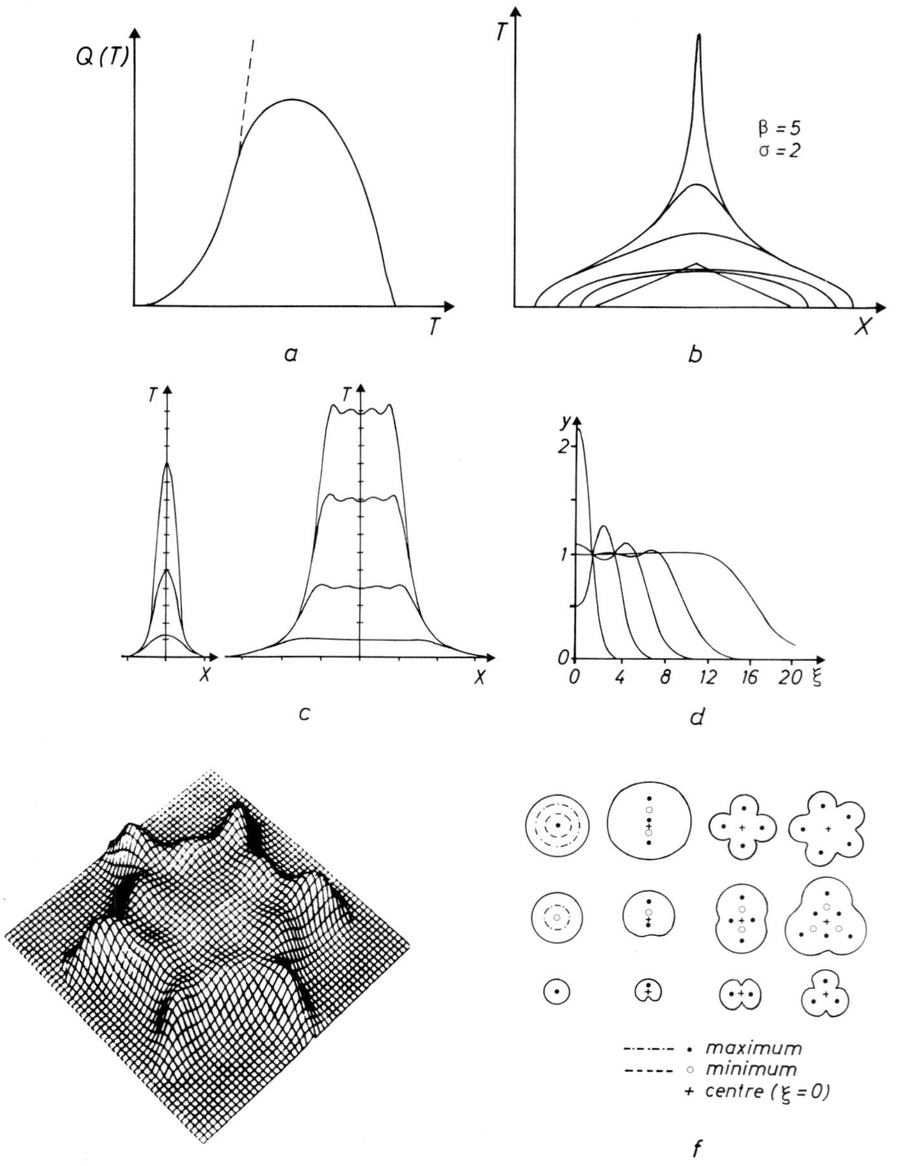

Fig. 1 see text

It can be seen in Fig. 1 b that at a developed stage (sufficiently close to t_f) the function T proves to be similar to itself. Calculations show that in this case it is described by the self-similar solution

$$T(x, t) = g(t) \cdot f(\xi), \quad \xi = x/\varphi(t). \tag{2}$$

Here g(t) determines a law of amplitude growth

$$g(t) = (t_f - t)^{\mu}, \qquad \mu = 1/(1 - \beta), \qquad g(t)_{t \to tf} \to \infty$$

$\varphi(t)$ is the halfwidth of the temperature profile

$$\varphi(t) = (t_f - t)^n, \qquad n = (\beta - \sigma - 1)/(2\beta - 2), \quad \varphi(t) \underset{t \to t_f}{\to} 0 \qquad ,$$

and f is a shape parameter of the profile. Convergence to this solution from different initial data permits consideration of the process as the formation of a nonstationary dissipative structure.

If arbitrary initial data are set, we shall have one or several localised structures with different life times t_f. Eventually only one structure with the minimum $t_f = t_{f\,min}$ will develop (in comparison with this, amplitudes of all the other structures will be negligible small when $t \to t_{f\,min}$). The formation of structures with a single maximum (we shall call them elementary), i. e. process of convergence to the self-similar solution, can be described by the dynamic system. This describes the laws of change in the halfwidth and amplitude and can be obtained by means of an averaging method (Kurdjumov, 1982).

We have considered, above, only elementary structures. The question arises: do more complex types of order (complex structures) exist in the medium that can be described by solutions which preserve their form and have several extrema? To answer the question one needs to study the whole set of self-similar solutions of type (2).

It has been shown that the amplitude and halfwidth of complex structures change according to the same law as for simple structures (Elenin et al., 1983). These structures describe thermal waves with increasing amplitude, converging to the centre (see Fig. 1 c).

The analysis performed revealed that in a medium with given β and σ (equation (1)) a finite set of different types of structures can exist. An example of the complete set of structures is depicted in Fig. 1 d for $\beta = 3.5, \sigma = 2$. (Replacing the variables allows k_0 and q_0 to be equal to 1). As all are symmetric only their right halves are shown.

The functions f determining the structures have been called the 'nonlinear medium eigenfunctions' (Kurdjumov, 1982; see above). In contrast to normal eigenfunctions, which arise when one solves linear problems in mathematical physics, they do not depend on the boundary conditions and describe localised processes. As the equation is nonlinear the functions cannot be combined to form a general solution. They determine all the types of organisation possible in the medium.

It can be seen that complex eigenfunctions (with several extrema) may be considered as several elementary structures with different age (t_f-values) and overlapping localisation areas (see Fig. 1 d). Setting the initial data according to the medium eigenfunctions has been called resonant perturbation of the system (Kurdjumov, 1982). Knowledge of the principles of structural organisation allows influence over the nonlinear medium in such a way that its evolution can be qualitatively different from that of the general case. Let us note that it is the form of the initial data rather than the amplitude that is important.

The architecture of eigenfunctions in the multidimensional case is of particular interest. It proves that a finite number of complex structures having different symmetry and spatial configurations can exist in a nonlinear medium (together with cylinder-symmetric solutions). One of the two-dimensional functions is depicted in Fig. 1 e. The complex structure here also determines a combustion wave converging to the centre of symmetry, its shape being constant.

An analysis of the two-dimensional analogue of equation (1) gives the following equation for nonlinear medium eigenfunctions

$$\alpha \Delta y - \frac{\beta-\sigma-1}{2} \xi (y^{\alpha})_{\xi} + y^{\alpha\beta} - y^{\alpha} = 0, \quad \xi = |\vec{\xi}|, \quad y = f^{\sigma+1}, \quad \alpha = \frac{1}{\sigma+1}. \qquad (3)$$

Attempts to solve the equation using standard numerical methods have failed. One needs *a priori* information about the solution. To gain it some analytical and approximation methods have been developed. They are based on 'matching' solutions for some linear problem with an asymptote describing y when $\xi \to \infty$. Study of a nonlinear elliptic problem allows determination of order parameters (they are closely related to the degree of symmetry of the structure under investigation). This proved to be very important. The availability of approximate solutions enabled us to numerically describe a large set of nonlinear medium eigenfunctions in two dimensions. Configurations of some of them are depicted in Fig. 1 f.

The study of model (1) yielded some interesting generalisations, rigorous mathematical results and some new methods (Samarskii, 1984; Galaktionov et al., 1984). But the possibility of discovering principles by which one can construct a set of localised complex structures in a simple nonlinear medium appears of great importance. Knowing these principles provides us with a key for effective control of processes in the system.

Principles of structure organisation in trigger media

Since the paper by Turing (1952) concerning morphogenesis and up until now systems of reaction-diffusion type equations

$$u_t = D_1 u_{xx} + Q_1(u, v, \lambda) \qquad (4)$$

$$v_t = D_2 v_{xx} + Q_2(u, v, \lambda)$$

have been used to simulate a wide class of different phenomena in biology. The right hand sides of these equations usually depend on the external parameter λ; change in it allowing influence over the course of processes in the system.

For example let us consider the following class of nonlinear media. Let equations without diffusion ($D_1 u_x = D_2 v_x = 0$) have two stable $(\bar{u}_1, \bar{v}_1)$, $(\bar{u}_2, \bar{v}_2)$ $(Q_i(u_j, v_j) = 0,\ i, j = 1, 2)$ and one unstable solutions $(0, 0)$ $(Q_i(0, 0) = 0,\ i = 1, 2)$. Such systems have been called bistable or trigger systems. An opportunity to use trigger media to simulate morphogenesis mathematically has been widely discussed in recent years.

For simplicity let us suppose that Q_1 and Q_2 are odd functions. Thus $(\bar{u}_2, \bar{v}_2) = (-\bar{u}_1, -\bar{v}_1)$ (from this point we shall omit the indices). The solutions $(\bar{u}, \bar{v})$, $(-\bar{u}, -\bar{v})$ we shall call the stable background and $(0, 0)$ the unstable background. Functions u and v can be interpreted as deviations from equilibrium of some concentrations of the substances in a chemical reaction, thus u and v may be both positive and negative.

Let us see what happens to small perturbations against an unstable background (Fig. 2, t_1, t_2, t_3). The wave of 'self-construction' propagates across the medium, dissipative structures arising successively in the neighbouring domains. There are transition regions between two stable backgrounds. The transition can be called an elementary structure. When $t > t_3$ they prove to be effectively localised (practically they do not influence each other and are independent of boundary conditions.[1])

In contrast to the model of thermal structures, localisation here is based on leaks rather than sources. (In the vicinity of stable backgrounds Q_1 and Q_2 behave like leaks). The availability of localised elementary structures permits us to create more complex types of order in the medium (Kurdjumov et al., 1981).

The solution of the two-dimensional analogue of equation (4) in a square domain with no flux at the boundaries is presented in Fig. 2. In the one-dimensional case structures arise in the same way. But further evolution of structures is entirely different in the two- and one-dimensional cases.

From equation (4) stationary solutions appear which are inhomogeneous in space (Malinetskii, 1981). In the two-dimensional problem structures which arise begin to transform slowly. If, for example, the structure has a circular configuration, then its radius slowly diminishes with time. The smaller its radius, the smaller is its life time.

Such behaviour is caused by the properties of diffusion processes. In the one-dimensional problem both diffusion directions (x and -x) are equally possible, but in two dimensions a particle at the boundary of the circle is more likely to leave it than to penetrate within.

---

1) Recall that in the thermal structure model rigorous localisation is possible, when $T(x, t) = 0$ outside of G_L. Here we deal with effective localisation; one can point out the region G_ε, outside of which one of the inequalities $(u-\bar{u})^2 + (v-\bar{v})^2 < \varepsilon$ or $(u + \bar{u})^2 + (v - \bar{v})^2 < \varepsilon$ is valid.

Superficially it appears that it is not possible to create stationary (i. e. time-independent) structures in 2D trigger media[2] but even here it is possible to apply some arguments apart symmetry.

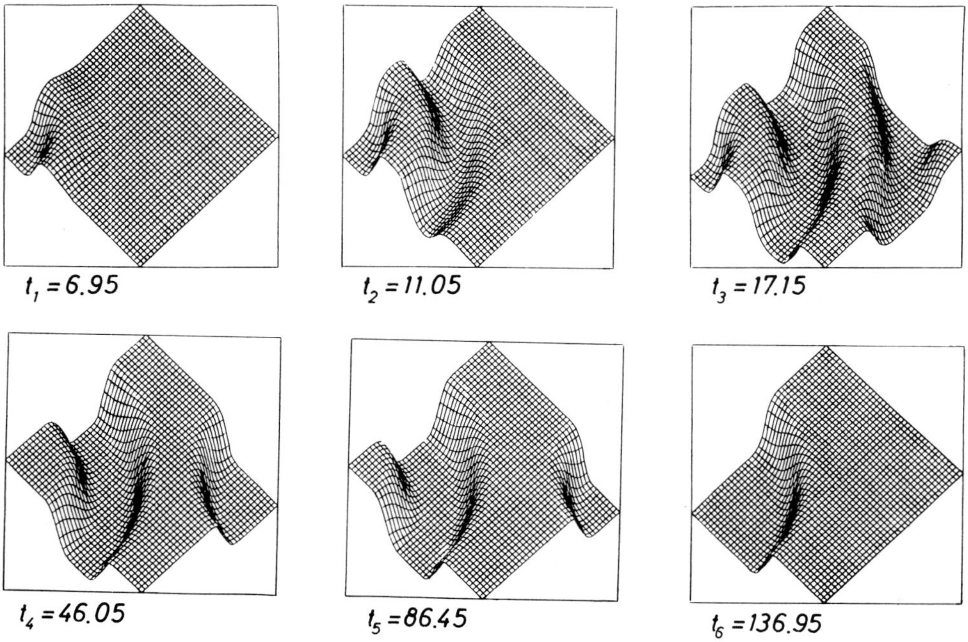

Fig. 2 Growth and subsequent disappearance of dissipative structures in the two-dimensional analogue of equation (4). The medium parameters are: $D_1 = 0.18$, $D_2 = 0.576$, $Q_1 = u - v + (4v - u)(u^2 + v^2)$, $Q^2 = u + v - (u + 4v)(u^2 + v^2)$.

The equation does not change when the spatial coordinates and functions change their sign: $(u, v) \rightarrow (-u, -v)$. Let us specify areas where $u > 0$ are black and where $u < 0$ are

2) Here we do not consider systems where diffusion instability develops against the background $(\bar{u}, \bar{v})$.

white. Then the evolution of the circular structure will appear as a diminishing white circle against a black background or a black circle against a white background.

Let us now put 'black' and 'white' into the same conditions, for example as depicted in Fig. 3 a. If the black squares diminish this would constitute a violation of the symmetry between 'black' and 'white', that exists in the equation.

Numerical calculations show that the 2D stationary structure, 'the Cross', can be constructed according to this principle (Fig. 3 a). The resonant perturbation, that results in the growth of stationary structures, is determined by symmetry while the specific details of the initial data are unimportant.

Other stationary structures also can be built up in this medium. For example, 'the Mill' and 'the Parquet' (Fig. 3 b, c). Note that the approximate methods of analysis developed for investigation of thermal structures prove to be efficient in the study of processes in trigger media. In particular they allow very accurate prediction of stationary distributions of components $u(x, y)$, $v(x, y)$ in configurations like 'the Cross'. Further studies of organisation laws for dissipative structures in trigger media are of great interest.

We have considered above two models that describe dissipative structures. It is possible that these will be useful in the creation of mathematical models of morphogenesis. In fact, features of self-organisation are reported in major papers in this field (Turing, 1952; Gierer, Meinhardt, 1972, 1974). Structures described by these models are closely related to the boundary conditions and the domain length l. While l increases, the number of structures may grow infinitely. But at the same time it is natural to expect that growth and development should stop at a definite spatial length characteristic of the given organism. This length must be defined by the internal properties of a nonlinear medium rather than by the boundary conditions which are characteristic of both the systems considered above.

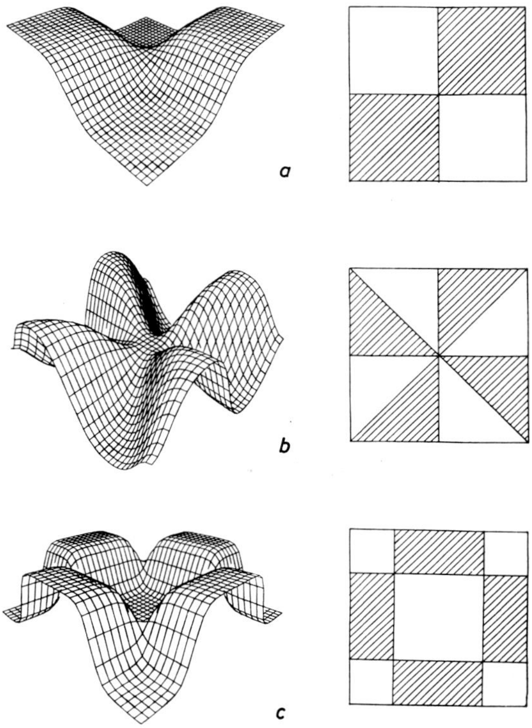

Fig. 3 Stationary structures in trigger medium. Parameters $D_1 = 2.5$, $D_2 = 0.5$, Q_1 and Q_2 are the same as in Fig. 2

Localised configurations and resonant perturbations in discrete systems

During the last few years discrete systems called cellular automata have been used to describe biological processes. A number of interesting applications have been proposed in the biology of growth (Alad'ev, 1981). Similar approaches are being developed for some ecological problems (Galitskii, 1985).

One of the most popular cellular automata is the game 'Life' proposed in the late Sixties by John H. Conway. This imitates growth, collapse and different changes in the population of living organisms. The rules of the game are as follows. An infinite flat lattice of square cells is considered. A cell may be alive or dead. Time in this model is discrete ($t = 1, 2, ...$). The change of cell state at the moment $t + 1$ is determined by the states of its neighbours (8 for each cell) at the moment t.

If the cell is dead at the moment t, it becomes alive at $t + 1$ when, and only when, three of its eight neighbours are alive at the same moment t.

If the cell is alive at the moment t, it dies at $t + 1$ when and only when fewer than two, or more than three, neighbour cells are alive at the moment t.

It has been shown that one can create arbitrary complex stationary and periodic configurations in this system. Definite combinations of cells in the system prove to be equivalent to a computer (Berlekamp et al., 1982). It was suggested that behaviour within simple automata may prove to be the key to understanding many complex natural phenomena (Wolfram, 1984; Lindenmayer, 1975).

Let us pay attention to two significant features of such systems - the existence of localised configurations of different types and an analogue of resonant perturbation.

As a rule, a randomly distributed set of live cells after a transition process becomes a set of localised configurations, that reproduce themselves at every step (for brevity let us call them stationary structures) or after n steps (let us call them n-cycles).

Convergence from different initial data to the same ordered configurations is a feature common to this system and to nonlinear dissipative media.

A set of simple stationary structures is depicted in Fig. 4 a. All of them are localised. Being separated by more than two cells they do not influence each other. It is apparent from the figure what the principles of construction of more complex stationary structures are that arise from these simple structures. In addition any set of localised stationars situated in different parts of the plane, some can be 'extended' in a diagonal direction or in a more complex fashion (Fig. 4 a).

In the discrete system discussed there is an interesting phenomenon - the spontaneous appearance of symmetry. Whilst developing a configuration may become spatially symmetric (for example with respect to a horizontal or vertical axis), and this symmetry does not vanish during evolution.

One can find several analogues of resonant perturbation in the game 'Life'. It may be an initial configuration which gives a permanently growing 'community' of cells (for example, 'the Glider Gun' (Berlekamp et al., 1982) or 'the Puffer-train' (Gosper, 1984)). It may be communities with complex evolution that have the fewest cells.

A resonant perturbation may be the result of collision between 'Gliders' - the simplest moving structures - and stationars (Fig. 4 b).

Collisions often result in disintegration of 'Glider' and 'target' or appearance of simple stationars. The resonant collision depicted in Fig. 4 b results in an increasing number of live cells, the appearance of several stationars of different types, and four 'Blinkers' ('the Blinker' is a figure of three cells in a line, at every step it turns through 90°).

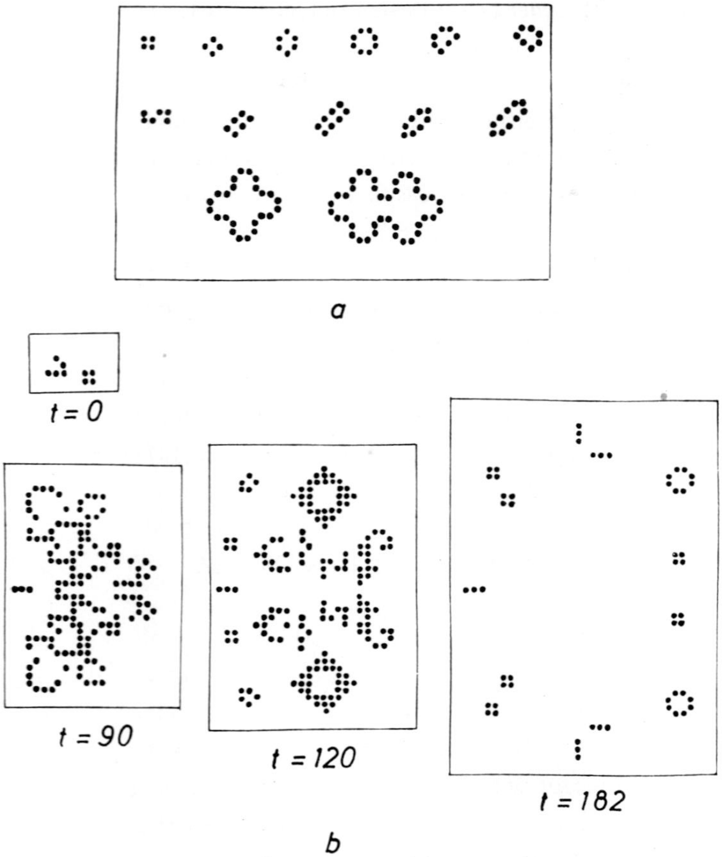

Fig. 4 see text

With only 'Gliders' and stationars one can make very complex orders. Thus the 'Life' game (and many other cellular automata) provides examples of localised structures, resonant perturbation, and examples of the appearance of complex structures on the basis of simple structures. Many profound properties are independent of whether we consider nonlinear dissipative media or different classes of discrete systems.

The complex temporal order and diffusion-induced chaos

Let us return to 'reaction-diffusion' systems. Studies of a number of biological phenomena led to the problem of the behaviour of solutions for different λ values. The following situation is a general case.

If $\lambda < \lambda_o$ equation (4) has a spatially homogeneous stationary solution that is often called the thermodynamic branch. When $\lambda > \lambda_o$ it loses stability and as a result stationary dissipative structures are formed or oscillations begin. To describe the behaviour of two-component systems (4) in the neighbourhood of the bifurcation point λ_o the Japanese scientists Y. Kuramoto and T. Tsuzuki (1975) proposed the equation

$$W_t = W + (1 + ic_1)\, W_{xx} - (1 + ic_2)\,|W|^2\, W, \qquad 0 \leqslant x \leqslant 1$$

$$W(x, 0) = W_0(x), \qquad W_x(0,t) = W_x(l,t) = 0 \qquad (5)$$

Here W is a complex funtion: $W(x,t) = u'(x,t) + iv'(x,t)$, its real and imaginary parts determine the behaviour of the solutions of the system (4). (For the simplicity of notation we shall omit primes.) An approach close to this has been developed in the study of models of morphogenesis (Haken, 1980). Investigation of equation (5) is related to the problem of classification of two-component systems (Akhromeyeva et al., 1984, 1986). Here the temporal order of processes rather than spatial configuration of dissipative structures is of interest.

The study of equation (5) for small l ($l \leqslant \pi$) led to the creation of a hierarchy of simplified models. Let us discuss some of them.

The calculations performed have shown that in small domains usually the zeroth and the first Fourier coefficients of the solution have the greatest amplitude while others are at least an order of magnitude less. In this case it is possible to propose a simplified model relating the amplitudes of the zeroth and the first harmonics ρ_0, ρ_1 and the phase shift between them:

$$\dot{\rho}_0 = \rho_0(1 - \rho_0^2 - \rho_1^2) - \rho_0\rho_1^2(\cos 2\psi + c_2 \sin 2\psi)/2$$

$$\dot{\rho}_1 = \rho_1(1 - \kappa^2 - 2\rho_0^2 - 3\rho_1^2/4) - \rho_0^2\rho_1(\cos 2\psi - c_2 \sin 2\psi) \tag{6}$$

$$\dot{\psi} = c_2(\rho_0^2 - \rho_1^2/4) + \sin 2\psi\,(\rho_0^2 + \rho_1^2/2) + c_1\kappa^2 + $$

$$+ c_2 \cos 2\psi\,(\rho_0^2 - \rho_1^2/2), \qquad \kappa = \pi/l$$

The behaviour of the solutions when $t \to \infty$ is determined by the so-called attracting sets or attractors. The simplest attractors are: the singular point ($\rho_0 \to$ const, $\rho_1 \to$ const, $\psi \to$ const) and the limit cycle (functions ρ_0, ρ_1, $\dot{\psi}$ have the same period). Let us denote the cycle that makes n revolutions per period on the plane (ρ_0, ρ_1) by S^n.

It is necessary to pay attention to the important differences between the system (6) and the well-known Lorenz model. Equations (6) have no special symmetries with respect to the signs of functions. They can reproduce both qualitative and a number of quantitative characteristics of the solutions of partial differential equations.

The system can have not only singular points and cycles, but also strange attractors - complex nonperiodical solutions. One of them is depicted in Fig. 5 a. This strange attractor has many interesting properties. It has been thoroughly studied as described in (Akhromeyeva, Malinetskii, 1985). Its probability dimension proved to be $D \approx 2.05$. This means that one can turn to a simpler description, related to one-dimensional maps,

$$M_{n+1} = f(M_n). \tag{7}$$

One can take for M_n the n-th local maximum of the $\varrho_0(t)$ function. Such models are useful in some other biological applications. It is important to emphasise that the analysis of the problem (5) also leads to one-dimensional maps like equations (6) do. Some maps obtained as a result of the numerical solution of the partial differential problem are shown in Fig. 5 b. The existence of those maps permits us to discover what is a scenario for transition from the ordered solutions to chaos as parameters change (Akhromeyeva et al., 1985).

The stochastic regime in equation (5) is of great interest. With no diffusion terms the system behaves regularly whilst with the presence of diffusion terms stochastic regimes result. This phenomenon has been called diffusion-induced chaos. It is interesting to investigate it in biological systems.

Different types of order, found in the simplified models, can also be sought in the original problem. For example, it can be proved that the analogue of a singular point in system (6) is the self-similar solution

$$W = R(x) \exp(iwt) + ia(x)). \tag{8}$$

This describes a periodic process where the function $R = (u^2 + v^2)^{1/2}$ does not depend on time. Simple and complex cycles S^n also have analogues in the original problem.

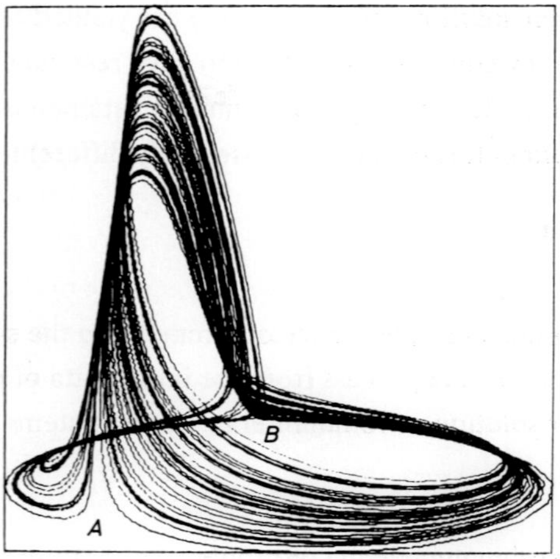

a

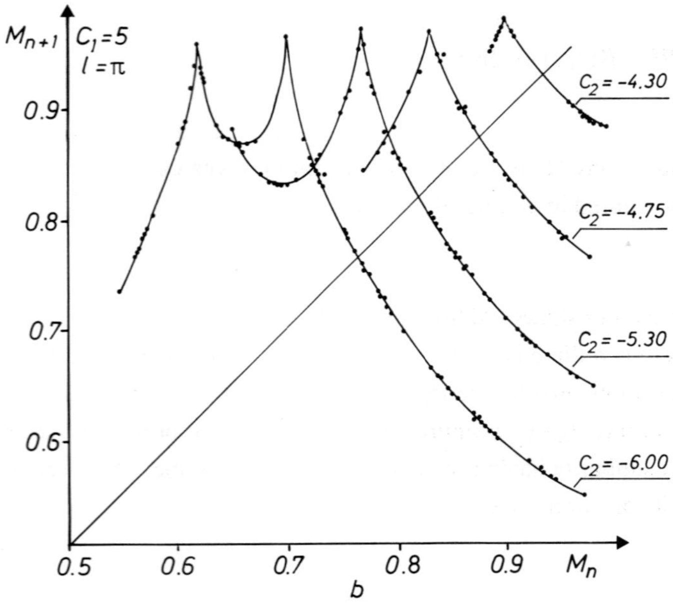

b

Fig. 5 a - Strange attractor in system (6) for $C_1 = 7$, $C_2 = -6$, $k = 1$
b - Change of one-dimensional maps in the partial differential problem as a function of parameter C_2, $C_1 = 5$, $l = \pi$.

The study of self-similar solutions of equation (5) has yielded two interesting results. The systems described by equation (5) proved to have resonant perturbations that were different classes of initial data with spatial symmetry retained during evolution. The behaviour of the system in this case can be essentially different from the general situation.

Along with this a spontaneous appearance of symmetry in the system can take place. In many cases when $c_1 \lessapprox 1$ the process from the initial data of general type converges to an even self-similar solution - two noninteracting subsystems arise.

Study of the two-dimensional analogue of problem (5) has unexpected results. It proves that in this case there are also self-similar solutions of the type

$$W = R(x, y) \exp(iwt + ia(x, y)). \tag{9}$$

For the proper functions R and a it describes spiral waves - an important type of order that is now studied in biological systems.

The analysis carried out shows (Akhromeyeva et al., 1986) that in addition to the self-similar solutions in two-dimensional dissipative systems more complex types of order must exist. For example, two-frequency regimes when $R(x, y, t)$ for one such solution is depicted in Fig. 6 (here $t_6 - t_1$ is approximately equal to the period of the function R). It is possible that similar behaviour can be observed in a number of experiments after the loss of stability of spiral waves.

'Forgetting' the initial data proves to be characteristic of equation (5) and many other systems where solutions converge to the attractor. According to the Second Law of Thermodynamics, entropy increases in closed systems, any order, present at the beginning, vanishing. In open systems the 'goal' of processes may be different:

dissipative structures form, complex periodic or stochastic regimes appear. Moreover, attractors of different types co-exist, and this means that evolution of the same system may be qualitatively different according to different initial data and other influences.

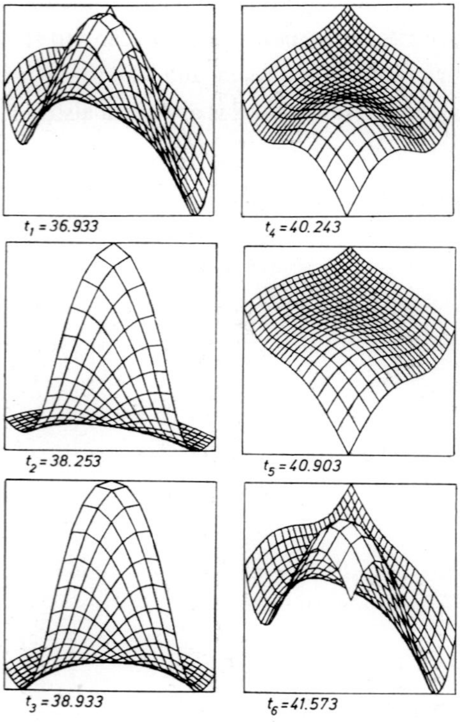

Fig. 6 Change of the function $R(X, Y, t)$ for $C_1 = 3$, $C_2 = 3$, $l = \pi$.

To 'create' definite organisation one may not 'give detailed instructions' to the system, but just transfer it to the domain of convergence of a corresponding attractor. The phenomenon of regeneration, methods of pattern recognition and intuition, and of other phenomena in biological systems may be closely related with this behaviour.

The investigation of the internal properties of nonlinear media, the principles of organisation of dissipative structures, new types of order and stochastic regimes prove to be very useful in many problems in physics, and lead to the formulation of profound mathematical problems. It is possible that this approach also will be useful for the study of biological systems.

Chirality Wave Propagation and the First Evolutionary Catastrophe

Ya. B. Zeldovich, A. S. Mikhailov

Life on earth is based upon 'left' molecules (L-enantiomers) of alpha-amino acids and 'right' (D) molecules of sugars. This homochiral biochemistry provides an important evolutionary advantage because of the kinetic efficiency and product economy of stereoselective single-enantiomer reactions. However, in the nonorganic nature the L- and D-enantiomers of these same substances are found in almost equal proportions (nonconservation of parity in weak interactions shifts the balance only to the extent of 10^{-17} (Kizel, 1985; Mason, Tranter, 1985; Zeldovich et al., 1977; Khriplovich, 1981). Selection of particular chiralities for sugars and alpha-amino acids was the first global symmetry breaking transition (or *catastrophe*) in the history of biological evolution.

Our direct knowledge of the early stages of biological and pre-biological evolution is still very poor. In some respects the situation is similar to that in cosmology. In an attempt to reconstruct the earliest evolutionary stages, all feasible scenarios must be thoroughly investigated and their consequences found.

The most radical hypothesis was proposed by P. Jordan (see Chubnikov, Koptsik, 1972). It asserts that all living organisms are the descendants of a single cell or a group of cells

Thermodynamics and Pattern Formation in Biology

brought to Earth by some miraculous event. Obviously, the assumption of unique origination is logically admissible and solves the whole problem. However, such a scenario seems to be extremely improbable. Evolutionary biologists believe (Oparin, 1968; Buvet, Ponnamperuma, 1971) that many sources of life were created simultaneously here on Earth, as a natural result of some complicated chain of chemical and physical processes in the primal ocean or in clays.

Chiral symmetry of an inorganic nature implies that sources of life, dependent on the opposite chiralities, should have been generated in equal proportions (a 10^{-17} asymmetry being negligible compared with statistical fluctuations). There were two principal factors that gave rise to competition between L and D forms of life and led to the final elemination of one of them in the history of evolution on Earth. First of all, there should be nonspecific competition for certain common achiral resources. Moreover, as modern biochemical data indicates (Kizel, 1985), the intermediate and final products of L species would be poisonous for D species, and *vice versa*. The combined effect of both factors is the mutual inhibition of L and D species.

The intensity of mutual inhibition in the earliest stages of evolution is not known, though such knowledge is of cardinal importance, because it can be used to make the choice between two different scenarios.

Weak mutual inhibition does not prevent coexistence of L and D species. This can be seen in the simplest phenomenological model (Frank, 1953) which takes into account the processes of self-replication and the effect of mutual inhibition

$$\dot{X}_L = k_1 X_L - k_2 X_L^2 - k_3 X_L X_D$$

$$\dot{X}_D = k_1 X_D - k_2 X_D^2 - k_3 X_D X_L \qquad (1)$$

Here X_L and X_D are the total numbers of L and D self-replicating objects, k_1 is the rate of self-replication, k_3 specifies the intensity of mutual inhibition. The nonlinear terms $- k_2 X_L^2$ and $- k_2 X_D^2$ are added to account for suppression of infinite exponential growth of L and D populations.

Stability analysis of the steady states of (1) indicates that for $k_3 > k_2$ coexistence is impossible and either the L or the D species disappears, depending on the initial ratio X_D/X_L. On the contrary, for $k_3 < k_2$ the symmetric state is stable, such that $X_L = X_D = k_1 / (k_2 + k_3)$, while the pure asymmetric states ($X_L = k_1 / k_2$, $X_D = 0$ or $X_L = 0$, $X_D = k_1 / k_2$) are unstable with respect to small fluctuations.

One set of scenarios (Morozov, 1979; Morozov et al., 1983; Kondepudi, Prigogine, 1981; Kondepudi, Nelson, 1983) assumes that, at the initial stage, mutual inhibition was sufficiently weak, and a spatially homogeneous symmetric state with coexisting L and D species was formed. Later, the environment gradually started to change and the effective intensity of inhibition increased; this forced a symmetry breaking transition (influenced, either by statistical fluctuations or by bias in the electroweak interactions) and this resulted, finally, in the extinction of the species that rely upon the opposite chirality of alpha-amino acids and sugar. Moreover, it is assumed that spatial homogeneity was preserved in the process of transition.

An alternative scenario is proposed in the present paper. It, too, is based on the idea of multiple, and approximately simultaneous, creation of life sources at opposite chiralities. We assume, however, that the intensity of mutual inhibition was sufficiently high to prevent homogeneous coexistence of L and D species, even at the initial stage of evolution. Under such conditions the system tends to become divided into separate domains, occupied by two different 'phases' with domination by L or D species. According to this scenario, two chiral forms of life coexisted for some period of time, occupying *separate* areas on the terrestrial surface. If this was so, the final selection of chirality can be seen to be a result of competition between L and D forms along the boundaries of contact between their areas (i. e. along the interphase boundaries).

The investigation, which is performed below, leads to the following principal conclusions:

a) When the rates of self-replication of L and D species are the same, the outcome of evolution depends on the initial geometry of the pattern, formed by the L and D areas on the terrestrial surface. In particular, we show that any closed area, completely surrounded by a region that is occupied by the form of life with the opposite chirality, would collapse with time. Generally, the border of contact between two areas tends to minimise its length.

b) The asymmetry of electroweak interaction results in the slightly higher rate of self-replication for L species. If the boundary between L and D areas is perfectly flat, it would slowly move into the region occupied by D species, so that the L area would expand. However, this effect is vanishingly small. The smallest curvature conceivable for the interphase boundary (e. g. a radius of curvature about that of the radius of the Earth) would completely destroy the effects of electroweak interactions.

We consider the following model that generalises (1) by including diffusion:

$$\begin{aligned} \dot{X}_L &= k_1 (1 + \varepsilon) X_L - k_2 X_L^2 - k_3 X_L X_D + D \Delta X_L \\ \dot{X}_D &= k_1 (1 - \varepsilon) X_D - k_2 X_D^2 - k_3 X_L X_D + D \Delta X_D \end{aligned} \qquad (2)$$

Here D is the diffusion constant; the same for both species; the small coefficient $\varepsilon \sim 10^{-17}$ takes into account the slight difference in replication rates for L and D species because of asymmetry of the electroweak interactions.

Quantitative estimates of the effects of these weak interactions are available only for the ground state energies of alpha-amino acid (Mason, Tranter, 1985). It seems plausible, however, to admit that the greater stability of L alpha-amino acids should result in higher replication rates for the L species. Moreover, since the effects are extremely small, the relative difference in replication rates would be of the same order of magnitude as the relative difference in the ground state energies.

It is convenient to introduce new dimensionless variables n and m (the population densities, t (time) and coordinates) as

$$n = (k_2/k_1)\, X_L , \qquad m = (k_2/k_1)\, X_D$$

$$r = (k_1/D)^{1/2} r', \qquad t = k_1 t'. \tag{3}$$

Equations (2) then take the form

$$\dot{n} = (1 + \varepsilon)n - n^2 - bnm + \Delta n,$$

$$\dot{m} = (1 - \varepsilon)m - m^2 - bnm + \Delta m, \tag{4}$$

where the coefficient $b = k_3/k_2$ characterises the intensity of mutual inhibition. Furthermore, we assume this inhibition to be very strong ($b >> 1$).

First we study the one-dimensional problem. Suppose that the rates of self-replication of both species are equal (i. e. $\varepsilon = 0$). Then the stationary solution of (1) can be found, such that the L species dominate in the region $x < 0$, while the region $x > 0$ is dominated by the species of opposite chirality. In the case of infinitely strong mutual inhibition ($b \to \infty$) densities n and m should vanish at the point $x = 0$ where the two populations meet (Fig. 1 a). The distribution of density satifies the equation

$$m'' + m - m^2 = 0 \tag{5}$$

with the boundary conditions $m(0) = 0$ and $m \to 1$ as $x \to \infty$. Its solution is

$$m_0 = 1/2 \, \frac{(\sqrt{3}\, \tanh(x/2) + 1)^2}{(1/\sqrt{3}\, \tanh(x/2) + 1)^2} - 1, \qquad x > 0 \tag{6}$$

It follows from (6), that $m'_0(0) = 1/\sqrt{3}$. Note also that $n_0(x) = m_0(-x)$.

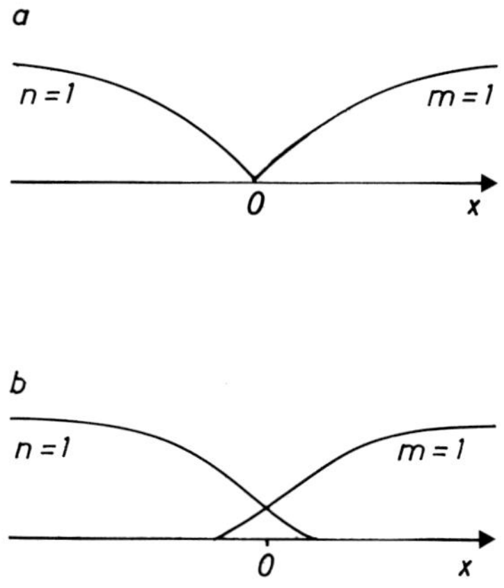

Fig. 1 a, b See text.

When b is finite, the two populations interpenetrate (Fig. 1 b). By using the scaling considerations it can be shown that the width of the interpenetration layer, where $n \sim m$, is $\delta x \sim b^{-1/3}$ and $n(0) = m(0) \sim b^{-1/3}$. It is interesting to note that almost the same estimate was found many years ago by one of the authors in a study of combustion of nonmixed gases (Zeldovich, 1949).

If the rate of replication of the L species is higher ($\varepsilon > 0$), the boundary between two populations would slowly move so that the region occupied by the more effective L species, would slowly grow at the expense of the area of the D species. We interpret the motion of this boundary as the propagation of the *chirality wave.*

The solution, in the form of a chirality wave, can be found by substitution of $n = n(x - st)$ and $m = m(x - st)$, where s is the propagation speed, into (4). Below we find that for small ε the propagation speed, s, is proportional to ε, and obtain an explicite expression for the speed of propagation of the chirality wave.

It is now useful to introduce the combinations $v = m - n$ and $w = m + n$. With these new variables (4) become

$$v'' + (1 - w)\, v = \varepsilon w - sv',$$

$$w'' + w - (1/2)\, b\, (w^2 - v^2) = \varepsilon v - sw', \tag{7}$$

where the derivatives are taken in respect to $\xi = x - st$. Since $\varepsilon << 1$, we can put $v = v_0 + v_1$, and $w = w_0 + w_1$, where $w_0(\xi)$ and $v_0(\xi)$ are the stationary solutions ($s = 0$) of (7) for $\varepsilon = 0$, and expect that $|v_1| << |v_o|$, $|w_1| << |w_o|$, $s << 1$.

In linearised form (7) can be written as

$$\hat{L}\phi = \Psi \tag{8}$$

with the two-component vectors

$$\phi = \begin{pmatrix} v_1 \\ w_1 \end{pmatrix} \qquad \Psi = \begin{pmatrix} \varepsilon w_0 - s v_0' \\ \varepsilon v_0 - s w_0' \end{pmatrix} \tag{9}$$

and the linear operator

$$\hat{L} = \begin{pmatrix} d^2/d\xi^2 + (1 - w_0), & -v_0 \\ b v_0 & , d^2/d\xi^2 + (1 - b w_0) \end{pmatrix} \tag{10}$$

Let us consider the eigenvalue problem

$$\hat{L}\phi_j = \lambda_j \phi_j \tag{11}$$

It can be noted that the same linear operator $\hat{L}$ describes the evolution of small perturbations of the stationary solution $v_0(\xi)$ and $w_0(\xi)$ for $\varepsilon = 0$. Since the position of the boundary between the two populations is arbitrary, the shift mode $\phi_0 = (v'_0; w'_0)$ should correspond to the eigenvalue $\lambda_o = 0$. But this implies that the decomposition of the vector Ψ in (8) cannot include the term which is proportional to ϕ_0 (see also Barenblatt, Zeldovich, 1957, 1959).

Since the operator $\hat{L}$ is not self-adjoint ($\hat{L}^+ \neq \hat{L}$), its eigenfunctions $\phi_j(\xi)$ cannot be orthogonal. However, if we take the eigenfunctions $\phi^j(\xi)$ of the adjoint operator $\hat{L}^+$,

$$\hat{L}^+ \phi^j = \lambda_j^* \phi^j, \tag{12}$$

then these would be orthogonal to $\phi_j(\xi)$, i. e. $(\phi^j, \phi_{j'}) \sim \delta_{jj'}$. Hence, if we multiply both sides of (8) by the eigenfunction ϕ^0, such that $\hat{L}^+ \phi^0 = 0$, we obtain

$$(\phi^0, \Psi) = 0. \tag{13}$$

This equation determines the speed of chirality wave propagation:

$$s = \frac{\varepsilon \int_{-\infty}^{\infty} (f w_o + g v_o)\, d\xi}{\int_{-\infty}^{\infty} (f v'_o + g w'_o)\, d\xi}, \tag{14}$$

where f and g are components of the vector ϕ^0.

Functions $f(\xi)$ and $g(\xi)$ satisfy equations

$$f'' + (1 - w_0)\, f + b\, v_0\, g = 0,$$

$$g'' + (1 - bw_0)\, g - v_0\, f = 0 \tag{15}$$

If we divide both equations by b, introduce the new function $\tilde{f} = f/b$, and take into account that $\min w_0 \sim b^{-1/3} >> b^{-1}$ for $b >> 1$, (15) can be transformed as

$$f'' + (1 - w_0) f + v_0 g = 0, \tag{16}$$

$$(1/b) g'' - w_0 g - v_0 f = 0.$$

When $b >> 1$ (16) includes two different length scales. The first of them is about one, while the second is about $b^{-1/3}$. For $|\xi| >> b^{-1/3}$ we have

$$\tilde{f} = \begin{cases} m_0'(\xi), & \xi > 0, \\ m_0'(-\xi), & \xi < 0, \end{cases} \qquad g = \begin{cases} \tilde{f}, & \xi > 0 \\ -\tilde{f}, & \xi < 0 \end{cases} \tag{17}$$

Inside the interval $|\xi| < b^{-1/3}$ the function $\tilde{f}(\xi)$ remains constant ($\tilde{f} = 1/\sqrt{3}$), while $g(\xi)$ varies from $+1/\sqrt{3}$ to $-1/\sqrt{3}$. Taking this into account, we find in the limit $b \to \infty$.

$$s = \varepsilon \int_0^\infty m_0 m_0' \, d\xi \Big/ \int_0^\infty (m_0')^2 \, d\xi \tag{18}$$

and, after simple integration, obtain

$$s = \frac{5\sqrt{3}}{2(3\sqrt{3} - 2)} \varepsilon \approx 1.3\,\varepsilon. \tag{19}$$

Next we investigate the two-dimensional problem. Here the boundary, separating two areas, can be curved and, as we shall see, the speed of propagation would depend on the curvature. If the curvature radius R is large ($R >> 1$), motion of the boundary is described by the equations

$$v'' + v'/R + (1 - w)\, v = \varepsilon w - s\, v',$$

$$w'' + w'/R + w - (1/2)\, b\, (w^2 - v^2) = \varepsilon\, v - s\, w', \tag{20}$$

that reduce to (7) after replacement $s \to s + 1/R$. Therefore, we can immediately obtain the final result[1)]

$$s = \frac{5\sqrt{3}}{2(3\sqrt{3} - 2)}\, \varepsilon - (1/R). \tag{21}$$

In the original dimensional units we have[2)]

$$s = \frac{5\sqrt{3}}{2(3\sqrt{3} - 2)} \sqrt{k_1 D \varepsilon} - (D/R). \tag{22}$$

First we estimate the effects of curvature. Suppose that we have a region of radius $R = 10^5$ cm, occupied by the L species and surrounded by the area of D species. If we take a diffusion constant $D = 10^{-5}$ cm^2/sec, which is typical for liquids and liquid solutions, (22) gives us $s \sim 10^{-10}$ cm/sec for $\varepsilon = 0$. This means that because of the effects of curvature such a region would collapse within the time interval $T \sim 10^{15}$ sec or about one million years.[3)] Although the corresponding speed of motion of the boundary between two areas is extremely small, the contribution toward the

---

1) Equation (21) is valid if the boundary is convex toward the area of D species, where $m = 1$. If the boundary is convex away from this region, the last term in (21) would be positive. Note, that if the speed s, given by (18), turns out to be negative, this means that the boundary moves into the area of L species.

2) Note that the same result would be valid for three-dimensional media, if we replace $1/R$ here by the sum $1/R_1 + 1/R_2$ where R_1 and R_2 are the prinicpal radii of curvature of the interphase surface.

3) This estimate holds if prebiological evolution had developed in clays. For the primal ocean turbulent mixing should have been important. Small-scale turbulent mixing results in the effective diffusion enhancement and can be taken into account by substitution into (22) of a renormalised diffusion constant.

propagation speed due to effects of weak interaction would be much smaller. Even if we assume that, in absence of competition, the populations double every second (i. e. if we put $k_1 = 1\ sec^{-1}$), the contribution from the first term in (22) would be $s \sim 10^{-19}$ cm/sec for $\varepsilon = 10^{-17}$. Smaller rates of replication lessen the contribution. It can be easily verified that for all reasonable radii of curvature R (up to a value comparable with the radius of the Earth globe) the effects of curvature grossly predominate over the consequences of parity violation in weak interactions.

Nonlinear Bioenergetics

A. S. Davydov

Introduction

The most active role in the cell bioenergetics is played by protein molecules. They are closely connected with the basic manifestations of life. All chemical processes in the cell take place with the participation of proteins, i. e. enzymes. Proteins transform chemical energy into mechanical energy and are responsible for cellular and intracellular movement. Proteins in complex association with lipids in cellular and intracellular membranes ensure the transport of substances and ions into and from the cell. They take part in the process of providing energy for all the needs of the living organism.

All proteins are polymeric molecules with a very high molecular weight. They are formed by polymerization of various amino acids. The polymerization is accompanied by the formation of water molecules; a hydrogen atom being separated from the amino group of one amino acid and the hydroxyl group from the carboxyl group of another one. At the same time the nitrogen and the carbon of amino acid residues join together to form a peptide bond. Given the appropriate enzymes and energy this process of

Thermodynamics and Pattern Formation in Biology

polymerization may be repeated producing long polypeptide chains (proteins) with repeated groups of four atoms HNCO called peptide groups (PG's).

Proteins take different spatial configurations determined by interactions between the elements of the peptide chain and the aqueous environment. One of the most important and interesting spatial protein structures is that of alpha-helix established by Pauling and Corey (1951).

Vibrational energy transport along protein molecules

It has been established that the universal unit of energy exploited by protein molecules for various kinds of work (movement, ion transfer, etc.) is the energy released under hydrolysis of adenosine triphosphate (ATP) molecules synthesised in specific organelles of the cell.

ATP hydrolysis is the universal intermediate step in an enormous number of energy transformations in the cells of all living organisms from the simplest bacteria to human beings. The hydrolysis takes place under the influence of enzymes and coenzymes (metal ions) at some sites in the protein molecule incorporated into ionic pumps, muscular fibres, etc. As a rule, the released energy is used in some other parts of these molecules. In this connection there arises an important problem concerning the mechanism of energy transfer of the hydrolysis of ATP molecules along large protein molecules.

The 1973 meeting of the New York Academy of Sciences was devoted to the elucidation of the mechanism of energy transfer in biological systems. The three main problems discussed were

1. Is there a crisis in bioenergetics?
2. What is the nature of the crisis?
3. How can the crisis be resolved?

The free energy released by the hydrolysis of ATP under normal physiological conditions is comparatively small - about 0.5 eV ($\sim$ 50 kJ). It is only twenty times greater than the average thermal energy. Consequently, this energy cannot be transferred by electronic excitation which requires 3 to 5 eV.

Recognising the small amounts of energy released by the hydrolysis reaction some participants at the conference (Geen, 1973) stuck to the view that the vibrational energy of carbon monoxide (C=0), incorporated into peptide groups of all proteins, is transferred along protein molecules in biological systems. These vibrations are called Amide 1.

Weber et al. disputed this opinion asserting that due to the short lifetime (of the order of 10^{-12} sec) of vibrational Amide 1 of individual peptide groups they cannot take part in the excitation transfer.

The supporters of the view that the energy is transferred by vibrational Amide 1 refuted this objection by arguing that a special structure of protein molecules considerably enhanced the lifetime of such excitations. However, they failed to supply convincing proofs to support their opinion. Thus, the problem of the crisis in bioenergetics remained open.

Theoretical investigations conducted at the Institute for Theoretical Physics, Academy of Sciences of the Ukrainian SSR (Davydov, Kislukha, 1973; Davydov, 1974, 1982 a, 1982 b, 1984) were directed toward a solution of this problem. It was shown that the energy of the hydrolysis of ATP can be transferred without loss along alpha-helical protein molecules as special collective excited states, i. e. solitons.

All protein molecules are formed by a linear polymerization of 20 different amino acids and represent very long polypeptide chains. In alpha-helical proteins the polypeptide chains are coiled into long helices. The structure is due to additional hydrogen bonding between the PG's in additon to the basic chemical link. There are three chains of hydrogen bonds in each helical protein molecule. The PG's are situated along the three chains at equal distances (a) between each other, forming periodic structures. The energy of a hydrogen bond is approximately an order of magnitude less than that of ordinary chemical bonds. Therefore, the displacement of PGs from their equilibrium positions along the chains of hydrogen bonds are much more easily fulfilled than are their displacements along the main peptide chain.

The quasi-periodic structure of an alpha-helical protein molecule provides the collectivization of vibration Amide 1 of individual PG's. This vibration has an energy about 0.21 eV and a comparatively large electric dipole moment (about 0.3 Debye) directed along the helix axis, ensuring a large resonance interaction between PG's. Owing to the dipole resonance interaction the vibrations Amide 1 of individual PG's are collectivized. As was shown by Davydov et al. (1973-1984) it is very important to take into account the displacements of equilibrium positions of PG's along the chain of hydrogen bonds in a theoretical description of collective excitation.

If we consider the independent excitation of each chain of PG's in protein molecules, then the collective excitation corresponding to the vibrations C=0 in PG's spreading along the chain with constant velocity V will be described by a nonlinear Schrödinger equation. This involves two types of collective excitations: excitons and solitons.

The excitons are collective excitations spreading along the chain with a group velocity exceeding the velocity V_0 of longitudinal sound waves. They are described by wave packets smeared with respect to time. Local deformation cannot follow the rapid motion of an exciton. Therefore, the exiton carries along the chain only the energy of an intrapeptide vibration which can be expressed by the formula

$$E_{ex}(V) = E_{ex}(0) + 1/2\, m_{ex} V^2. \qquad (1)$$

Here $E_{ex}(0)$ is the internal energy of an exciton which is slightly smaller than the energy of an isolated vibration Amide 1. The second term in (1) determines the exciton kinetic energy which is proportional to the square of the velocity. The proportionality coefficient, m_{ex}, is called an effective exciton mass. This is inversely proportional to the energy of resonance interaction, J, between the neighbouring PG's in the chain. Whilst in movement the excitons generate phonons, therefore, they are rapidly decelerated.

The solitons are collective excitations formed as a result of the combination of intrapeptide vibrations Amide 1 with a local deformation of the chain. They always move with a velocity V less than that, V_o, of sound. The energy of a soliton travelling along the chain with velocity V ($< V_o$) is expressed by the formula

$$E_{sol}(V) = E_{sol}(0) + 1/2\, m_{sol} V^2. \qquad (2)$$

It is very important that the internal soliton energy $E_{sol}(0)$ is less than the internal exciton energy $E_{ex}(0)$:

$$E_{sol}(0) = E_{ex}(0) - \chi^{4}/3\, \varkappa^{2} J. \qquad (3)$$

Here the parameter χ determines the interaction of the Amide 1 vibrations with the displacement of equilibrium positions of PG's in the deformation region; $\varkappa$ is the chain longitudinal elasticity coefficient. Hence because of the connection between intrapeptide excitations (Amide 1) and local deformation of the chain their total energy decreases.

The multiplier m_{sol} entering expression (2) can be called the effective mass of a soliton

$$m_{sol} = m_{ex}\,(1 + 2\,a^2 \cdot \chi^{4}/3\ \kappa^{3}\ \hbar^2) \tag{4}$$

This mass in soft chains (small κ) significantly exceeds that of an exciton mass, m_{ex}, since soliton motion is accompanied by a local deformation of the chain. The soliton, with a large mass can transfer great kinetic energy at small velocities.

The probability distribution of the vibrational excitation on PG's is expressed by a bell-like function

$$W(x, t) = aQ/\{2 \cosh^2 [Q(x - x_o - Vt)]\} \tag{5}$$

where

$$Q = \chi^2 (2\,a^2\,\kappa\,J)^{-1}. \tag{6}$$

According to (5), the maximal probability corresponds to the excitation of a peptide group with the coordinate $x - x_o - Vt$, travelling along the chain with velocity V. The region of the chain embraced by an excitation travelling with velocity V has the approximate size $x = 2\pi/Q$, which does not change with time. In the region embraced by an excitation the distances between PG's decrease by the values

$$\Delta a(x, t) = a\,\chi Q\,\{\kappa \cosh^2 [Q(x - x_o - Vt)]\}^{-1}. \tag{7}$$

The presence of the energy gap (3) in the spectrum of the excited states between exciton and soliton proves to be one of the reasons for very high stability of solitons. In order to destroy a soliton, i. e. to split it into a free exciton and a local deformation which then relaxes into thermal motion, it is necessary to expend an energy $E_{ex}(0) - E_{sol}(0)$.

Since solitons always move with velocity less than that of sound along the chain they do not emit phonons. In other words, their kinetic energy is not transformed into the energy of thermal motion. This is the second important characteristic that ensures the high stability of solitons in soft molecular chains of the type of alpha-helical protein molecules.

Finally, the third causative factor in the stability of solitons is their topological stability. When solitons move along the chain, a displacement of the equilibrium position of PG's takes place. To the right of a soliton $x > x_o + Vt$ all PG's are in undisplaced positions, na , and to the left ($x < x_o - Vt$) they are displaced by the value $4 \chi / \kappa$. In order to destroy a soliton it is necessary that all the left-hand PG's of the chain must be returned to their initial positions.

As is known, at the moment of light absorption by molecular systems the coordinates of heavy particles cannot be displaced (the Franck-Condon principle). Since the formation of a soliton is associated with the displacement of equilibrium positions of heavy PG's, the probability of soliton excitation by light is very small. Solitons may be excited by local effects. One such effect is a chemical reaction. The probability of the excitation is highest when such a local effect takes place at the end of a molecular chain.

Solitons in real alpha-helical protein molecules

In the formation of the alpha-helical structure of protein molecules there appear three chains of PG's. Therefore in a mere detailed theoretical description of excited states of the molecule it is necessary to take into account the resonance interaction between PG's of different chains. This more complex mathematical model has been described by Davydov (1982 b, 1985).

The solutions of the equations that account for the resonance interaction between neighbouring PG's in a single chain and between neighbouring chains and changes in the pitch and diameter of the helix have shown that intrapeptide vibration Amide 1 corresponds to a total of three types of stationary solutions:

a) symmetrical solutions (excitations propagate along the three chains in a plane)
b) two kinds of asymmetric solitons.

In the case of symmetric solitons in the region of excitation the pitch of the helix decreases and the diameter increases. In the case of asymmetric solitons the excitations propagate along the three chains with a slight shift in phase. The diameter of the helix, then, increases and a local 'bond' in the molecule occurs.

In order to obtain analytical results models of infinite chain lenghts are usually used in the continuum approximation. An investigation of limited discrete chains can be made only through use of modern computers. Numerical calculations of the soliton excitations of protein molecules have greatly contributed to a more complete understanding of these states.

The first numerical integration of the discrete differential equations proposed in the work of Davydov (1978) was performed by Hyman et al. (1979). They arrived at the conclusion that 'numerical studies of Davydov's nonlinear dynamic model of alpha-helical proteins confirm his prediction of the formation of solitons'. This conclusion is very important for the authors have used values of the model parameters which correspond to real protein molecules. Their work is also interesting in the respect that the process of formation of solitons from the defined initial state is investigated and the role of chain discreteness clarified.

Eilbeck (1979) in Meriot-Watt University (Edinburgh, Scotland) made a computer based 16 mm film that demonstrates the propagation of an internal vibrational excitation Amide 1 of an edge group along a PG chain. This film vividly shows that the excitation propagates along the protein molecule in the form of a soliton, i. e. in the form of a distinct local pulse which has the width of several PG's. Its shape and width remain constant during the motion.

Eilbeck's film and the numerical calculations of Hyman et al. (1979) show that the soliton is formed at the very beginning of the chain. Therefore, solitons can arise within comparatively short parts of the alpha-helical protein molecule.

The numerical calculations performed by Lomdahl et al.(1984) confirmed the particle-like properties of solitons. When two peptide groups are excited at the opposite ends of the polypeptide chain there appear two solitons moving towards each other with the velocity constituting three-eights of that of longitudinal sound. It turned out that after collision, they pass through each other retaining their velocity and form. Thus the collision of solitons resembles the elastic collision of two particles.

Some examples where soliton-like dynamics may well be operating

The problem of Muscle contraction

One of the most interesting problems in bioenergetics is to explain, at the molecular level, how the chemical energy of the hydrolysis of ATP molecules is transformed into mechanical energy in different intra- and intercellular movement processes in living organisms. The best studied movement of this kind is that caused by transversely striated muscles.

Based on theoretical studies of solitons in helical proteins Davydov (1974, 1982 b) proposed a new hypothesis to describe the mechanism of muscle contraction at the molecular level.

Skeleton muscles in vertebrates form a tissue consisting of millions of muscle fibres. Most of the volume of muscle fibres is occupied by contractive elements, called myofibrils. Each muscle fibre is surrounded by a membrane; the sarcoplasmic membrane. The myofibrils consist of identical parts arranged periodically along the fibre. They are sarcomeres.

Electron microscope studies have shown that each sarcomere consists of a large number of parallel running threads (filaments) of two types: thick and thin. The thin filaments are made up of one type of protein, myosin. The thin filaments lie between the thick. They are made up of three proteins: actin, tropomyosin and troponin.

Contraction of the muscle fibre occurs at a time, when, under the influence of a nerve pulse, calcium ions enter the sarcoplasm from special microtubes. The calcium ions stimulate the process of hydrolysis of ATP molecules at locations where thick and thin protein strands touch.

With the help of electron microscopic studies in 1954 it was established that contraction of a sarcomer is associated with the penetration of thin filaments into the space between the thick filaments without contraction of the filaments themselves. The experimentally established model of slipping raised a very important problem for researchers: to explain the mechanism of slipping at the molecular level. What forces cause the slipping of thin strands relative to thick ones? How does the energy at hydrolysis of ATP transform into slipping energy?

At present biologists believe that slipping of thin strands relative to thick ones is due to the active motion of the 'head' of myosin molecules which make up thick strands. It is proposed that with hydrolysis of an ATP molecule, attached to the head of myosin molecule the latter stretches, forming a bond ('bridge') with the actin molecule, obtained in the thin strand. Then the head rotates moving the thin strand towards the centre of the sarcomere and, finally, separates from the actin molecules returning to the previous size and position in the thick strand.

The idea of bridges between the thick and thin strands, which are joined, pull the fibres causing their displacement and then break, does not elucidate the molecular nature of the phenomenon of muscle contraction. The following questions remain unsolved: How is ATP hydrolysis energy used in the elongation, the cross-bridge formation, the pull force and the cross-bridge breaking? What is the molecular mechanism of change in the myosin molecule head that leads to these phenomena?

And, finally, why does only the head of the enormous myosin molecule (molecular weight - 500 000 which in addition to the head, has a long 'tail' formed by an alpha-helical protein) take an active part in the contraction?

According to a new hypothesis of the mechanism of muscle contraction proposed by Davydov, the calcium ions upon reaching the first series of the myosin molecule heads at the ends of the thick filaments initiates the hydrolysis of the ATP molecules attached to them. The energy released generates solitons in the long helical section of the myosin molecules. These move from the heads of myosin molecules to their ends situated in the region of the sarcomere centre. The motion of the solitons along the helical section of myosin molecule is accompanied by a local bending and an increase in the diameter of the molecule. The motion of this swollen region of a thick filament from its end to the centre causes the displacement of thin protein filaments to the sarcomere centre. Analogous phenomena take place at the other end of thick filaments. The motion of thin filaments in opposite directions causes the shortening in the length of each sarcomere and, consequently, the contraction of the length of muscle fibre.

Intracellular dynamics and solitons

Alpha-helical structures are quite common in proteins, and, in particular, they are present where energy appears to be transported from one end of a protein to the other or where two processes appear to be coupled by proteins.

The dominant configuration of cytochromes is alpha-helical. Helical proteins span the inner membrane (of mitochondria). Some people think that soliton-like mechanisms in these proteins may have something to do with the stabilisation of electron transport and its connection to ATP synthesis.

Recent studies of the spatial structure of cells have shown that the whole cellular interior is spanned by a network of protein microfilaments and tubules which retain the organelles of the cell in definite sites, determine the form of the cell, its change and all the movements inside the cell. This network of microfilaments and microtubules has been called the 'cytoskeleton'.

The protein molecules incorporated into the cytoskeleton determine all the internal movements in the cell, the energy transport and the internal bond. All these actions are due to the energy released by hydrolysis of ATP. The process of hydrolysis is, apparently, controlled by the change in the concentration of calcium ions as in muscle fibres.

The presence of contractile actin, myosin and tropomyosin, tubulin and duenin in nonmuscular cells confirms the hypothesis of a general way of converting the chemical energy of ATP hydrolysis into mechanical motion. Such a process could be realised by a sliding mechanism which differs in some details from that in muscular fibres.

In the muscular fibres thick filaments lie parallel to one another in the sarcomere. One end of the thin actin filaments is attached to z-plates separating one sarcomere from another. They enter the space between the thick filaments. In nonmuscular cells myosin filaments have no fixed position. They 'float' in the cytoplasma.

Since helical sections constitute a considerable part of the protein structure of the cell cytoskeleton, they can transfer energy and information from one part of the cell to another with the help of propagating soliton excitations.

In the vital activity of the cell an important role is played by transmembrane glycoprotein. Glycoproteins represent a covalent combination of protein molecules with a remified chain of carbohydrate residues, namely polysaccharides. Long protein fractions are in the alpha-helical conformation. Such a protein penetrates the entire thickness of the outer membranes of the cell to form the membrane channel. A remified polysaccharide portion is present on the cell surface. Some glycoproteins are closely associated with the microfilaments and tubules of the cytoskeleton of the cell. Thus glycoproteins connect the inside and the outside of the cell.

Glycoproteins are implicated in cellular identity, cellular adhesion, intercellular communication and transmembrane signalling. They can convey the signals that originate on the cellular exterior through the binding of hormones, neurons, immunoglobulins and other molecules. Thus, the nonlinear dynamics of solitons may provide a key for understanding the mechanism of transmission of information from the outside to the inside of the cell.

Solitons and molecular mechanism of the action of radiation on living cells

One of the essential pecularities of the nonheating effect of radiation on living cells is its sharp resonance. A new hypothesis for the mechanism of such resonance interaction was suggested by Eremko (1984) on the basis of the concept of an active participation of solitons in the vital activity of the cell.

As we have described above, solitons are ideal carriers of the energy of ATP hydrolysis along protein molecules. Eremko (1984) has shown that under the influence of electromagnetic radiation solitons can be transformed into a metastable excitonic-type state. Excitons are decelerated rapidly returning their energy to the environment through the heat of vibration. Hence, the transformation of solitons into excitons breaks the effectiveness of the transiton of energy and information. This process of transmission has a sharp resonance at a wave length $\lambda = 4.6$ to 8.8 nm. The Franck-Condon principle shows that the probability of the process is very small.

Possible mechanisms for anesthesia

Molecules from a wide class of intravenous anesthetics form hydrogen bonds with protein molecules situated on the inside and outside of the cytoplasmatic membrane. Such an attachment of anesthetic molecules to protein inhibits the normal functioning of the cell. One of the possible inhibitory mechanisms during attachment of anesthetic molecules (barbiturates) was proposed by Layne (1984) on the basis of the soliton model.

The barbiturate molecules contain groups of atoms (H-N-C-O) very similar to the peptide groups present in proteins. When this type of molecule is attached to a protein molecule the hydrogen bond between neighbouring PG's in a protein molecule is broken, since a new hydrogen bond between the O and H atoms of PG's with the atoms of the (H-N-C-O) group of barbiturates is formed. When the hydrogen bond between PG's is ruptured the distance between them in the protein molecule increases. Owing to this a soliton propagating along the PG's chain stops or decelerates and, consequently, the energy and information transfer along a protein molecule is violated.

In numerical calculations performed by Layne (1984), the PG chain perturbation in the region of three PG's was assessed. The numerical calculations have revealed the soliton to be well shaped before entering the perturbation region, however, having transversed through the perturbation it significantly degrades emitting phonons. These phonons move at the velocity of sound exceeding the soliton velocity. Hence, the attachment of an anesthetic molecule to protein decreases the ability to transfer the energy and information.

Layne suggests that the decrease in efficiency of energy and information transfer by solitons would be important in two cellular structures:

a) in the helical protein of the inner mitochondrial membranes which participate in ATP synthesis;

b) in membrane helical proteins which are responsible for chemical reception and signal transduction.

Electron transfer along protein molecules

Many biological phenomena are associated with electron transfer from donor to acceptor molecule. One of the possible explanations of electron transfer over large distances is connected with the assumption that this process is facilitated by the participation of protein molecules situated between the donor and acceptor of electrons.

The investigations performed by Davydov (1978 b, 1984) have shown that the processes of excess electron (caught by protein from donor) transfer along alpha-helical molecules are determined by their specific properties. Peptide groups, because of asymmetric distribution of electric charges, have a comparatively large constant electric dipole moment; about 4 Debye. Therefore they form potential wells capable of preventing outer electrons from entering from donor.

The overlap of wave functions of an excess electron in the chain with periodically arranged dipole electric moments leads to exchange interaction and collectivization of electron states. Strong electron interaction with the PG chain described by the deformation potential leads to a change in intrapeptide distances in the region where an electron is located. Therefore the energy states of an electron in the chain have a complicated character due to the interaction between the electron and the local deformation of the chain. The electron motion is described by a system of nonlinear differential equations. It has been shown that such a system has soliton-like-stable states. In these states the electron is associated with local chain deformation and moves with it as a whole with a velocity less than the velocity of sound V_0 in the chain.

The electron in this state will be called the electrosoliton. It transfers along the chain of electron electric charge and energy, including that of the deformation. An electrosoliton moving with velocity V has the energy

$$E_{sol}(V) = E_{sol}(0) + 1/2\, M_{sol} V^2.$$

Here $E_{sol}(0)$ is the energy of a soliton at rest, M_{sol} is the electrosoliton effective mass. It exceeds the mass of an electron which does not interact with the local deformation.

With coupling of the electron and the local chain deformation energy is released. This energy is proportional to the fourth power of the parameter of deformational potential and is inversely proportional to the square of the elastic coefficient of the chain.

The stability of electrosolitons is due to the intercompensation inherent in nonlinearity and dispersion effects. The exceptional electrosoliton stability makes alpha-helical sections of protein molecules the ideal guides for transporting electrons from donor to acceptor molecules.

Selection Between Aging Species

W. Ebeling, A. Engel, V. G. Mazenko

Introduction

In the early stages of biological evolution simple cells were the units of life. They were able to adapt to their environment and to improve their porperties under selection pressure (Eigen, Schuster, 1977; Ebeling, Feistel, 1982: Conrad, 1983). In principle simple cell organisms are already complete immediately after their birth, they do not develop in time, i. e. they do not show aging. With increasing selection pressure and increasing complexity species developed a special life-cycle consisting of several periods as e. g. birth, youth, period of learning and training, self-reproduction, taking care and teaching of the following generation, illness and death. Typically a more complex organism is not the result of a simple act of reproduction, the copying of a given template, but more likely the result of a developmental process which includes morphogenesis and learning. The development of the complete organism through the course of its life cycle is typical for more complex cells as e. g. yeast cells and especially for multi-cellular organisms. It should be stressed that development and aging are great achievements of evolution. Species could increase their fitness by developing good developmental and aging strategies. Therefore the better understanding of this important aspect of evolution should be a special goal of a general theory. Let us note

Thermodynamics and Pattern Formation in Biology

that many models of aging populations go back to the classical models of McKendrick and von Foerster (Poluektov, 1974; Busenberg, Ianelli, 1985; Hadeler, 1985).

In this paper aging populations under selection pressure are investigated. Several general models of development and aging under selection pressure have been discussed in earlier communications (Ebeling et al., 1986 a, b). Here we shall concentrate on the mathematical analysis of a more general model which includes aging and selection. Selection pressure is modelled by the condition that the total number of individuals in the population is constant. Therefore species with better aging strategies may develop only at the expense of species with worse aging strategies. It will be shown how this strategic game performs explicitely assuming simple functions for the age-dependence of birth and death rates. In this way the meaning of fitness will be clarified.

The model

The classical model of age-dependent population dynamics is given by (see e. g Poluektov, 1974)

$$
\begin{aligned}
\partial x/\partial \tau + \partial x/\partial t &= -d(\tau)\cdot x(\tau,t) && \tau, t > 0 \\
x(0,t) &= \int_0^\infty b(\tau)\, x(\tau,t)\, d\tau && t \geq 0 \qquad (1)\\
x(\tau,0) &= \varphi(\tau) && \tau \geq 0
\end{aligned}
$$

where $x(\tau, t)$ is the concentration of individuals of age τ at time t. The first equation in (1) describes the decrease in the population due to death, $d(\tau)$ is the age-dependent mortality rate. The second equation gives the concentration of newborns with age $\tau = 0$; $b(\tau)$ is called fecundity rate.

The parameters of (1) have to satisfy the following conditions:
$d(\tau)$ and $b(\tau)$ are continuous, nonnegative functions for $\tau \in [0, \infty)$, moreover $\int_0^\infty d(\tau)\, d\tau = \infty$.
The initial condition satisfies

$$\varphi(\tau) \in C^1[0, \infty) \cap L^1[0, \infty)$$

$$\varphi(\tau) \geq 0 \qquad \text{for} \quad \tau \in [0, \infty)$$

$$\varphi(0) = \int_0^\infty b(\tau)\varphi(\tau)\, d\tau \neq 0; \quad \partial\varphi/\partial\tau \in L^1[0, \infty)$$

$$\partial\varphi/\partial\tau(0) + d(0)\varphi(0) = \int_0^\infty b(\tau)[\partial\varphi/\partial\tau + d(\tau)\varphi(\tau)]\, d\tau.$$

Under these conditions (1) possesses a unique nonnegative solution $x(\tau, t)$, $\tau \in [0, \infty)$ $t \in [0, \infty)$.

Concentrating here on the question of the existence and uniqueness of a stationary solution we introduce a new parameter H which combines both the birth and death rate and is given by

$$H = \int_0^\infty b(\tau) \exp\left\{-\int_0^\tau d(\xi)\, d\xi\right\} d\tau. \tag{2}$$

The value of H determines the stationary solution and its stability. It is well known that

for $H > 1$ $\quad \|x(\tau, t)\| \to \infty \quad$ if $\quad t \to \infty$

for $H < 1$ $\quad \|x(\tau, t)\| \to 0 \quad$ if $\quad t \to \infty \quad$ and

for $H = 1$ $\quad$ there are infinite numbers of stationary solutions of the form

$$x^0(\tau) = n_0 \exp\left\{-\int_0^\tau d(\xi)\, d\xi\right\}$$

with $n_0 \in [0, \infty)$.

The norm can be defined here by

$$\|x(\tau,t)\| = \max_{\tau} |x(\tau,t)| \quad \text{or} \quad \|x(\tau,t)\| = \int_0^\infty x^2(\tau,t)\,d\tau \,.$$

The linear model (1) has been well studied (Poluektov, 1974; Busenberg, Ianelli, 1985; Mazenko, Rubanovskii, 1983). Moreover, a number of papers report study of model (1) with different assumptions about the character of the death and birth processes (Hadeler, 1985).

Here, however, we will study another important question, namely the relation between age-structure and selection processes. If several species compete there should be some influence of the selection pressure on the internal age-structure and, on the other hand, the birth and death rates of a species should determine their fitness in the selection process (Ebeling et al., 1986 a, b).

A simple model for selection processes between n species not taking into account the age-structure was proposed by Eigen and has been much studied since then (Eigen, 1971; Ebeling, Feistel, 1982). It is given by

$$dn_i/dt = (\lambda_i - \langle \lambda \rangle)\, n_i$$

$$n_i(0) = n_{i0} \qquad i = 1, \ldots, n.$$

Here $n_i(t)$ is the overall concentration of species i, λ_i its fitness and

$$\langle \lambda \rangle = \sum \lambda_i n_i(t) / \sum n_i(t).$$

As is well known in this model the species m with $\lambda_m = \max_i \lambda_i$ wins the competition (survival of the fittest), i. e.

$$n_m(t) \to C = \sum_i n_{io}, \quad n_i(t) \to 0 \text{ for } i \neq m \text{ if } t \to \infty .$$

We now generalise model (1) to a set of competing species $i = 1, ..., n$ with population densities $x_i(\tau,t)$. Similar to the Eigen model the selection pressure is introduced via the condition of constant overall organisation

$$\sum_i \int_0^\infty d\tau \, x_i(\tau,t) = \sum \int_0^\infty \varphi_i(\tau) d\tau = C = \text{const.} \tag{3}$$

In preceding papers (Ebeling et al., 1986 a, b) we have studied cases where condition (3) is satisfied by either altering the fecundity rate alone or the mortality rate alone. Here we investigate a model where both rates are influenced by the selection process which gives rise to new, interesting phenomena. To this end we introduce the index i and two, so far unkown functions $\phi(t, x_1...x_n)$ and $\psi(t, x_1...x_n)$ into (1) and obtain

$$\begin{aligned} \partial x_i/\partial\tau + \partial x_i/\partial t &= -[d_i(\tau) + \phi(t, x_1, .., x_n)]\, x_i \qquad && \tau, t > 0 \\ x_i(0,t) &= \int_0^\infty [b_i(\tau) + \psi(t, x_1, .., x_n)]\, x_i \, d\tau && t \geq 0 \\ x_i(\tau,0) &= \varphi_i(\tau) \end{aligned} \tag{4}$$

We note that $\phi(.)$ and $\psi(.)$ neither depend on τ nor on i. We assume that the functions $b_i(\tau)$, $d_i(\tau)$ and $\varphi_i(\tau)$ satisfy conditions similar to those described for $b(\tau)$, $d(\tau)$ and $\varphi(\tau)$ at the beginning of this section.

Differentiating (3) by t and using (4) we find

$$\begin{aligned} 0 &= \sum_i \int_0^\infty \partial x_i/\partial t \, d\tau = -\sum_i \int_0^\infty [\partial x_i/\partial\tau + d_i(\tau) x_i + \phi x_i]\, d\tau = \\ &= \sum_i \int_0^\infty b_i(\tau) x_i d\tau + \psi \sum_i \int_0^\infty x_i d\tau - \sum_i \int_0^\infty d_i(\tau) x_i d\tau - \phi \sum_i \int_0^\infty x_i d\tau . \end{aligned}$$

Therefore

$$\phi - \psi = \langle b \rangle - \langle d \rangle \tag{5}$$

where the social average $< \;\; >$ is defined as

$$< b > = \sum_i \int_0^\infty b_i(\tau)\, x_i(\tau,t)\, d\tau \Big/ \sum_i \int_0^\infty x_i(\tau,t)\, d\tau .$$

The general solution of (5) can be written as

$$\phi = \alpha < b > - \beta < d > , \quad \psi = (\alpha - 1) < b > - (\beta - 1) < d >$$

with $\alpha, \beta \in \mathbb{R}$. However, the interpretation of (4) is very much more transparent if both ϕ and ψ are nonnegative. Then $\phi(t)$ is simply an overall dilution flux as in the Eigen-Model and $\psi(t)$ is an overall input of newbornes. Negative ϕ would correspond to offspring with age $\tau > 0$ and negative ψ could entail negative $x_i(0, t)$. One can show that the only choice avoiding these problems is $\alpha = 1$, $\beta = 0$, i. e. $\phi(t) = < b >$, $\psi(t) = < d >$.

Therefore our model for age-dependent selection processes reads

$$\begin{aligned} \partial x_i/\partial\tau + \partial x_i/\partial t &= - D_i(\tau,t)\, x_i \\ x_i(0,t) &= \int_0^\infty B_i(\tau,t)\, x_i\, d\tau \qquad (6) \\ x_i(\tau,0) &= \varphi_i(\tau) \end{aligned}$$

For nonnegative $x_i(\tau, t)$, $\tau \in [0, \infty)$, $t \in [0, \infty)$ the effective death rates $D_i(\tau, t) = d_i(\tau) + < b >$ and birth rates $B_i(\tau, t) = b_i(\tau) + < d >$ are also non-negative and (6) possesses a unique non-negative solution.

Ecological picture

It is convenient to write $x_i(\tau, t)$ in the form

$$x_i(\tau, t) = n_i(t)\, \varrho_i(\tau, t) \qquad (7)$$

with $n_i(t) = \int_0^\infty d\tau\, x_i(\tau, t)$

implying $$\int_0^\infty d\tau\, \varrho_i(\tau, t) = 1 . \tag{8}$$

$n_i(t)$ gives then the number of individuals of species i without specification of the age and $\varrho_i(\tau, t)$ describes the normalised age-structure within species i. Putting (7) into (6) we obtain

$$\begin{aligned} dn_i(t)/dt &= \left[\int_0^\infty [b_i(\tau) - d_i(\tau)]\varrho_i(\tau, t)d\tau - \langle b - d\rangle\right] n_i \\ n_i(0) &= \int_0^\infty d\tau\, \varphi_i(\tau) \end{aligned} \tag{9}$$

and

$$\begin{aligned} d\varrho_i/d\tau + \partial\varrho_i/\partial t &= -\left[d_i(\tau) + \int_0^\infty (b_i(\tau) - d_i(\tau))\varrho_i(\tau, t)d\tau + \langle d\rangle\right]\varrho_i \\ \varrho_i(0, t) &= \int_0^\infty b_i(\tau)\varrho_i(\tau, t)d\tau + \langle d\rangle \\ \varrho_i(\tau, 0) &= \varphi_i(\tau) \Big/ \int_0^\infty \varphi_i(\tau)d\tau . \end{aligned} \tag{10}$$

Contrary to the cases studied before (Ebeling et al., 1986 a, b) the dynamics of the overall number of individuals n_i and of the inner age-structure ϱ_i are coupled to each other as specified by the appearance of $\varrho_i(\tau, t)$ in (9) and of $\langle d \rangle$ in (10) respectively.

Stationary solutions

We denote the stationary solution of (9) and (10) by n_i^o and $\varrho_i^o(\tau)$. From (9) we get

$$(\lambda_i - <\lambda>)\, n_i^o = 0 \tag{11}$$

with the fitness

$$\lambda_i = \int_0^\infty [b_i(\tau) - d_i(\tau)]\, \varrho_i^o(\tau)\, d\tau \tag{12}$$

depending functionally on the age-structure $\varrho_i^o(\tau)$. Similar to the Eigen-model (11) possesses n stationary solutions of the form

$$\begin{matrix} n_1^o, & 0, & \cdots\cdots, & 0 \\ 0, & n_2^o, & \cdots\cdots, & 0 \\ \vdots & \vdots & & \vdots \\ 0, & 0, & \cdots\cdots, & n_n^o \end{matrix} \tag{13}$$

For the stationary solution 0,...., n_i, ..., 0 we find

$$<d> = \int_0^\infty d_i(\tau)\, \varrho_i^o(\tau)\, d\tau$$

therefore $\varrho_i^o(\tau)$ is the solution of

$$d\varrho_i^o/d\tau = -\left[d_i(\tau) + \int_0^\infty b_i(\tau)\, \varrho_i^o(\tau)\, d\tau\right] \varrho_i^o \tag{14}$$

$$\varrho_i^o(0) = \int_0^\infty b_i(\tau)\, \varrho_i^o(\tau)\, d\tau + \int_0^\infty d_i(\tau)\, \varrho_i^o(\tau)\, d\tau$$

$$\int_0^\infty \varrho_i^\circ(\tau)\,d\tau = 1 . \tag{15}$$

Using

$$c_i = \int_0^\infty b_i(\tau)\,\varrho_i^\circ(\tau)\,d\tau \tag{16}$$

we find

$$\varrho_i^\circ(\tau) = \varrho_i^\circ(0)\,\Lambda_i(\tau)\exp(-c_i\tau) \tag{17}$$

with

$$\Lambda_i(\tau) = \exp\{-\int_0^\tau d_i(\xi)\,d\xi\} . \tag{18}$$

c_i and $\varrho_i^\circ(o)$ can now be obtained from equations (14) and (15):

$$\begin{aligned} 1 &= \int_0^\infty (K_i(\tau) + D_i(\tau))\exp(-c_i\tau)\,d\tau \\ \varrho_i^\circ(0) &= \Big[\int_0^\infty \Lambda_i(\tau)\exp(-c_i\tau)\,d\tau\Big]^{-1} \end{aligned} \tag{19}$$

with $K_i(\tau) = b_i(\tau)\Lambda_i(\tau)$, $D_i(\tau) = d_i(\tau)\Lambda_i(\tau)$

We note that (16) follows from equation (19). Since

$$\begin{aligned} &\int_0^\infty D_i(\tau)\exp(-c_i\tau)\,d\tau = -\int_0^\infty \exp(-c_i\tau)\,d\exp\{-\int_0^\tau d_i(\xi)\,d\xi\} = \\ &= 1 - c_i\int_0^\infty \Lambda_i(\tau)\exp(-c_i\tau)\,d\tau = 1 - c_i/\varrho_i^\circ(0) \end{aligned} \tag{20}$$

we have $c_i = \varrho_i^\circ(0)\int_0^\infty K_i(\tau)\exp(-c_i\tau) = \varrho_i^\circ(0)\big[1 - \int_0^\infty D_i(\tau)\exp(-c_i\tau)\,d\tau\big] =$

$$= \varrho_i^\circ(0)\,(1 - (1 - c_i/\varrho_i^\circ(0))) = c_i .$$

Every equation (19) has a unique positive solution c_i since the function

$$\mathcal{U}_i(c) = \int_0^\infty (K_i(\tau) + D_i(\tau)) \exp(-c\tau)\, d\tau$$

is monotonically decreasing ($\mathcal{U}'_i(c) < 0, \quad c \in [0, \infty)$) and

$$\mathcal{U}_i(0) = \int_0^\infty (K_i(\tau) + D_i(\tau))\, d\tau = 1 + \int_0^\infty K_i(\tau)\, d\tau > 1 .$$

Therefore the stationary solution is completely determined for every $i = 1, \ldots, n$.

Stability of the stationary solutions

First we consider the stability of $\varrho_i^\circ(\tau)$. To this end we write (10) in the form

$$\partial\varrho_i/\partial\tau + \partial\varrho_i/\partial t = -\left[d_i(\tau) - \int_0^\infty d_i(\tau)\varrho_i(\tau,t)\,d\tau + \varrho_i(0,t)\right]\varrho_i(\tau,t) . \tag{21}$$

The analogous equation for $\varrho_i^\circ(\tau)$ reads

$$\partial\varrho_i^\circ/\partial\tau = -\left[d_i(\tau) - \int_0^\infty d_i(\tau)\varrho_i^\circ(\tau)\,d\tau + \varrho_i^\circ(0)\right]\varrho_i^\circ(\tau) . \tag{22}$$

Putting $\varrho_i(\tau, t) = \varrho_i^\circ(\tau) + \xi_i(\tau, t)$ in (21) and using (22) we get a first order equation in

$$\begin{aligned} \partial\xi_i/\partial\tau + \partial\xi_i/\partial t = &- \left[d_i(\tau) + \varrho_i^\circ(0) - \int_0^\infty d_i(\tau)\varrho_i^\circ(\tau)\,d\tau\right]\xi_i(\tau,t) \\ &+ \varrho_i^\circ(\tau)\int_0^\infty d_i(\tau)\xi_i(\tau,t)\,d\tau - \varrho_i^\circ(\tau)\xi_i(0,t) . \end{aligned} \tag{23}$$

Defining

$$\alpha_i = \varrho_i^\circ(0) - \int_0^\infty d_i(\tau)\varrho_i^\circ(\tau)d\tau$$

we find

$$\alpha_i = \varrho_i^\circ(0)[1 - \int_0^\infty D_i(\tau)\exp(-c_i\tau)d\tau =$$

$$= \varrho_i^\circ(0)\int_0^\infty K_i(\tau)\exp(-c_i\tau)d\tau > 0.$$

We look for the solution of (23) in the form

$$\xi_i(\tau,t) = Q_i(\tau)\exp\{(t-\tau)z_i\}. \tag{24}$$

From (23) then follows a linear equation for the function $Q_i(\tau)$ of the form

$$dQ_i/d\tau = -[d_i(\tau)+\alpha_i]Q_i + q_i\exp(z_i\tau)\varrho_i^\circ(\tau) \tag{25}$$

where $$q_i = \int_0^\infty d_i(\tau)Q_i(\tau)\exp(-z_i\tau)d\tau - Q_i(0).$$

Solving (25) we get

$$Q_i(\tau) = [Q_i(0) + q_i\varrho_i^\circ(0)\int_0^\tau \exp(z_i\tau)d\tau]\exp\{-\int_0^\tau d_i(\xi)d\xi - \alpha_i\tau\}.$$

For the existence of the solution $Q_i(\tau)$ we must have $z_i < 0$ which in turn ensures $\xi_i(\tau, t) \to 0$ if $t \to \infty$. Consequently the stationary normalised age-structure $\varrho_i^\circ(\tau)$ is stable. In order to investigate the stability of the stationary solutions n_i^0 given by (13), i. e. to determine the species surviving the selection process, we have to determine, according to (11) and (12),

$$\max_i \lambda_i = \max_i \varrho_i^\circ(0) \int_0^\infty (b_i(\tau) - d_i(\tau)) \Lambda_i(\tau) \exp(-c_i\tau)\, d\tau$$

under the condition (19).

From (19) and (16) we find

$$\max_i \Lambda_i = \max_i (c_i - \varrho_i^\circ(0) \int_0^\infty D_i(\tau) \exp(-c_i\tau)\, d\tau) =$$

$$= \max_i (c_i - \varrho_i^\circ(0) [1 - \int_0^\infty K_i(\tau) \exp(-c_i\tau)\, d\tau]) =$$

$$= \max_i (c_i - \varrho_i^\circ(0)(1 - c_i/\varrho_i^\circ(0)) = \max_i (2c_i - \varrho_i^\circ(0)) =$$

$$= \max_i \frac{2c_i \int_0^\infty \Lambda_i(\tau) \exp(-c_i\tau)\, d\tau - 1}{\int_0^\infty \Lambda_i(\tau) \exp(-c_i\tau)\, d\tau}$$

Furthermore, since $\frac{d}{dc}(\int_0^\infty D_i(\tau) e^{-c\tau} d\tau) < 0$, (20) yields

$$c \int_0^\infty \Lambda_i(\tau) \exp(-c\tau)\, d\tau \to 1 \qquad \text{if } c \to \infty$$

Consequently the function

$$\frac{2c \int_0^\infty \Lambda_i(\tau) \exp(-c\tau)\, d\tau - 1}{\int_0^\infty \Lambda_i(\tau) \exp(-c\tau)\, d\tau}$$

is monotonously increasing on the set of c_i satisfying

$$2c_i > \varrho_i^0(0)$$

Therefore $\max_i(2c_i - \varrho_i^\circ(0))$ is realised by the same i as $\max_i c_i$, where c_i, $\varrho_i^\circ(0)$ are given by (19). Hence there is only one stable stationary solution among (15), namely $0,\ldots, n_i^o, 0,\ldots,0$ with $c_i > c_j \; \forall j \neq i$. The surviving species is characterised by the maximal value of the parameter c_i. Using equation (19) this can be interpreted biologically: In order to survive the competition a species should have a small mortality rate in the youth of an individual (small τ) and the fecundity rate should be large, i. e. self-reproduction should occur early in life.

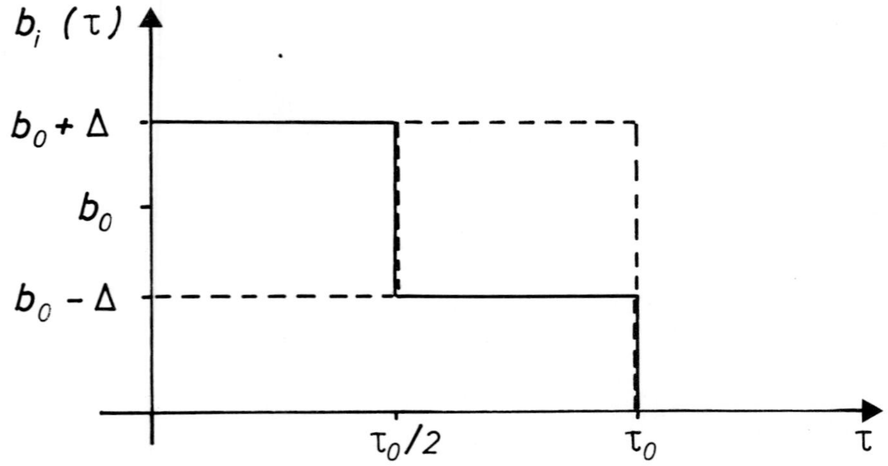

Fig. 1 Qualitative plot of the birth-rates $b_1(\tau)$ (full line) and $b_2(\tau)$ (dotted line) as given by (27).

Example

Finally we discuss a simple example which shows the importance of the age-structure in selection. Consider two species $i = 1,2$ with

$$d_1(\tau) = d_2(\tau) \equiv d, \tag{26}$$

$$b_i(\tau) = \begin{cases} b_0 \pm \Delta & \text{if } 0 \leq \tau \leq \tau_0/2 \\ b_0 \mp \Delta & \text{if } \tau_0/2 < \tau \leq \tau_0 \\ 0 & \text{otherwise.} \end{cases} \tag{27}$$

Here and in the following the upper sign is valid for $i = 1$ and the lower one for $i = 2$ (see Fig. 1). We ask which species survives in the selection process? Neglecting the age-structure one would characterize both species by the same average birth-rate b_0, hence both would survive. Taking into account the detailed age-dependence of $b_i(\tau)$, however, we have to proceed according to the last section and to determine the species with the higher value c_i as given by (19). Now from (26) follows

$$\Lambda_i(\tau) = \exp(-d\tau)$$

and (18) can be written as

$$c_i = (c_i + d) \int_0^\infty b_i(\tau) \exp\{-(c_i + d)\tau\} d\tau . \tag{28}$$

Using (27) we find

$$c_i = b_0 [1 - \exp\{-(c_i + d)\tau_0\}] \pm \Delta [1 + \exp\{-(c_i + d)\tau_0\} - \\ - 2 \exp\{-(c_i + d)\tau_0/2\}] . \tag{29}$$

We substitute $x_i = \exp\{-(c_i + d)\tau_0/2\}$ and note that $c_i(x_i)$ is monotonically decreasing. (29) now reads

$$-\frac{2}{\tau_0} \ln x_i - d + b_0 (x_i^2 - 1) = \pm \Delta (1 - x_i)^2 \tag{30}$$

which is solved graphically in Fig. 2. We infer $x_1 < x_2$ and hence $c_1 > c_2$, i. e. species 1 survives whereas species 2 dies out. The advantage the first species has in its youth cannot be compensated by the advantage of the second species for large τ. This is intimately connected with the fact that the stationary age-structure $\overset{\circ}{\varrho}_i(\tau)$ as given by (17) is monotonically decreasing in τ.

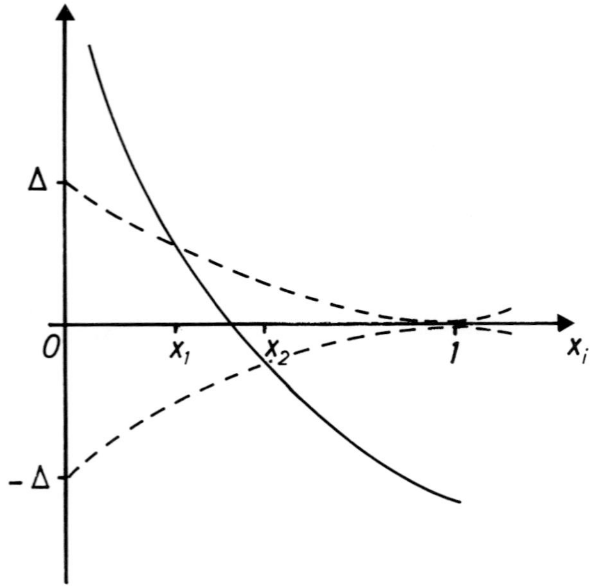

Fig. 2 Graphical solution of equation (30). The full line represents the l. h. s., the dotted line the r. h. s.

Conclusion

In this paper a rather general mathematical model for a biological selection process taking into account the age-structure of the competing species has been proposed and analysed. The possible stationary solutions and their stability have also been obtained. Thereby it is possible to determine from the age-dependent mortality- and fecundity-rates the species, that wins the competition and survives. It was shown that advantageous properties of a species are most valuable if they occur in the youth of the individuals. Favourable properties in old age have only minor importance for the fitness of the species. Several additional interesting questions, as, e. g. the change in the stationary age-structure of a species due to competition with other species, can also be studied with the help of the model.

Pattern Formation in Chemical and Biochemical Systems

Pattern Formation and Wave Propagation in Chemical Systems

I. R. Epstein

Introduction

Pattern formation in biological sytems is a fascinating and important phenomenon. It is also an extremely complex phenomenon whose origins, as the other chapters in this volume illustrate, we are only now beginning to comprehend. Despite the complexity and the multiplicity of beautiful and intricate pattern formation phenomena that occur in nature, the processes that give rise to spatial organisation are few in number: reaction, diffusion and, in some cases, heat conduction and convection. Nonliving systems are not inherently more complex than living ones with respect to their transport properties (except, perhaps, where specialised structures such as membranes may provide distinctive boundary conditions). They do, however, offer a more intricate set of chemical reactions from which to generate spatial structure. It makes good sense, therefore, if one wishes to understand biological pattern formation at a fundamental level, to examine relatively simple chemical systems for evidence of similar phenomena and, if such phenomena do occur, first to seek an understanding of "chemical" pattern formation and then to try to extrapolate that knowledge to systems of biological significance.

Thermodynamics and Pattern Formation in Biology

Spatial patterning has been of interest to chemists for nearly a century. Perhaps the first phenomenon to be studied systematically was the periodic precipitation discovered by Liesegang in 1896 (Hedges, 1932). In a typical Liesegang ring experiment, two solutions containing a pair of ions, such as magnesium and hydroxide, that form a sparingly soluble compound, are introduced at opposite ends of a tube containing a gel. As the ions diffuse toward one another, a precipitate is deposited not, as one might expect, in a continuous fashion, but in sharp bands. Liesegang rings and theoretical mechanisms for their origin have provided models for a variety of geological patterning phenomena (Ortoleva, Schmidt, 1985, see also chapter 8).

More significant from a biological point of view is another discovery dating from the same period. Luther (1906), in a lecture delivered to the Deutsche Bunsengesellschaft für Angewandte Physikalische Chemie in Dresden, suggested that chemically reacting systems could propagate waves of chemical activity just as nerve cells could conduct waves of electrical activity. He developed this analogy in some detail, discussed the relative velocities of the two types of waves, and even gave a recipe for a permanganate-oxalate solution that would generate such chemical waves (see Arnold et al., 1987, for a recent interpretation of Luther's work).

The modern study of chemical wave phenomena begins with Belousov's work in the 1950's on the cerium-catalyzed reaction of bromate ion with citric acid (Belousov, 1985). This system was later modified by Zhabotinskii (1964) to become the prototypical chemical oscillator, the BZ reaction. If one adds the redox indicator ferrous phenanthroline to an appropiately prepared BZ solution, and pours an initially homogeneous mixture in a thin (~1 mm) layer in a Petri dish, then one soon observes a stunning phenomenon (Zhabotinskii, 1967). In the red solution, one or more blue dots appear. Each dot expands to form a blue disk. The centre of the disk returns to its original red state, leaving a blue ring, and then a new dot appears at the centre of the ring and also begins to expand. Ultimately a pattern of concentric rings is established. Where two rings come together, they annihilate and the solution goes back to red.

As has been the case with temporal oscillation, the vast majority of studies of spatial pattern formation in chemical systems have been carried out on the BZ reaction. In this review, we shall discuss BZ systems, but shall also treat wave propagation and pattern formation in other, more recently discovered reaction systems. Additional, more comprehensive reviews on chemical waves and patterns may be found in Field and Burger (1985), Winfree (1974, 1978, 1980) and Showalter (1985).

In an attempt to give the reader a feeling for the variety of spatial phenomena that have been seen in chemically reacting systems, we begin with a survey of experimental results. We then discuss efforts at modelling some of these phenomena. We conclude with a discussion of the significance of spatial pattern formation and wave propagation for biological systems and some speculations on future prospects in this field.

Experimental Results

Chemical pattern formation and wave propagation may be studied in a variety of geometries. Perhaps the simplest is a pseudo-one-dimensional arrangement in which the reaction occurs in a cylindrical tube whose length significantly exceeds its radius. A second, commonly employed configuration is a thin layer of solution in a circular Petri dish, an essentially two-dimensional experiment. Observations of waves and patterns in systems with three comparable dimensions are relatively rare, though we shall mention two examples.

One-dimensional waves

In one dimension, the phenomenon of interest is generally the propagation of a wave or a pulse of chemical reactivity through a solution. Most often, the reaction involved is autocatalytic, and the basic mechanism of propagation is the diffusion of the autocatalytic product in front of the wave causing the reaction rate to increase and thereby carry the wave further. Initiation may be spontaneous, resulting from inhomogeneous fluctuations within the solution, or may be induced by external means such as introducing a drop of autocatalyst, heating the solution or changing the potential of an electrode at the desired point of initiation. Velocities of propagation tend to be on the order of mm/min. Table 1 gives a summary of reactions in which such waves have been observed.

Table 1.

Systems Exhibiting Trigger Wave Propagation

System	Range of measured velocities (mm/min)	Reference
BZ	5 - 10	Field and Noyes (1974)
BZ	3 - 5	Showalter et al. (1979)
BZ	1 - 4	Jorne (1980)
bromate - ferroin	1 - 22	Showalter (1981)
iodate - arsenite	1 - 5	Gribschaw et al. (1981)
HNO_3 - NH_2OH	0.02 - 9	Gowland and Stedman (1983)
chlorite - iodide	0.3 - 3	Weitz and Epstein (1984)
HNO_3 - Fe(II)	30 - 115[a]	Bazsa and Epstein (1985)
	8- 20[b]	
	0.7[c]	
ClO_2^- - $S_2O_3^{2-}$	2-25	Nagypál et al. (1986)

a. Downward in vertical tube.
b Upward in vertical tube.
c Isotropic velocity with added silical gel.

It is important to distinguish these waves, often referred to as *trigger waves* from another, less interesting, type of wave phenomenon, the phase or kinematic wave. A trigger wave involves both reaction and diffusion, and travels with a constant velocity determined by the rate and diffusion constants of the reacting species. Since diffusion is essential, trigger waves cannot penetrate barriers impermeable to key reactants or intermediates. Phase waves, in contrast, depend not upon diffusion, but upon differences in reaction rate due to initial inhomogeneities in reactant concentrations. As a result of these differences, different points in the medium reach any given stage in the reaction (concentration versus time) profile at different times, and the reaction appears to propagate through the solution, as in a trigger wave. However, since diffusion is not necessary for propagation (in fact, diffusion tends to destroy the wave by obliterating the concentration gradient), phase waves can travel *through* impermeable barriers. Their velocities are determined not by the chemistry of the reaction, but by the magnitude of the gradient. For a more detailed discussion of the differences between trigger and phase waves, see Winfree (1980).

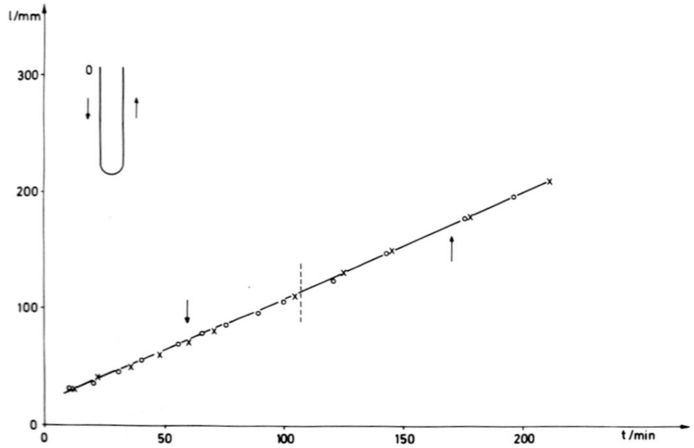

Fig. 1 Position as a function of time of chemical wave in a mixture consisting initially of 3 M HNO_3, 0.2 M Fe(II) and 1×10^{-4} M N_2H_4. Tube is filled with silica gel to prevent convection. Sketch in upper left indicates shape of tube, and arrows show direction of wave propagation with respect to vertical. Inner diameter of tube: x, 2.5 mm, o, 1.3 mm (Bazsa, Epstein, 1985).

Interest in one-dimensional wave propagation has largely focused on measuring the velocity of the wave as a function of system parameters such as reactant concentration and temperature and on explaining the parameter dependence of that velocity. Under carefully controlled conditions, the wave velocity can be remarkably constant as illustrated in Figure 1. Recent experiments (Bazsa, Epstein, 1985; Nagypál et al., 1986), however, have shown that one must be exceedingly careful to avoid convective effects. Figure 2 illustrates ferrous-nitric acid waves in a vertical tube bent into three segments. We note that the wave travels at essentially constant velocity within each segment, but the velocity when the wave moves downward is roughly six times greater than that of the upward travelling wave! Figure 3 illustrates an oppositely directed *gravity effect*. The polar coordinate plot of wave velocity versus direction for chlorite-thiosulfate waves shows more rapid propagation in the upward direction. These surprising results will be explained below.

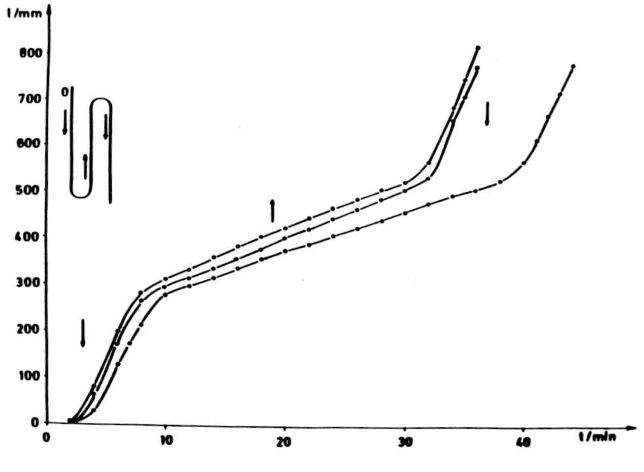

Fig. 2 Similar to Figure 1, but without added silical gel in a three-segment tube of inner diameter 2.5 mm. The three curves represent three replicate runs. (Bazsa, Epstein, 1985).

Another unusual wave phenomenon is that found by Gowland and Stedman (1983) in the hydroxylamine-nitric acid reaction.Here, as the reaction proceeds the stoichiometry undergoes a transition from scavenging of nitrous acid, HNO_2, by hydroxylamine:

$$NH_2OH + HNO_2 \rightarrow N_2O + 2\,H_2O \qquad (1)$$

to autocatalytic production of HNO_2:

$$NH_2OH + 2\,HNO_3 \rightarrow 3\ HNO_2 + H_2O \qquad (2)$$

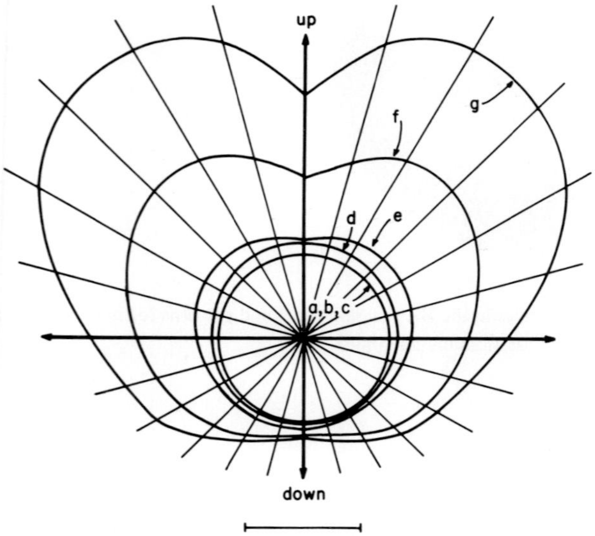

Fig. 3 Velocity of chemical wave as a function of propagation direction and internal tube diameter (i.d.) in the chlorite-thiosulfate system. [$ClO2^-$] = 0.05 M, [$S_2O_3^{2-}$] = 0.025 M, [OH^-] = 0.025 M. Notations: a-c, intramedic tubing, i.d.(mm) = 0.381 (a), 0.580 (b), 0.764 (c); d-g, tygon tubing, i. d. = 0.794 (d), 1.584 (e), 2.381 (f), 3.175 (g) (Nagypál et al., 1986)

An initially homogeneous solution approximately 0.1 M in NH_2OH and 4 M in HNO_3 can spontaneously form two layers, both of which are colourless, but which can be distinguished by their unequal indices of refraction. The upper layer is rich in nitrous acid and depleted of hydroxylamine, while the lower one resembles the starting mixture, containing almost no nitrous acid, but essentially 0.1 M hydroxylamine. The boundary between the two layers moves steadily downward until all the hydroxylamine is consumed.

Fig. 4 Spiral chemical waves in the BZ reaction (left) and patterns formed by the aggregating slime mould Dictyostelium discoideum (right)

One type of experiment with potential analogies to waves in biological systems involves the propagation of chemical waves in the presence of applied electric fields. Schmidt and Ortoleva (1977) suggested that, since ionic species play key roles in wave-producing reactions such as the BZ, it should be possible to influence the wave behaviour by applying an external field. Their experiments (Feeney et al., 1981), conducted mainly in a two-dimensional geometry, showed that circular waves were indeed distorted into crescent-shaped waves. More elaborate one-dimensional experiments have been performed by Sevcikova and Marek (1983, 1984). They studied the effects of applied fields on BZ and arsenite-iodate waves travelling in a thin tube, and observed a variety of new phenomena. These include the stopping of the wave at a sufficiently high field, and *wave splitting*, the generation of several oppositely directed secondary waves from a single intitial wave. The fields required to produce these phenomena were of the order of 10 - 30 V/cm.

Waves and patterns in two dimensions

While the one-dimensional geometry has, thanks to its computational simplicity, received the lion's share of attention in theoretical studies, most experimental investigations of chemical waves and patterns have been directed toward two-dimensional configurations. Some of the phenomena observed are indeed striking, particularly in their resemblance to biological structures. An example is shown in Figure 4, where chemical waves in the BZ reaction are juxtaposed with the pattern formed during aggregation of the slime mold *Dictyostelium discoideum*.

Two-dimensional experiments have sought to address several questions. These include

- How do the velocities of propagating waves depend upon reactant concentrations?
- What types of patterns are possible, and what determines which of several possible patterns will arise?
- How are waves initiated, and what is the nature of the central pacemaking region of circular and spiral waves?
- Are truly stationary patterns possible, and if so, what forms do they take?

Studies of wave velocities in two dimensions are very similar to those of one-dimensional waves. One induces initiation at a convenient point and then observes the position of the wave as a function of time. Wave velocities tend to be nearly constant as long as the wave is not too close to the edge of the container. Experiments of this type have been carried out on the BZ (Field, Noyes, 1974 a; Showalter et al., 1979; Jorne, 1980), bromate-ferroin (Showalter, 1981), iodate-arsenite (Gribschaw et al., 1981), chlorite-iodide (Weitz and Epstein, 1984) and ferrous-nitric acid (Bazsa, Epstein, 1985) reactions. Showalter et al. (1979) give an excellent discussion of methods to initiate waves. Wood and Ross (1985) have used laser and digital imaging techniques to generate detailed concentration profiles of waves in the BZ reaction.

The simplest two-dimensional waves consist of *target* patterns, concentric circles about one or more central points or *organising centres*. Figure 5 contains an example of one of the few non-BZ reactions to exhibit target patterns (de Kepper et al., 1982). Note the regions where two circles overlap and the waves annihilate. Winfree (1972) has shown that target patterns may easily be converted to spiral waves by tilting the vessel to shear the propagating wave fronts. The broken fronts soon wind around themselves, producing spirals such as those shown in Figure 4. Spiral waves are always created and destroyed in pairs such that the number of left- and right-handed spirals remains equal. Spirals may also be generated chemically by appropriately crossed concentration gradients (Winfree, 1978). More elaborate patterns, such as multi-armed spirals (Agladze, Krinskii, 1982) may be generated by more complex manipulations of the reacting medium.

The nature of pacemakers, the points at the origin of concentric circular waves, has been a subject of some controversy among investigators in this field. The analogous behaviour in biological systems can be generated by ectopic foci, i. e. patches of tissue irritated chemically or by abrasion, in heart muscle, or by special cells with an anomalously short oscillatory period in slime molds. Rastogi et al. (1977) have pointed out the role of dust particles and defects in the container surface in generating waves in the BZ reaction. However, both theory (e. g. Zaikin, Kawczynski, 1977) and experiment suggest that chemical waves may arise spontaneously, most likely from intrinsic concentration fluctuations. Müller et al. (1986) have carried out a remarkable study of circular and spiral waves in the BZ reaction in which computerized digital imaging was used to provide high resolution (in time and space) concentration profiles at the center of the wave. One sees clearly the marked difference in behaviour between the central core and the rest of the wave. Similar techniques were also applied to waves resulting from glycolytic oscillations in yeast cell extracts.

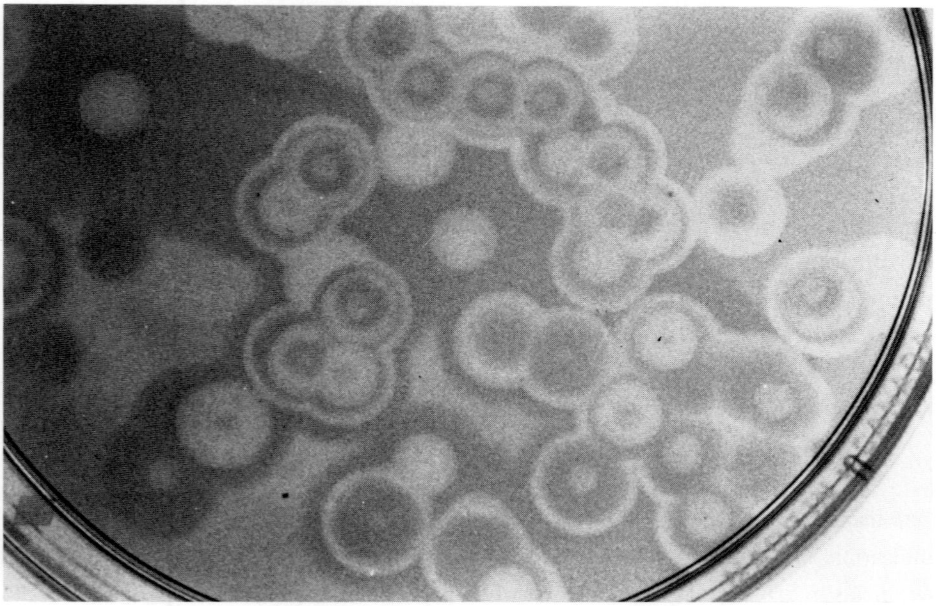

Fig. 5 Target patterns in the chlorite-iodide-malonic acid reaction (de Kepper et al., 1982).

Stationary reaction-diffusion patterns in chemical systems have been predicted from simple models (Nicolis, Prigogine, 1977), and there have been several reports of such phenomena. Showalter (1980) and Orbán (1980) have observed stationary patterns in the ferroin-catalysed and uncatalysed BZ reactions respectively. Avnir and coworkers (e. g. Avnir, Kagan, 1983; Avnir et al., 1983), expanding on an earlier discovery by

Möckel (1977), have shown that a wide variety of chemically reacting systems produce mosaic patterns at interfaces. In all cases, however, controversy exists over whether or not the observed structures are generated solely by reaction and diffusion or whether convection, too, plays an essential role.

Three-dimensional waves

A typical two-dimensional wave propagation experiment involves a layer of solution approximately 1 mm high in depth. Bazsa and Epstein (1985) have studied wave propagation in the ferrous-nitric acid system as a function of layer depth. They find that the general form of the wave remains the same, but that both the velocity and the thickness of the wave increase sharply as the layer depth is changed from 1 mm to 4 mm. They attribute this behaviour primarily to hydrodynamic effects.

If one increases the depth of the solution still further, so that the medium is truly three-dimensional, more elaborate waveforms can be generated. Winfree (1973) has shown how spirals in two dimensions become scroll-shaped waves in three dimensions. Welsh et al. (1983) have modified the usual experimental conditions for BZ waves so that the medium will support propagation in three dimensions. They find such new structures as toroidal and spheroidal waves. Both works contain impressive photographs of three-dimensional wave structures.

Theoretical approaches

The theory of spatial structure, i. e. waves and patterns, in chemical systems is at a far more primitive stage than the corresponding theory of temporal structure, i. e. oscillations, chaos, etc. Nevertheless, a great deal of progress has been made in recent years by applying a wide variety of techniques. These range from detailed numerical calculations of wave velocities in a particular system to qualitative theories about the topological properties of allowed spatial structures in any reaction-diffusion system. In this section we give a brief overview of some of the theoretical methods that have been utilized, again, though occasionally somewhat arbitrarily, dividing our treatment into one-, two- and three-dimensional approaches. There is a voluminous literature on the theory of chemical waves and patterns, and what follows is, of necessity, a rather superficial sampling designed to whet the reader's appetite rather than an attempt at comprehensive coverage.

One-dimensional waves

Perhaps the single most important theoretical work in the field of chemical waves and pattern formation is a paper by Turing (1952) entitled 'The Chemical Basis of Morphogenesis'. In this remarkable work, Turing introduces the notion of symmetry breaking, the idea that an initially homogeneous system 'may later develop a pattern or structure due to an instability of the homogeneous equilibrium, which is triggered off by random disturbances'. He gives a detailed example, a set of seven elementary steps, and shows that this model when coupled to diffusion does indeed produce stationary spatial structures. He discusses a variety of forms that symmetry breaking may take and suggests that this phenomenon may provide the route to morphogenesis, gastrulation, phyllotaxis and a variety of patterns such as those of tentacles in *Hydra*. Turing's paper remains today a fruitful source of insight into pattern formation in both chemistry and biology.

More recent work has tended to utilize either simpler models than those of Turing or models derived from actual chemical reactions that exhibit the behaviour of interest. In the former category, Prigogine and coworkers (Nicolis, Prigogine, 1977) have done some very thorough analyses of the Brusselator model (Prigogine, Lefever, 1968):

$$\mathrm{A} \longrightarrow \mathrm{X} \qquad \text{(B1)}$$
$$\mathrm{B} + \mathrm{X} \longrightarrow \mathrm{Y} + \mathrm{D} \qquad \text{(B2)}$$
$$2\,\mathrm{X} + \mathrm{Y} \longrightarrow 3\,\mathrm{X} \qquad \text{(B3)}$$
$$\mathrm{X} \longrightarrow \mathrm{E} \qquad \text{(B4)}$$

The concentrations of A and B are held fixed, so the Brusselator is a two-variable model. After the addition of diffusion terms, the stability analysis can be done analytically, and a variety of both time- and space-periodic solutions exist. The bifurcations that give rise to these *dissipative structures* have been elucidated in considerable detail.

The other model system that has received considerable attention is the Oregonator (Field, Noyes, 1974 a), a three-variable model derived from the full Field-Körös-Noyes (1972) mechanism for the BZ reaction:

$$\mathrm{A} + \mathrm{Y} \longrightarrow \mathrm{X} \qquad \text{(O1)}$$
$$\mathrm{X} + \mathrm{Y} \longrightarrow \mathrm{P} \qquad \text{(O2)}$$
$$\mathrm{B} + \mathrm{X} \longrightarrow 2\,\mathrm{X} + \mathrm{P} \qquad \text{(O3)}$$
$$2\,\mathrm{X} \longrightarrow \mathrm{Q} \qquad \text{(O4)}$$
$$\mathrm{Z} \longrightarrow f\,\mathrm{Y} \qquad \text{(O5)}$$

The concentrations of A and B are held fixed.

With the Oregonator, the focus has been on understanding the travelling wave seen in the BZ reaction rather than on modelling the stationary patterns found in the Turing and Brusselator models. Troy (1985) gives a proof that wave solutions exist to the equations generated by (O1)-(O5) in the presence of diffusion, and the numerical results of Reusser and Field (1979) suggest that these solutions are in fact stable. Field and Noyes (1974 b) showed that the wave velocity should be proportional to $([H^+][BrO_3^-])^{1/2}$, a result in agreement with experiment (Field, Noyes, 1974 b; Showalter, 1981), though reconciliation between the calculated and measured constants of proportionality requires some reassessment of the rate constants (Tyson, 1985). A great deal more work has been done on modelling wave propagation in the BZ reaction, nearly all of it with the Oregonator model. We refer the interested reader to the reviews listed in the references. Here we mention only the recent numerical calculation by Becker and Field (1985) in which large amplitude stationary structures are found in the Oregonator.

Weitz and Epstein (1984) simulated waves in the chlorite-iodide reaction using an approach similar to that taken by Reusser and Field (1979) in their calculation on the Oregonator. A three-variable (I^-, I_2, ClO_2^-) model was employed for the chemistry, and the reaction-diffusion equations were integrated numerically using the method of lines, in which the spatial region is divided into n compartments and the m partial differential equations for the m independent chemical species are converted into $m\,n$ ordinary differential equations by replacing the spatial derivatives resulting from diffusion by difference quotients. Qualitative agreement was obtained between the simulations and the experimentally determined wave velocities and profiles.

Perhaps the most detailed theoretical results on travelling waves in a chemical system are those obtained by Showalter and collaborators on the arsenite-iodate reaction. The chemistry of this system is described by two overall stoichiometric processes:

$$IO_3^- + 5\,I^- + 6\,H^+ \rightarrow 3\,I_2 + 3\,H_2O \qquad (3)$$

$$H_3AsO_3 + I_2 + H_2O \longrightarrow H_3AsO_4 + 2\,I^- + 2\,H^+ \qquad (4)$$

With arsenite in excess, reaction (3) is rate determining and the net reaction is (3) + 3 x (4), i. e. the autocatalytic production of iodide

$$IO_3^- + 3\,H_3AsO_3 \longrightarrow I^- + 3\,H_3AsO_4 \qquad (5)$$

according to the empirical rate law (Liebhafsky, Roe, 1979)

$$d[I^-]/dt = R_3 = (k_a + k_b[I^-])\ [I^-]\ [IO_3^-][H^+]^2. \qquad (6)$$

If one works in buffered solution and assumes (Hanna et al., 1982) that iodide and iodate are the only stoichiometrically significant iodine-containing species and that both of these species have the same diffusion constant D, then on adding the diffusion term, (6) yields a one-variable reaction-diffusion equation:

$$\partial c/\partial t = D\,\partial^2 c/\partial x^2 + (k_A + k_B c)\,c\,(A - c) \qquad (7)$$

where $c = [I^-]$, $A = [IO_3^-]_o$, $k_A = k_a[H^+]^2$ and $k_B = k_b[H^+]^2$. On introducing the ansatz of a wave solution travelling with constant velocity v:

$$c(x, t) = c(x - v\,t) \qquad (8)$$

one obtains the ordinary differential equation

$$Dc'' + vc' + (k_A + k_B c)\, c\, (A - c) = 0$$

which can be solved exactly to yield

$$c(x, t) = A/(1 + B\, e^{kA(x - v\,t)}) \quad (9)$$

where B is an arbitrary constant, $k = A(k_B/2D)^{1/2}$ and

$$v = (Dk_b/2)^{1/2}[H^+]_o\, [IO_3^-]_o + k_1(2D/k_b)^{1/2}[H^+]_o \quad (10)$$

The analytical solution for the waveform (9) and the explicit expression for v as a function of the parameters (10) make possible a detailed comparison between theory and experiment. The results are quite good. It is also possible to treat the case of excess iodate, though a complete analytic solution is not available. The reader is referred to Saul and Showalter (1985) for further details.

Successful theoretical treatments have also been carried out of more complex one-dimensional wave phenomena. Periodic precipitation phenomena lie beyond the scope of this chapter, but interested readers will find thorough discussions of recent developments in the theory of Liesegang rings and related phenomena in Kai et al. (1982) and in Ortoleva and Schmidt (1985) and in chapter 8 by Müller. We conclude this section with a look at the qualitative theory (Nagypál et al., 1986) developed to explain the puzzling gravity effects shown in Figures 2 and 3.

In order to understand why wave velocities in the ferrous-nitric acid and chlorite-thiosulfate reactions vary with the direction of propagation, one must take into account several factors. First, the reactions are significantly exothermic; the temperature rises by several degrees as the wave front passes (Bazsa, Epstein, 1985). A second, less obvious, factor is that the isothermal density, i. e. after correction for the heating and resultant thermal expansion, of the reaction mixture increases during the course of the reaction as a result of the changing composition of the solution. This latter fact was established (Nagypál et al., 1986) via careful measurements of the temperature and volume of the reaction mixture. Finally, excess heat can leave the solution through the walls of the tube.

Whether the wave travels more rapidly in the upward or the downward direction depends upon a balance between the thermal expansion and the *chemical contraction*. Figure 6 illustrates the model in a case where the contraction is dominant so that the maximum wave velocity occurs for downward travelling waves. Suppose that the reaction is initiated in a small sphere (Figure 6 a). The temperature rises within this sphere because of the heat of reaction. The temperature in a larger sphere surrounding the reacted sphere at this moment is equal to that of the bulk solution, and the density of the reacted sphere is less than the density of the environment. Heat exchange takes place between the reacted sphere and its unreacted environment, leading to the relations in Figure 6 b. Since the isothermal density of the reacted sphere is greater than that of the unreacted environment, the reacted sphere is denser than its surroundings. Thus, the warmer but heavier reacted sphere drops down (Figure 6 c). These processes - together with diffusion and chemical reaction - occur simultaneously, without a sharp break. The observable net effect is a downward extension of the reacted sphere, as illustrated in Figure 6 d.

The model implies that by changing the conditions, e. g. by using tubes with thicker walls that transfer heat less efficiently or by choosing reactant concentrations that produce a greater temperature rise, it should be possible to revers the preferred direction of propagation. That this is indeed the case is illustrated in Figure 3.

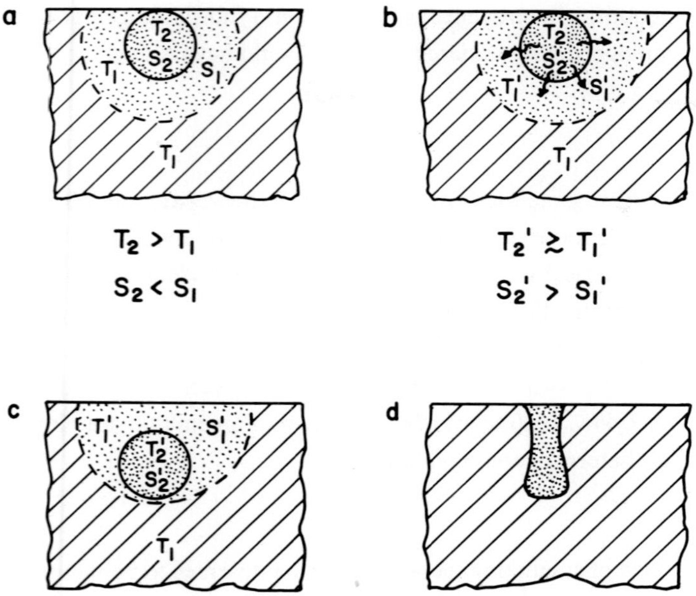

Fig.6 Schematic model of wave propagation in system where waves move most rapidly in downward direction. High concentrations of dots indicate unreacted solution immediately surrounding reacted sphere, diagonal lines show bulk unreacted solution, and arrows indicate heat dissipation. T - temperature; S - density (Nagypál et al., 1986).

Two-dimensional waves and patterns

Numerical calculations of wave propagation in two dimensions are conceptually equivalent to, though computationally more elaborate than, those in one dimension. More interesting are the analytical efforts aimed at understanding the origins of target patterns and spiral waves. Tyson and Fife (1980) have used a two-variable approximation to the Oregonator to show how target patterns might develop from inhomogeneous nuclei in the BZ reaction. Other theories (e. g. Zaikin, Kawczynski, 1977; Kopell, Howard, 1981) are able to account for target waves without invoking spatial heterogeneity. The primary difference between the predictions of these theories is that patterns requiring heterogeneous nuclei should show a range of oscillatory periods, while homogeneously initiated patterns should have a single value of the period (Tyson, 1985).

Spiral waves have been generated from a number of theoretical models, with biological applications predating those of chemical relevance. Gul'ko and Petrov (1972) obtained spiral waves in a numerical study of a reaction-diffusion model of excitable neuromuscular tissue. Winfree (1974) appears to have been the first to model spiral waves in a *chemical* system, using a pair of reaction-diffusion equations with an idealised piecewise linear kinetics in his numerical simulations. Several analytical studies (e. g. Greenberg, 1980; Zykov, 1980) exist of spiral waves in media that are either oscillatory or excitable under heterogeneous conditions.

One other approach of simulating these phenomena deserves to be mentioned here. It is the use of the cellular automaton theory. In this technique, one divides the medium into cells, each of which may be in any of several states (e. g. oxidized, reduced, refractory). The time is also discretized. Rules are specified by which a cell in a given state at time t_i either remains in that state or undergoes transition to another state at time t_{i+1}. The rules involve the state of the cell under consideration and possibly those of neighbouring states as well. Examples of such calculations in which spiral waves and other patterns of interest are produced may be found in Greenberg and Hastings (1978) and Fraser and Kapral (1986).

Three-dimensional waves

Owing to the computational difficulties, numerical simulations of three-dimensional waves are exceedingly rare. Strogatz et al. (1984) have presented some beautiful colour computer-generated displays of what such objects as twisted scroll-waves might look like, but these are based upon the topological rather than the chemical properties of theses waves. Winfree and Strogatz (1983 a, b, c; 1984) in a series of papers have developed a topological theory of what types of structures may be found for three-dimensional reaction-diffusion structures. By applying a reasonable set of chemical and topological constraints (e. g. infinite concentration gradients are forbidden; far from organising centers, concentrations must approach quiescent uniformity), they are able to derive an *exclusion principle* that limits the numbers and connections of rings, twists and knots that any pattern may have. They proceed to classify spatial structures according to a set of quantum numbers which specify their complexity.

Significance and the future

Are the patterns and waves found in relatively simple, usually inorganic, chemical systems likely to be helpful in understanding pattern formation in biology, or are they simply curiosities that make for pretty pictures and dramatic lecture demonstrations? It seems clear to this author that these phenomena are of considerable significance to an elucidation of the mechanisms of biological pattern formation. Unless one invokes some vitalistic principle, the factors that give rise to the phenomena discussed above: reaction, diffusion and, in some instances, heat conduction and electrical or gravitational fields, are the only ones available to generate the structures that give living organisms their form and behaviour. The insights of Turing (1952) into the *chemical* basis of morphogenesis are all the more valid today as we have gained deeper knowledge of the chemistry of genetics and development.

What can one predict about the future direction of studies in this field? It seems clear that technological advances will make major contributions on the experimental side. Experiments like that of Müller et al. (1986) show that it is becoming possible to obtain high resolution information about even the most detailed substructures at the core of a wave. It should thus become possible not only to formulate, but also to test theories about this most difficult aspect of pattern formation. One may expect the number of systems in which waves and patterns are observed to increase rapidly, just as (Epstein et al., 1983) the range of systems showing interesting temporal behaviour has been extended well beyond the BZ reaction. Advances in numerical methods, particularly those for treating systems of partial differential equations, as well as new computer hardware should make possible numerical studies with more realistic chemical mechanisms rather than the simplified models that have generally been employed thus far. One may expect profitable interactions between those studying spatially distributed phenomena and those investigating the properties of coupled chemical oscillators, since diffusion is really nothing more than a linear, continuous coupling of elements in space. The prospects for breakthroughs of sufficient magnitude to enable chemical wave and pattern studies to have a real impact on biology have never been brighter.

Spatial Ordering Processes in Chemical Reactions

S. C. Müller

Introduction

Macroscopic spatial structures evolving from an initially homogeneous state are frequently observed in open chemical or biochemical systems in which the reactions occur far from thermodynamic equilibrium. Reports on the spontaneous formation of beautiful patterns can be found throughout the literature over the past century. Unifying views of the underlying principles and mechanisms could be developed only recently following the maturation of nonequilibrium thermodynamics and progress in understanding nonlinear system dynamics (Glansdorff, Prigogine, 1971; Haken, 1983). In the last two decades extensive theoretical and experimental work has been performed on this type of symmetry-breaking in space in which the highly nonlinear coupling of complex reaction kinetics with transport processes such as diffusion and convection is investigated (Vidal, Pacault, 1984; Zierep, Örtel, 1984; Field, Burger, 1985). Although research has achieved an unprecedented level of systematic methodology and quantification, it should be noted that most of the extensively studied chemical reactions exhibiting spatial order were discovered quite unexpectedly. Despite today's accumulated knowledge an element of chance still plays a crucial role and this may lead to the detection of still unknown types of ordering processes.

Thermodynamics and Pattern Formation in Biology

In this article a survey of various patterns forming in chemical and biochemical solutions is given with emphasis on recent experimental studies. Without being at all complete the selection is considered to be representative for the large variety of spatial and spatio-temporal organisation observed in predominantly liquid-phase (bio)chemical systems. Within this scope the presentation of experimental results is restricted to a mostly qualitative discussion. However, application of an experimental method which allows a quantitative description of the spatial distribution of reactants in a thin solution layer by means of a two-dimensional digital spectrophotometer (a computerised video apparatus combined with an appropriate optical setup) is discussed in some detail.

Experimental

The experiments required the preparation of two or more chemical solutions in water, in a gel-forming medium, or in an organic solvent. Reagent-grade chemicals and doubly-distilled water were used throughout. The solutions were filtered through a 0.44 um Millipore filter. The investigations were performed at room termperature, usually 24 $\pm$ 1 ^{o}C. Specific preparation procedures are given in the subsequent sections and the corresponding literature. The Petri dishes for thin solution layers were constructed from an optically flat bottom plate to which a shallow cylinder was glued. They were siliconised in order to minimise the influence of boundaries and microscopic scratches.

In all cases visual observations and photography are an important first step in pattern characterisation. For further analysis suitable, non-invasive optical techniques are used such as photometry, light scattering, beam deflection, and Schlieren-optics. In a few cases chemical analysis and measurements of electrochemical potential differences are performed. Solution layers are investigated with a two-dimensional (2D) spectrophotometer (Müller et al., 1985). This methodology is based on an apparatus which consists of precision optics for homogeneous illumination and imaging purposes, a video equipment for the ultraviolet and visible range, a video frame buffer, and a fast, large memory computer for the storage of digital images and their further processing. The raster solution is 512 x 512 picture elements each of which measures one out of 256 possible grey levels. Acquisition, and storage of the 2D data are feasible at a frequency of 30 frames per minute. Elaborate software routines have been developed for further evaluation of the image date (Müller et al., 1986).

The chemical and biochemical patterns

Runge paintings

An early example of the investigation of spatially inhomogeneous reactions is the work of Runge around 1850 (Runge, 1855). After impregnating a piece of filter paper with a reactant, he placed a drop or a sequence of drops of an appropriately chosen chemical counterpart on this paper at regular time intervals. He thus discovered a large number of pairs of solutions for which this procedure results, upon spreading of the drops into the surrounding area, in quite spectacular and colourful spatial structures. The example in Fig. 1 was prepared according to one of his recipes. (Runge provided a set of such 'paintings' from his laboratory as illustrations for each volume of his book 'Der Bildungstrieb der Stoffe' (Runge, 1855). The chemical reactions combined with the influence of the physical nature of the filter paper used lead to complicated interactions which cannot be discussed in this framework (Harsch, Bursenas, 1985).

Whereas the importance of Runge‘s methodological contribution to the development of chromatography is generally accepted, his more artistic approach to these ‘paintings‘ has not been equally appreciated by the scientists. However, the importance of his work as an essential forerunner of later, more systematic and mechanistic research on patterns in chemistry should be emphasised. In fact, he was the first to observe ringshaped regions of precipitated reaction product, many years before Liesegang rediscovered this phenomenon in 1896 and to whom the first systematic study is credited (Liesegang, 1896).

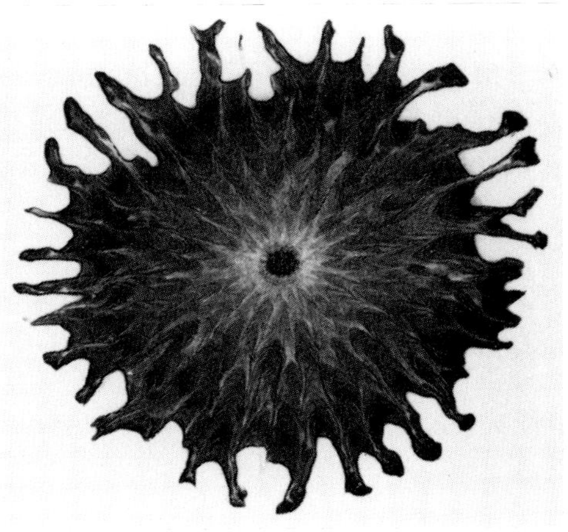

Fig. 1 'Painting' according to a recipe by Runge, produced by repetetively placing drops of a solution of potassium hydroxide on a filter paper impregnated with manganese and copper sulfate. Diameter approximately 10 cm.

Periodic precipitation

The periodic precipitation process known as Liesegang ring formation has been investigated over the past 90 years. In the first such experiments a drop of silver nitrate was placed on a gelatin layer containing potassium dichromate. The subsequent process of diffusion led to the deposition of silver dichromate in concentric rings (Liesegang, 1896). More generally: if a soluble electrolyte is placed in contact with a second electrolyte and, on interdiffusion, both react to form a poorly soluble salt, a metastable solution can be produced. For many combinations of electrolytes at suitable concentrations the subsequent precipitation process does not occur homogeneously in the volume. Instead, well separated bands of precipitate appear parallel to the diffusion front. They develop sequentially in a period of hours to several days. Usually, a gel-forming material is added to the solutions to prevent sedimentation and convection, but structures can be also observed without a gel in very thin layers (films) or capillary tubes. Excellent reviews (Hedges, Liesegang, 1954) and bibliographies (Stern, 1967) exist concerning this subject.

In a typical two-dimensional Liesegang experiment, the reaction is initiated in the centre of a Petri dish. The resulting pattern consists of a set of concentric rings. When the two reagents are placed in a test tube the direction of diffusion is confined to essentially one spatial coordinate and the precipitate is deposited in a set of parallel bands. Examples of lead iodide bands in agar gel are shown in Fig. 2. The bands in the thin tubes to the left (Fig. 2 a) differ in spacing and sharpness. This is a consequence of different initial concentrations of the two reacting salts, in this case potassium iodide and lead nitrate, the latter compound having a much lower concentration than the former one. As long as the initial difference between the electrolyte concentrations as well as the concentrations themselves are chosen to be sufficiently high, one finds that numerous bands appear at reproducible distances x_n from the origin of the imposed gradient. Spacing is usually according to x_{n+1}/x_n = const., but deviations have been reported (Packter, 1955; Müller et al., 1982). A stochastic element prevails when the initial concentrations of the two electrolytes are comparable (Müller et al., 1982).

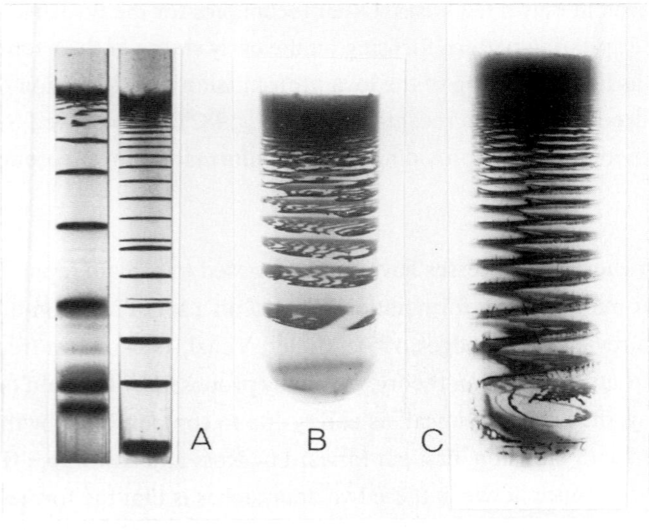

Fig. 2 Periodic precipitation patterns (Liesegang bands) of lead iodide in agar gel. (a) Parallel bands with large and small spacing, (b) sharp bands with regular structures within the bands, (c) helicoidal band. Inner tube diameter: 5.5 mm (a), 13 mm (b, c). (Picture(c) after Kai, Müller, 1985).

In tubes of large diameter additional structure formation tends to occur within each band as illustrated in Fig. 2 b. Since concentration gradients can be assumed to be very small or vanishing in a plane perpendicular to the tube axis, this phenomenon is probably related to pattern formation in initially homogeneous supersaturated layers of lead iodide in agar gel as reported in (Feinn et al., 1978). More complex structures are found in such precipitating systems, e. g. two-armed spirals instead of concentric rings, dislocations between rings (Kai, Müller, 1985) or a helical band instead of a set of parallel bands (Fig. 2 c). The detailed initial conditions required to produce such complex features are not known.

The number of phenomenological descriptions of precipitation phenomena is large, while the obviously important quantification of spatial concentration distributions has remained sparse. An early study yielded concentration data along a tube after performing a chemical analysis of thin sections of the gel column (Fricke, Suwelack, 1926). Surprisingly, in the subsequent literature such local experimental procedures have been reported in only a few cases. Other techniques for the investigation of local properties of precipitation pattern focusing on the early stages of their temporal evolution, such as light scattering of the local transmission and deflection of a thin beam of laser light, were applied quite recently (Kai et al., 1982). They proved to be efficient tools for experimental analysis providing relevant information for theoretical modelling.

Various physico-chemical processes have been proposed to explain repetitive band formation. Their mathematical formulation is based on a set of reaction-diffusion equations. For a review see (Hedges, 1954; Müller, Venzl, 1984). Most frequently discussed are the supersaturation theory, originally proposed by Ostwald (Wagner, 1950), and the hypothesis of chemical instability due to competitive growth of colloidal particles coupled with diffusion, first put forward by Ross and coworkers (Lovett et al., 1978). A major difference between these two approaches is that the former theory postulates pattern formation to occur at the nucleation stage, while the latter assumes uniform nucleation and subsequent patterning due to the autocatalytic nature of colloidal particle growth, meaning that larger particles grow at the expense of smaller ones. Recent investigations seem to indicate that a common basis may be found for both approaches which could solve the controversy of whether the patterns evolve from a pre- or a postnucleation phenomenon (LeVan, Ross, preprint).

Chemical waves

While the first liquid-phase oscillator, the decomposition of acidic hydrogen peroxide in the presence of iodate, was described by Bray in 1921 (Bray, 1921), it was only after the discovery of Belousov in the 1950s and the subsequent work of Zhabotinskii (Belousov, 1959; Zhabotinskii, 1964) that systematic research on chemical oscillations began. In the so-called Belousov-Zhabotinskii (BZ) reaction malonic acid is oxidized and decarboxylised by bromate in the presence of a metal ion catalyst, e. g. cerium or ferroin. Many other oscillators are known today (Epstein, 1984), but for the investigation of chemical waves the BZ reaction is still the most frequently studied system.

The propagation of such waves of chemical activity is perhaps the most striking example of pattern formation due to the coupling of reaction and diffusion (Field, Burger, 1985; Field, Noyes, 1974). It is usually investigated in thin layers, in a Petri dish, of a ferroin-catalysed BZ reaction. Patterns may evolve in an oscillating or in a quiescent, excitable medium. Here we only describe socalled *trigger waves* in an excitable reaction mixture.

The most common geometry is a circular wave or a set of concentric waves (*target* patterns). These are triggered by, e. g., a hot platinum wire or they evolve around impurities or small CO_2 bubbles. Because of the excitability of the surrounding regions this perturbation then travels outwards, while excited solution volumes gradually return - through a refractory period of time - back to a quiescent and newly excitable state. Until recently, quantitative data on geometric and kinematic properties and the detailed concentration distribution of wave patterns have been rare (Field, Burger, 1985; Field, Noyes, 1974; Showalter et al., 1979; Smoes, 1980). A recent investigation yielded quantitative descriptions of concentration profiles of circular waves by recording light absorption on a one-dimensional photodiode array (Wood, Ross, 1985). In further extension of such measurement spectrophotometric measurement in two dimensions was applied (see above) and results for circular waves are given in (Müller, Plesser, 1986). The application of such modern techniques is most useful in the analysis of chemical patterns, because a large amount of information is contained in the digitised image of a pattern which can be processed by computer software routines.

Spiral-shaped waves in an excitable medium have been the focus of our space-resolved investigations (Müller et al., 1985). These can be produced, for instance, by disrupting the front of an expanding circular wave with a gentle blast of air ejected from a pipette. The irregularly shaped open ends of the wave are the starting points of a quite rapid evolution towards spiral structures with highly regular geometry. Spirals are frequently arranged in pairs but more complex systems of many spirals may be easily observed, as illustrated in the digital image of Fig. 3. Each tip runs inward around a rotation centre with a period of about 18 s, while the fronts propagate away from this centre. Note that the collision of two fronts leads to their mutual annihilation without any interference phenomenon as is seen in many other physical systems.

Fig. 3 Digital image of transmitted light showing a set of spiral-shaped waves of chemical activity propagating in an excitable layer (thickness 1 mm) of the Belousov-Zhabotinskii reaction. Observation wavelength 490 nm (maximum ferroin absorption). Image area: 10 x 10 mm^2.

The shape of the spirals is Archimedian outside a certain region around the centre - the spiral core. This was proved within experimental error by fitting simple mathematical functions to iso-concentration lines extracted from such digital images (Müller et al., 1985). Furthermore, the location of the rotation centre could be determined precisely by applying an overlay technique to a sequence of spiral images covering one or several revolutions (Fig. 4). In addition, this procedure reveals the detailed structure of the core which is shown in Fig. 5 by a three-dimentional display technique and resembles that of a chemical 'tornado'. One finds that inside the core (diameter ≈0.7 mm) a transition takes place from the rotation centre, a singular site (diameter ⩽ 30 µm) at which the concentration of the catalyst remains quasi-stationary, to the surrounding area where waves attain their full amplitude between maximum oxidation and reduction of the catalyst.

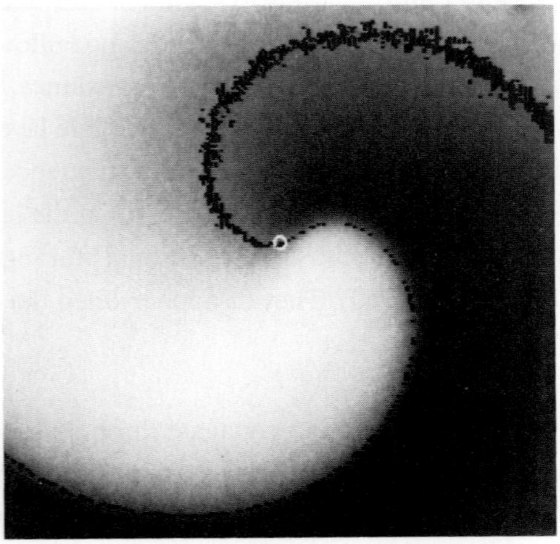

Fig. 4 The tip of the sprial wave recorded with high spatial resolution. The 1 x 1 mm^2 area of the image is obtained from an image of a 2 x 2 mm^2 section by twofold digital expansion. The white circle indicates the centre around which the wave rotates. The intensity level passing through the centre is presented in black.

For modelling of chemical waves, in particular spirals, we refer to the literature (Field, Burger, 1985; Keener, Tyson, 1986). It is clear that, with the new experimental data at hand, the analysis of the proposed partial differential reaction-diffusion equations will continue on a more quantitative basis than was possible previously.

Convection induced patterns in an excitable medium

Usually the sample dish containing a BZ reagent is covered by a glass plate in order to minimise evaporation at the layer surface. If, however, the cover is removed and the layer thickness exceeds 1 mm, a variety of spatial effects due to evaporative cooling are observed. The sensitivity of our photometric technique allows for the detection of several structural details (Müller et al., 1986). For instance, stationary patterns of high regularity ('mosaic' patterns) may evolve in an excitable layer that are independent of travelling chemical waves (Fig. 6 a). There is some evidence that these patterns are correlated with a network of convection cells set up in the layer due to temperature- or concentration-induced gradients of surface tension (Marangoni convection) (Platten, Legros, 1984; Müller et al., 1985). They disappear when the dish is covered.

In the circumstance that wave fronts propagate through the detection area it is difficult to observe such stationary patterns, because, usually, the difference in transmitted light between the fronts and the tails of waves is considerably larger than that between the bright and dark areas. However, with a specific choice of initial concentration the coexistance of waves and line patterns indicating the boundaries of convection cells can be visualised. Wave fronts tend to be distorted at the points of intersection with the line network (Müller et al., 1987).

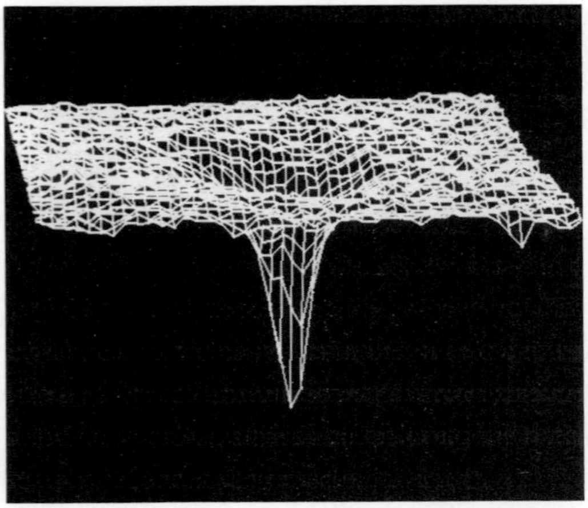

Fig. 5 The spiral core in three-dimensional perspective. The grid image represents the time-averaged level of maximum ferriin concentration (maximum oxidation) in a square section of 1.5 x 1.5 mm^2. It is derived from an overlay of an image sequence covering four spiral revolutions.

The interaction of natural convection due to evaporative cooling with a reaction-diffusion pattern of a given wavelength frequently results in a highly disordered structure as shown in Fig. 6 b. A strong distortion of a front may be followed by a complete wave disruption thus giving rise to order-disorder transitions, which were discussed and predicted on theoretical grounds for the case of controlled Bénard convection (Walgraef, 1984). Recently, image sequences of the decomposition of spiral waves were treated by image analysis techniques with the goal of finding numerical criteria for the characterisation of disorder and order in such chemical patterns (Markus et al., 1987). In this context, the question as to whether spatial chaos occurs is most interesting.

Convection in a non-reactive solution

In many of the experiments performed in reactive pattern-forming solution layers the influence of convective currents plays an important role, as shown above for the BZ reaction. In general, it is difficult to distinguish between the coupling of reaction to diffusive or to convective transport. It is important to note that even without reaction, spatial ordering of a chemical compound may occur due to natural convection. Experimental evidence to support this exists for a nonreactive biochemical mixture:

reduced nicotinamide adenine dinucleotide (NADH) in water or in buffer solution (Müller, Plesser, 1984).

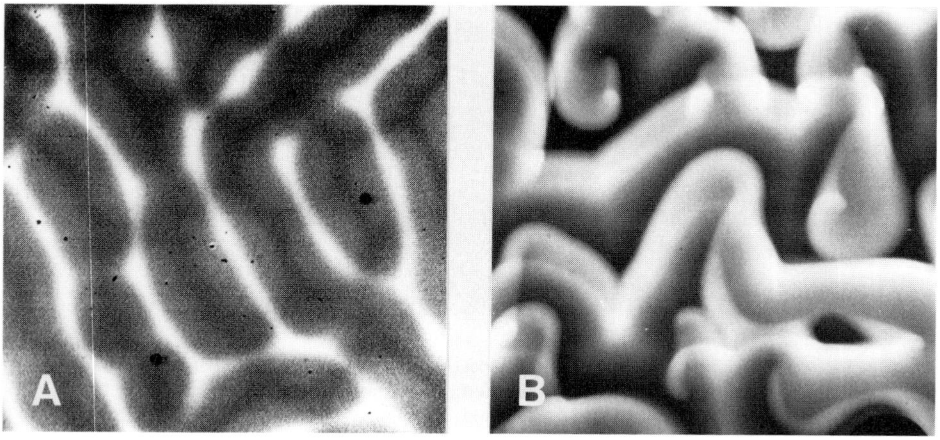

Fig. 6 Convective effects in open layers (thickness 1.5 mm) of an excitable Belousov-Zhabotinskii reaction. (a) Stationary ('mosaic') structure without wave propagation; (b) distortion and partial decomposition of a previously regular pattern of chemical waves. Image section: 9 x 9 mm^2. (Picture (a) after Müller et al, 1987).

About four minutes after pouring a concentrated solution of typically 5 mM NADH in a Petri dish measurements with the 2D spectrophotometer in the ultraviolet (380 nm) leads to the detection of faint rod-like structures. These are shown in Fig. 7 after conversion of intensities of transmitted light to concentrations of NADH. The structures arise only in an open dish in which temperature gradients up to 0.5 $^{o}C/cm$ are measured in the horizontal direction. They fail to develop when the dish is covered, thus preventing the evolution of temperature gradients due to evaporative cooling of the liquid/gas interface. Other nonreactive solutions exhibit similar structuring phenomena, but a critical threshold of concentration may have to be exceeded in order to produce spatial inhomogeneities (Müller et al., 1985).

The importance of the critical thermal and solutal Marangoni numbers for the onset of structure formation has been pointed out (Platten, Legros, 1984; Müller et al., 1985). For these parameters the variation of surface tension with temperature and chemical composition plays a crucial role. No detailed mechanism for such two-component solute-solvent systems has been put forward yet, but it is likely that the fundamental conditions for spatial symmetry-breaking are to be found in the effect of unbalanced forces at the solution interface.

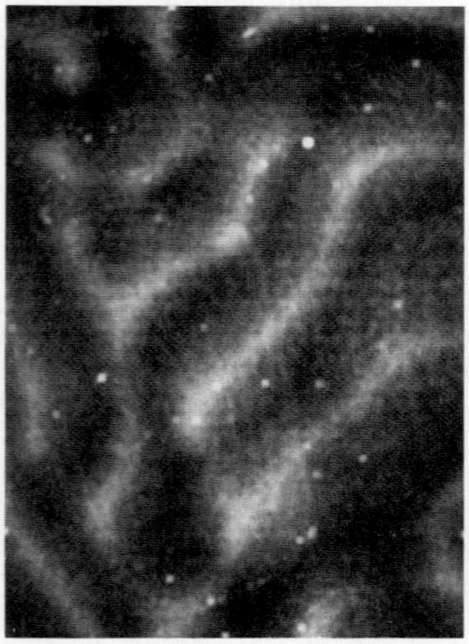

Fig. 7 Roll-shaped pattern in an open layer (thickness 1.8 mm) of a 5 mM solution of NADH, observed at 380 nm. The intensity of transmitted light is converted into absorbance and subsequently contrast-enhanced by digital methods (after Müller, Plesser, 1984).

Patterns in oscillating glycolysis

Glycolysis in yeast cells or in cytoplasm extracted from these cells is the most extensively studied example of an oscillating biochemical reaction (Hess et al., 1964; Hess, Boiteux, 1968). The protein content (about 50 mg/ml) of the extracted medium, in which most investigations are performed, corresponds to roughly one third of the concentration in the cytoplasm of an intact cell. If the medium is prepared according to (Hess, Boiteux, 1968) the metabolic degradation of glycolytic substrates such as glycogen is performed in an oscillatory manner with periods of the order of several minutes.

In thin open layers of the cytoplasm one observes the repetitive formation of regular structures (length scale 1 mm) at a wavelength specific for the absorption of NADH (Müller et al., 1985). More insight into the coupling of biochemical metabolism and transport processes is achieved by using the UV-sensitive 2 D photometer. A digital image of a typical pattern, taken at the moment of maximum intensity contrast, is shown in Fig. 8 a. While the structures tend to have predominantly rod-like shape during the first one or two phases of pattern occurrence, they break up into smaller patches at later times of an experiment (picture a).

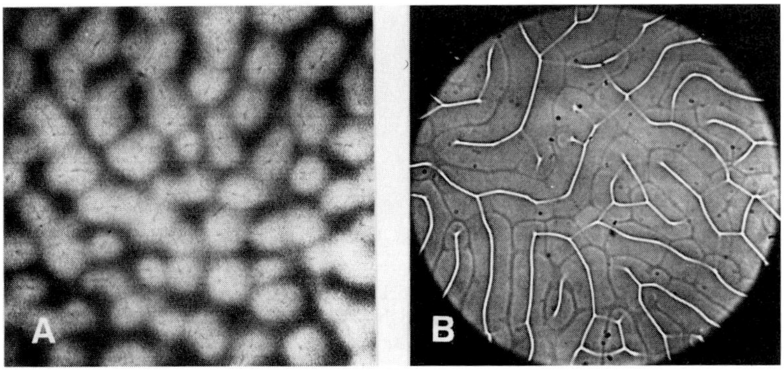

Fig. 8 Patterns in an open layer of yeast cytoplasm (thickness 1.8 mm) with oscillatory metabolic turnover. (a) Repetitively occurring patches as detected with transmitted diffuse light in the NADH absorption band (at 370 nm); (b) stationary line pattern as detected with a parallel light beam in the visible (at 560 nm). Image section: 14 x 14 mm^2. (Picture (a) after Müller et al., 1986).

The pattern morphology is mainly determined by a network of convection cells which, after a short induction period, is permanently present in the sample layer (Müller et al., 1985). As in the previous sections, there exist gradients in surface tension due to inhomogeneities in temperature and/or chemical composition. In Fig. 8 a the cell network is reflected by thin dark lines at the centre of the bright patches. These are shadows cast by pronounced gradients in refractive index that exist at the cell boundaries and are directly related to the distribution of temperature and chemical composition in the liquid. While Fig. 8 a was observed in a diffuse light field, one can observe these boundaries in a much more pronounced manner by using a parallel beam of light (Fig. 8 b). In this image, taken at a nonabsorbing wavelength in the visible, one detects the patch-like convective pattern on the basis of changes in refractive index.

A temporal correlation of the glycolytic oscillations in the bulk of the layer with the time intervals of pattern formation is established by evaluating a time sequence of digital images (Fig. 9). The spatial average of the measured intensity values (Fig. 9 a) reflects the oscillations. The peaks in the standard deviation (Fig. 9 b), which is a heuristic measure of structure formation, indicate that patterns occur twice per glycolytic cycle at those times when the metabolic turnover is especially high (Müller et al., 1985). A comparison of the time course of local changes in NADH concentration at two adjacent sites, that exhibit maximum contrast in a pattern as in Fig. 8 a, reveals that twice per period there are characteristic differences in the time course of local oscillations which point to differences in the enzymatic activities at the corresponding sites (Fig. 9 c). They cannot be explained by changes of heat production by the reaction alone, because these were measured to have the same period as the metabolic oscillations (Plesser et al., 1985).

The glycolytic system thus generates oscillatory reaction-convection patterns in which symmetry is broken by an interfacial instability. Observations of chemical patterns of related nature have been reported in (Micheau et al., 1983).

Transport coupled with enzymatic activity

Transport processes in an enzymatic reaction can be demonstrated in a Petri dish by the enzyme lactate dehydrogenase (LDH) that oxidises NADH in the presence of pyruvate.

The experiments are performed so that a small drop of enzyme solution is pipeted into a solution layer containing the substrates NADH and pyruvate at almost equal concentrations. With the 2 D photometric technique, local changes of NADH concentration are detected at 370 nm. The changes are caused by dilution resulting from the addition of the drop and by the following oxidation reaction.

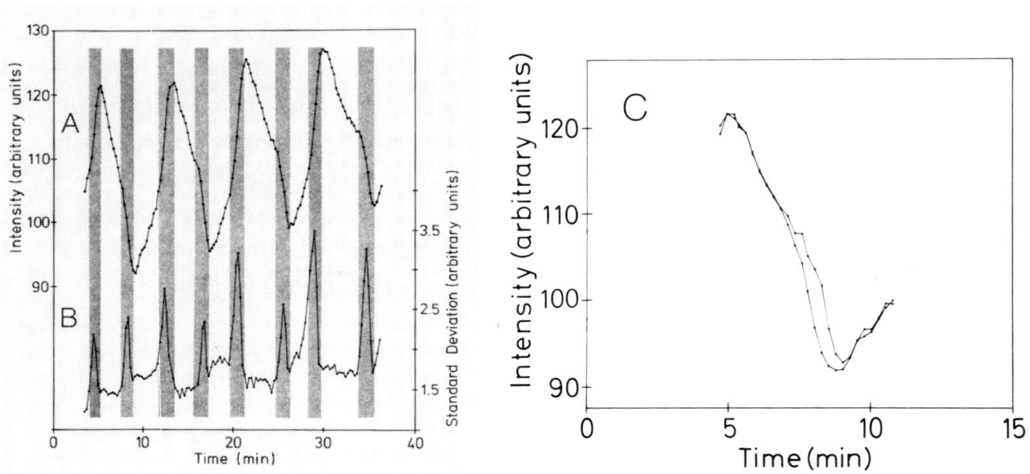

Fig. 9 Temporal relation between glycolytic oscillations and the formation of patterns. The average of transmitted light (a) and its standard deviation (b) are plotted at intervals of 15 s. The peaks in (b) correspond to those time intervals during which patterns are visually detectable on the observation screen (shaded stripes). (c) shows the transmitted light intensity at two nearby 10 x 10 pixel areas during one half-period of the oscillations (after Müller et al., 1985).

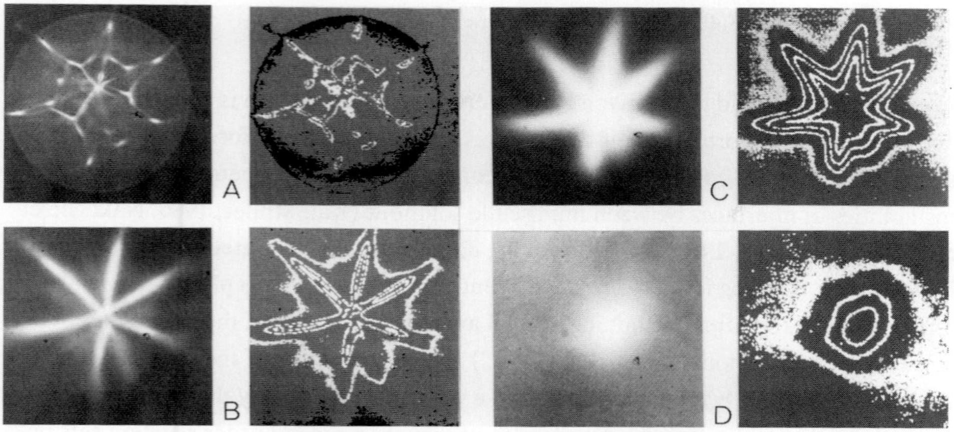

Fig. 10 Spreading of enzymatic activity after addition of a drop of LDH to a layer of NADH/pyruvate solution, observed at 370 nm. The initial geometry is to a large extent determined by differences in surface tension between drop and substrate medium. The left column contains a time sequence of original images recorded 5, 10, 15, and 25 min after drop addition (a to d). In the right column these images are transformed into contour maps. Observations area: 9 x 9 mm^2.

If the surface tension of the drop is adjusted to that of the layer, the disturbances at the moment of drop addition are minimised and the spreading of oxidation is mainly governed by diffusion processes leading to a smooth concentration distribution (Müller et al., 1986). If, however, the surface tension of the drop differs from that of the solution, a quite different pattern is produced. Right after the layer drop touches the layer surface, it 'explodes' and the enzyme material is instantaneously spread out in a repoducibly star-like fashion (Fig. 10 a). The detailed features of the initial geometry becomes blurred in the course of time (Fig. 10 b, c) because diffusion overtakes the scene and completely dominates in the last picture of the sequence (Fig. 10 d). The contour lines in the right picture column of Fig. 10 represent specific iso-intensity lines and emphasise structural details.

Dynamical motion at an oil-water interface

Convection in a liquid layer driven by gradients of surface tension is always accompanied by deformations of the free surface. While these deformations remain small in the cases discussed above, they become much larger in chemically-driven instabilities at interfaces between immiscible solutions (Kai, Müller, 1985; Nakache et al., 1983; Sackmann, 1985). As an example, if nitrobenzene saturated with potassium iodide and containing iodine at small concentrations (≈10 mM) is placed in a beaker and an aqueous solution containing a small amount (≈10 mM) of the surfactant trimethylstearylammonium chloride (TSAC) is pipetted on top of the organic solvent, a system consisting of two liquid phases with a well-defined, initially flat interface is formed. Immediately after the start of the experiment a second order reaction sets in that removes the surfactant dissolved in the aqueous phase from the interface by forming a precipitating yellow reaction product. This chemical process induces local variations in surface tension, and thus leads to an interfacial instability with the potential of creating a large variety of spatio-temporal patterns of macroscopic dimensions.

Fig. 11 Snapshot of a twofold wave-like distortion of the interface between an aqueous solution containing a small amount of surfactant and a nitrobenzene solution containing potassium iodide and iodine. The solutions were placed on top of each other in the space between the walls of two concentric cylindrical columns (double column container). Outer diameter: 50 mm. Courtesy: S. Kai, Fukuoka, Japan.

Single or multiple interfacial waves having amplitudes of typically 5 mm and rotating with a constant velocity of the order of 10 cm/s along the glass wall of the cylindrical container belong to the most spectacular examples of dynamical behaviour (Fig. 11). Depending on the solute concentrations, the container walls, and the details of initial conditions, many types of wave forms can be observed. Some of these are drawn schematically in Fig. 12. From a large number of measurements with different combinations of TSAC and iodine concentrations a phase diagram was established in the concentration plane consisting of several areas in which certain modes of motions predominate (Kai, Müller, 1985). In addition to periodic wave motion there are areas with unsteady, intermittent rotation, stretching or breather motion from the centre to the outer boundary of the vessel, nonperiodic (possibly chaotic) motion with relatively small amplitude and quiescent areas. The process is accompanied by the formation of a microscopic emulsion at the interface.

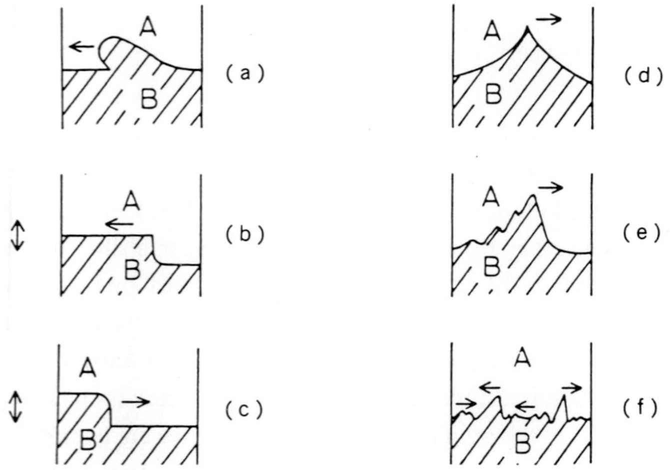

Fig. 12 Schematic drawing of typical wave forms at the interface between two immiscible liquids with similar chemical composition as in Fig. 11, placed in a cylindrical (single column) container with diameter 19 mm (after Kai, Müller, 1985).

A more quantitative characterisation of the modes of motion is the measurement of the electrochemical potential difference between platinum electrodes located close to the interface as a function of time (Kai, Müller, 1985). Although a considerable amount of experimental knowledge has been accumulated, no comprehensive theoretical model of this interfacial phenomenon has been proposed (Nakache et al., 1983; Sackmann, 1985).

There is an interesting variation of the experiment described above. When a small drop of the organic solution is surrounded by a large volume of the aqueous phase, the drop moves horizontally over the bottom of the container or along the vertical wall boundaries. The quite erratic motion is similar to the movements of an amoeba. The interfacial instability has therefore been considered as a model for the conversion of chemical into mechanical energy with relevance for biological systems (Sackmann, 1985).

The beating mercury heart

To conclude the list of spatio-temporal patterns in reactive systems we return to the last century. The first report on the periodic geometric distortion of the interface of a mercury drop is due to Lippmann in 1873 (Lippmann, 1873), although the original oscillator was observed by Kühne who apparently never published his results. The phenomenon belongs to the large family of electrochemical oscillators and is performed as follows: A mercury drop is placed in a watch glass and submerged in a pool of an acidic solution of an oxidiser, e. g. potassium dichromate in sulfuric acid. An iron needle is positioned such that it just touches the side of the mercury drop. This procedure results in pulsations of the drop on a 1 Hz scale lasting for an hour or longer. Depending on the geometry of the setup, the initial concentration, and the ageing of the chemical composition the oscillations are isotropic or the shape of the drop becomes triangular as illustrated by the two examples of Fig. 13. We observed more complex modes of motion with, for instance, four- or five-fold symmetry.

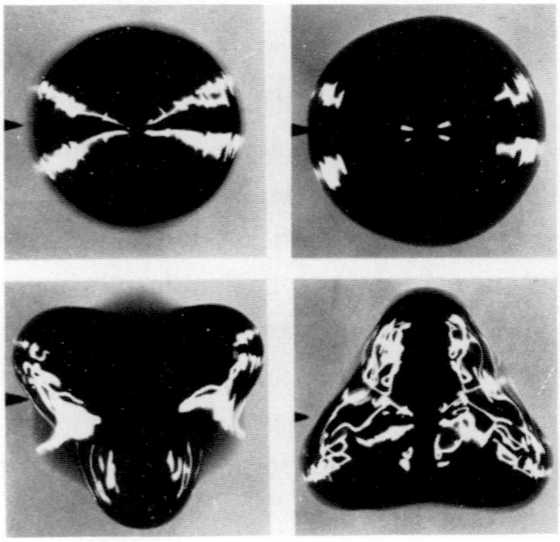

Fig. 13 Two modes of motion of the 'beating mercury heart'. Upper row: periodic isotropic contraction and expansion; lower row: oscillation between two triangular shapes. The tip of the iron needle can be seen at the left margin of each picture. Drop diameter approximately 1 cm.

The mechanism involves a redox chemical reaction, surface tension, and gravity. Oxidation of the drop leads to positive charges on the mercury and reduces its surface tension. Consequently the shape becomes flatter and broader in the gravitational field. Upon contact with the iron needle electrons can flow to the mercury, surface tension increases again, the drop contracts, and the metal contact is interrupted. A new cycle begins.

The electrochemical-mechanical oscillator has often been treated as a curiosity, but a number of related experimental systems have been found and more in-depth work for the elucidation of the molecular mechanism has put this phenomenon into the broader context of spatio-temporal order in chemical far-from-equilibrium systems (Keizer et al., 1979).

Concluding remarks

The examples of patterns presented here give some insight into the richness and complexity of spatio-temporal ordering processes in chemical and biochemical non-equilibrium systems. The combination of classical observation and measuring techniques with today's sophisticated technology significantly improves our experimental knowledge. The extension of space-resolved photometric methods to the ultraviolet and to the infrared may soon lead to the discovery of new classes of pattern-forming systems. First steps in this direction have been undertaken by investigating 'invisible' chemical waves in the ultraviolet (Ungvarai-Nagy, Müller, 1987).

The findings in the field of chemical organisation are confidently expected to fertilise other disciplines for which nonlinear behaviour plays a crucial role. In this context biological patterns on the subcellular and cellular level are of particular interest. These display an immense variety of shapes, forms and colours, but many of the structural elements can already be produced in a chemical or biochemical solution. There is a greater chance of explaining the ordering processes in these chemical systems and, once the relevant mechanisms are elucidated, to improve our understanding of the more intricate mechanisms acting in biology.

Evolution of Populations of Biothermodynamic Systems Including Birth and Death-Processes

J. U. Keller

Biothermodynamic Systems with limited lifetime (BTS)

We consider a piece of matter $\Sigma_{\tau\sigma}$ which during the time interval

$$\tau < t < \sigma$$

has clearly defined boundaries dividing its interior from its surroundings or environmental system Σ^*. The system may start its evolutionary process in a well defined 'birth' or initial state $\Sigma_{\tau\sigma}(\tau)$ at time τ and finalise it in another well defined 'death' or final state $\Sigma_{\tau\sigma}(\sigma)$ at time $\sigma > \tau$.

For times $t < \tau$ and $\sigma < t$ the system $\Sigma_{\tau\sigma}$ may exist in a physical sense - or not. In any case it is not considered to be a system itself but rather as being part of other systems or of the environmental system Σ^*.

Thermodynamics and Pattern Formation in Biology

We do not need to know the nature of the pre-birth- and the post-death-period of the system, since we here confine ourselves to the application of the laws of thermodynamics, which is possible when one can define, unambiguously, a birth- and a final- or death-state of the system.

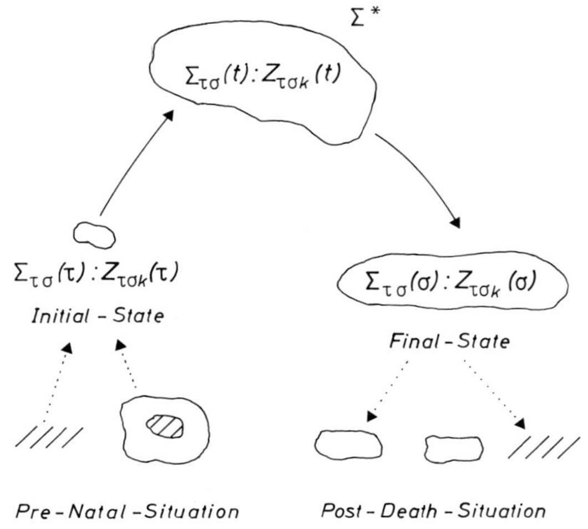

Fig. 1 Evolution of a Biothermodynamic System (BTS) from its initial or birth state ($\Sigma_{\tau\sigma}(\tau)$) to its final or death state ($\Sigma_{\tau\sigma}(\sigma)$), ($\tau < \sigma$). The symbols $Z_{\tau\sigma k}(t)$, k = 1,2 ... indicate extensive quantities like mass, energy, etc. of $\Sigma_{\tau\sigma}$.

The system $\Sigma_{\tau\sigma}$ in $\tau < t < \sigma$ may

- exchange heat, mechanical work and mass with its surroundings
- reproduce itself or any other kind of system.

Such a system we call a Biothermodynamic System with limited lifetime (BTS) (Keller, 1984). Obviously, bacteria are examples of BTSs. However, purely thermodynamic systems such as snow-flakes which grow in cold supersaturated humid air and again may dissolve or sublime in another environment, or the droplets of a liquid within an aerosol can also be considered as examples of such systems.

It should be noted that the symbols τ , σ have twofold meaning. Firstly, they indicate the instant of time when the BTS is considered to be created (τ) and annihilated (σ) respectively. Secondly, they serve as an identification code for the BTS itself to distinguish it from other systems. This 'code' may be represented as a pair of integers by multiplying τ , σ by appropriate powers of 10.

A single BTS (Σ) can be described thermodynamically by a set of extensive parameters

$$Z_k: U, V, m_1 \dots m_N \dots , \tag{1}$$

i. e. the internal energy, volume, masses of different components etc., which are defined uniquely (and are measurable) in any equilibrium- and non-equilibrium state. Isolating the system at a certain time t, it will approach an equilibrium state with unique values of the thermodynamic parameters conjugated to (1), i. e.

$$Z_k: T, P, \mu_1 \dots \mu_N \dots , \tag{2}$$

the temperature, pressure and chemical potentials of the various components. These parameters are the so-called 'accompanying intensive parameters' of the system. They only are realised within the system if it is in an equilibrium state. However, they serve as auxiliary quantities in any non-equilibrium state since they indicate average values of the (locally varying) temperature, pressure and chemical potential within the system.

The extensive and intensive parameters are related by Gibbs' equation (Schottky, 1973),

$$dS = \frac{1}{T} dU + \frac{P}{T} dV - \sum_i^N \frac{\mu_i}{T} dm_i + \dots \tag{3}$$

which for sake of brevity can be written as

$$dS = \sum_k Z_k \, d\, Z_k, \tag{4}$$

indicating that the intensive parameters in principle can be calculated from the 'accompanying thermostatic entropy'

$$S = S\,(Z_k, \qquad k = 1 \ldots K), \tag{5}$$

via the relations

$$Z_k = \left(\frac{\partial S}{\partial Z_k}\right)_{Z_1} \quad k = 1 \ldots K \tag{6}$$

Exchange processes of work, heat and mass between a BTS and its environmental system Σ^*, cause a change in the extensive variables which, according to the balance equations, can be described by

$$\dot{Z}_k = J_k + P_k \tag{7}$$

The flow terms J_k, and the production terms P_k are assumed to be functionals (or functions) of the intensive variables of the system Z_k, and of those of the environmental system Z_k^*, viz

$$J_k = \phi_k(z_1, z_1^*), \qquad k, 1 = 1 \ldots K, \tag{8}$$

$$P_k = \pi_k (z_l, z_l^*), \qquad k, l = 1 \ldots K \tag{9}$$

This is a consequence of the Second Law of Thermodynamics, i. e. Clausius' inequality (Keller, 1984). Given the functionals (8, 9) and the entropy function (5), the evolutionary process of the BTS can be calculated by iterative numerical integration of equations (5 - 9) starting from a given initial state.

Populations of Biothermodynamic Systems

Let us consider a thermodynamic system (Schottky, 1973) which includes a set of BTSs $\Sigma_{\tau\sigma}$, τ = ..., σ = ..., all of which belong to a single type, and these may be in various different states of their individual evolutionary process. Assuming all BTSs to being separate from each other, i. e.

$$\Sigma_{\tau\sigma} \cap \Sigma_{\mu\nu} = 0 \ldots \text{all } \tau, \sigma, \mu, \nu \tag{10}$$

with either $\tau \neq \mu$ or $\sigma \neq \nu$,
we formally have the additivity relation

$$\Sigma^{\circ} = \sum_{\tau\sigma} \Sigma_{\tau\sigma} + \Sigma^{*} \tag{11}$$

Σ^{*} being the environmental system.

The set of BTSs is called a population, its evolution being described by a so-called frequency distributions function

$$\nu = \nu(\tau, \sigma) > 0, \tau \leq \sigma. \tag{12}$$

The quantity $\nu \, d\tau \, d\sigma$ is the number of BTSs which are both created in $(\tau, \tau + d\tau)$ and annihilated in $(\sigma, \sigma + d\sigma)$. Since $\tau < \sigma$, we have

$$\nu(\tau, \sigma) = 0 \quad \text{for } \tau > \sigma. \tag{13}$$

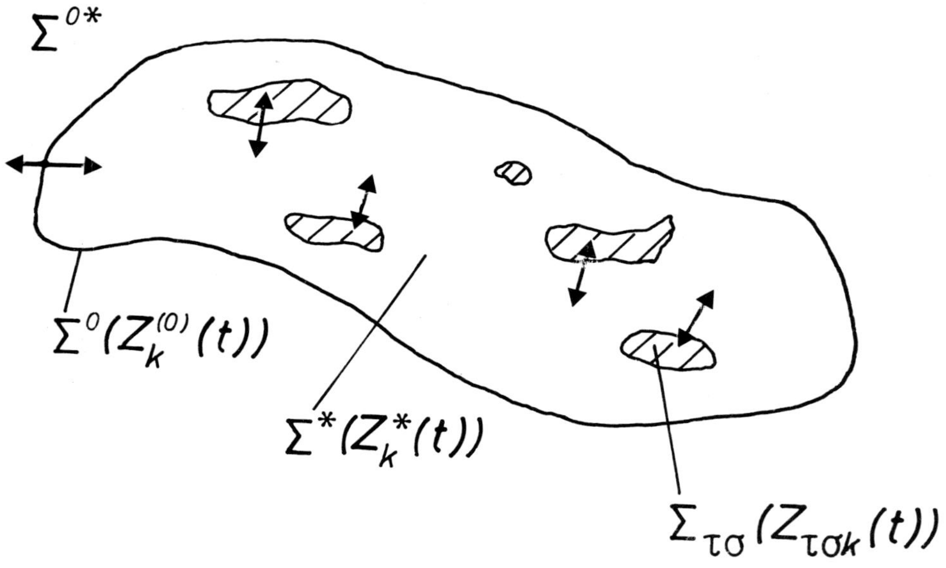

Fig. 2 A population of Biothermodynamic Systems $\sum_{\tau\sigma} \Sigma_{\tau\sigma}$ which together with its surroundings Σ^{*} forms the entire system Σ°.
..... exchange processes of heat, work and mass.

The number n(t) of BTSs within the population at time t is given by

$$n(t) = \int_{-\infty}^{t} d\tau \int_{t}^{\infty} d\sigma \; \nu(\tau, \sigma) . \tag{14}$$

This can be rewritten as

$$n(t) = \int_{-\infty}^{t} \nu^{+}(\tau, \infty)\, d\tau - \int_{-\infty}^{t} \nu^{+}(\tau, t)\, d\tau , \tag{15}$$

with

$$\nu^{+}(\tau, t) = \int_{\tau}^{t} \nu(\tau, \sigma)\, d\sigma \quad \ldots \quad t \geq \tau , \tag{16}$$

indicating the number of BTSs being created at time τ per unit of time <u>and</u> being annihilated in the time interval (τ , t).

The number of BTSs with a duration of existence (lifetime) within (t, t + dt) is given by N(t)dt ; the frequency distribution function being

$$N(t) = \int_{\tau=-\infty}^{\infty} \nu(\tau, \tau + t) . \tag{17}$$

The average lifetime of a system within the population is

$$\bar{t} = \lim_{t \to \infty} \frac{1}{N_t} \int_{0}^{t} t' \, N(t')\, dt' ,$$

with

$$N_t = \int_{-\infty}^{t} d\tau \int_{\sigma=\tau}^{\infty} d\sigma \; \nu(\tau, \sigma), \tag{18}$$

indicating the total number of systems which have been created in $(-\infty, t)$.

The time derivative of (14) reads

$$\dot{n}(t) = \int_{t}^{\infty} d\sigma \; \nu(t, \sigma) - \int_{-\infty}^{t} \nu(\tau, t)\, d\tau \gtrless 0, \tag{19}$$

implying that the net growth rate of the population is the difference between the number of BTSs which, per unit of time, are created at time t and those which, per unit of time, are annihilated at time t .

An integral equation for the distribution function ν of a population of BTSs

In this section we are going to derive a general evolution equation for the distribution function $\nu(\tau, \sigma)$ from the extensivity relation (11) and the laws of thermodynamics. In view of (10, 11) for any extensive quantity Z_k, $k = 1 \ldots K$, the additivity relation

$$Z_k^{(0)}(t) = \int_{-\infty}^{t} d\tau \int_{t}^{\infty} d\sigma \; Z_{\tau\sigma k}(t)\, \nu(\tau, \sigma) + Z_k^{*}(t) \tag{20}$$

must hold (cp. Fig. 2). The time derivative of (20) reads

$$\dot{Z}_k^{(0)}(t) = \int_t^\infty d\sigma\, Z_{t\sigma k}(t)\, v(t,\sigma) - \int_{-\infty}^t d\tau\, Z_{\tau t k}(t)\, v(\tau,t) + \\ + \int_{-\infty}^t d\tau \int_t^\infty d\sigma\, v(\tau,\sigma)\, \partial_t Z_{\tau\sigma k}(t) + \dot{Z}_k^*(t), \quad k = 1,2,\dots,K \tag{21}$$

Specifying (21) for the entropy S which obviously is an extensive quantity too, we get

$$\dot{S}^{(0)}(t) = \int_t^\infty d\sigma\, S_{t\sigma}(t)\, v(t,\sigma) - \int_{-\infty}^t d\tau\, S_{\tau t}(t)\, v(\tau,t) + \\ + \int_{-\infty}^t d\tau \int_t^\infty d\sigma\, v(\tau,\sigma)\, \partial_t S_{\tau\sigma}(t) + \dot{S}^*(t). \tag{22}$$

Let us introduce Gibbs' equation for the various systems. Using the notation introduced in (4) we have

$$\Sigma^0(t) : \dot{S}^{(0)} = \sum_k^K z_k^{(0)} \dot{Z}_k^{(0)}, \tag{23}$$

$$\Sigma_{\tau\sigma}(t) : \dot{S}_{\tau\sigma} = \sum_k z_{\tau\sigma k} \dot{Z}_{\tau\sigma k} \qquad \tau = \dots, \sigma = \dots \tag{24}$$

$$\Sigma^*(t) : \dot{S}^* = \sum_k z_k^* \dot{Z}_k^* . \tag{23*}$$

In additon we utilise the Gibbs-Duhem-relations for the initial and the final state of $\Sigma_{\tau\sigma}$, viz.

$$\Sigma_{\tau\sigma}(t=\tau) : \quad S_{t\sigma} = \sum_k^K z_{t\sigma k} \dot{Z}_{t\sigma k}, \tag{25}$$

$$\Sigma_{\tau\sigma}\,(t=\sigma): \quad S_{\tau t} = \sum_{k} z_{\tau t k}\,\dot{Z}_{\tau t k}. \tag{26}$$

Inserting (23 - 26) in (22) and substracting the resulting equation from (21) (which itself has been multiplied by the intensive variables of the environmental system Σ_k^* and summed over $k = 1, 2 \ldots K$) we can eliminate the extensive variables Z_k^* of the environmental system to get

$$\int_t^{\infty} M(t,\sigma)\,\nu(t,\sigma)\,d\sigma - \int_{-\infty}^{t} N(\tau,t)\,\nu(\tau,t)\,d\tau = \\ = L(t) + \int_{-\infty}^{t} d\tau \int_t^{\infty} d\sigma\, K(\tau,\sigma,t)\,\nu(\tau,\sigma). \tag{27}$$

Here we have introduced the abbreviations

$$M(t,\sigma) = \sum_{k=1}^{K} (z_k^*(t) - z_{t\sigma k}(t))\,\dot{Z}_{t\sigma k}(t), \tag{28}$$

$$N(\tau,\sigma) = \sum_{k} (z_k^*(t) - z_{\tau t k}(t))\,\dot{Z}_{\tau t k}(t), \tag{29}$$

$$L(t) = \sum_{k} (z_k^*(t) - z_k^{(0)}(t))\,\dot{Z}_k^{(0)}(t), \tag{30}$$

$$K(\tau,\sigma,t) = \sum_{k} (z_{\tau\sigma k}(t) - z_k^*(t))\,\dot{Z}_{\tau\sigma k}(t). \tag{31}$$

Equation (27) is a modified balance equation of the distribution function $\nu(\tau,\sigma)$. This is made clear by comparing the l. h. s. of (27) with (19) noting that the auxiliary quantities M and N are connected to the initial and the final state of the BTSs. Therefore, the l. h. s. of (27) can be interpreted as a generalised (or weighted) storage term, whereas on the r. h. s. of (27), L indicates a flow term which includes external influences on the overall system, Σ^0 and K is the kernel function of a production term and this includes because of its definition, (31) and (7 - 9), the evolutionary process of a single BTS and environmental parameters.

Given the entropy function (5), the constitutive equations (8), (9), well defined initial and final states of the BTSs considered and the external influences on the overall system Σ^0 (cp. (30)), the auxiliary quantities M, N, L, K can be calculated. Then (27) becomes an integral equation from which the distribution function $\nu(\tau, \sigma)$ of a population can, in principle, be calculated once initial conditions are given. This equation is a generalisation of an evolution equation which has been derived and discussed recently (Keller, 1984). Equation (27) appears to be linear in $\nu(\tau, \sigma)$. However, generally this will not be true. Simple examples prove that the auxiliary quantities L, K, internally, may again depend on ν which in nonlinear terms are introduced into (27) (Keller, 1987). Irrespective of this, it should be emphasised that the evolutionary process of a single BTS and the evolution of a population are generally not independent of each other, but the latter can be inferred from the former via (27). Since we have used only purely thermodynamic arguments, this seems to us to be a surprising result which should be discussed in the light of realistic examples.

Some special cases of the evolution equation (27)

In this section we would like to consider special cases of the general evolution equation (27).

Populations in large and/or closed evironmental systems.

If the environmental system Σ^* is very large compared to the whole population of BTSs, i. e. if symbolically the inequality

$$\Sigma^* \gg \sum_{\tau,\sigma} \Sigma_{\tau\sigma} \tag{32}$$

holds, the intensive parameters of the environmental and the overall system will be nearly the same, i. e. we have

$$z_k^* = z_k^0, \quad k = 1...K, \tag{33}$$

which implies (cp. (30))

$$L(t) = 0. \tag{34}$$

This relation also holds if the overall system Σ^0 is isolated from its surroundings Σ^{0*}, i. e. if

$$Z_k^{(o)} = \text{const}, \quad k = 1...K. \tag{35}$$

Population of zero-systems

Let us introduce the concept of a zero-system. By this we understand a BTS which starts and finishes its evolutionary process with zero mass, i. e.

$$m_{\tau\sigma}(t = \tau) = m_{\tau\sigma}(t = \sigma) = 0. \tag{36}$$

This implies that

$$Z_{t\sigma k}(t) = 0, \quad k = 1...K, \tag{37}$$

$$Z_{\tau t k}(t) = 0 \tag{38}$$

and, due to (28, 29)

$$M = 0 \tag{39}$$

$$N = 0. \tag{40}$$

Hence the evolution equation (27) reduces to

$$0 = L(t) + \int_t^\infty d\sigma \; K(\tau, \sigma, t) \; v(\tau, \sigma), \tag{41}$$

which after taking the time derivative can be rewritten as

$$\int_{-\infty}^{t} d\tau\, K(\tau,t,t)\, \nu(\tau,t) = \dot{L}(t) + \int_{-\infty}^{t} d\tau \int_{t}^{\infty} d\sigma\, (\partial_t K(\tau,\sigma,t))\, \nu(\tau,\sigma). \tag{42}$$

This relation again can be considered as a balance equation for the distribution function ν since the l. h. s may be interpreted as a modified (or weighted) storage term and in the r. h. s. a flow and a production term again occur. Zero systems can serve as useful idealisations of real BTSs. In view of (36 - 38) the initial- and the final states of such systems have to be defined by the <u>intensive</u> variables $z_{\tau\sigma k}(\tau)$, $z_{\tau\sigma k}(\sigma)$ of the system, the values of which normally will be different from those of the environmental system (z_k^*). This can be caused locally by spontaneous thermodynamic fluctuations, and/or by germs, as is the case for droplets within supersaturated water vapor.

Populations of systems with constant lifetime

Assume all BTSs within a population to be of the same kind and to anticipate the same, constant lifetime t_0. Then the distribution function can be written as

$$\sigma(\tau,\sigma) = \hat{\nu}(\tau)\, \delta(\sigma - (\tau + t_0)). \tag{43}$$

Inserting this in (27) we get the relation

$$M(t, t+t_0)\, \hat{\nu}(t) - N(t-t_0, t)\, \hat{\nu}(t-t_0) = \tag{44}$$

$$= L(t) + \int_{t-t_0}^{t} K(\tau, \tau+t_0, t)\, \hat{\nu}(\tau)\, d\tau,$$

from which the function $\hat{\nu}(\tau)$ can be determined. It should be noted that the number n(t) of BTSs present in the population at time t because of (14), (43) is related to $\hat{\nu}$ by

$$n(t) = \int_{t-t_0}^{t} d\tau\, \hat{\nu}(\tau) . \tag{45}$$

In addition we have, for the frequency distribution N of the lifetimes of the BTSs from (17), (43)

$$N(t) = N_{\infty}\, \delta(t - t_0), \tag{46}$$

with

$$N_{\infty} = \int_{-\infty}^{\infty} d\tau\, \hat{\nu}(\tau) . \tag{47}$$

Finally we would like to mention that for $t_0 \to \infty$ (44) becomes

$$\Lambda \hat{\nu} = L + \int_{-\infty}^{t} \hat{K}(\tau, t)\, \hat{\nu}(\tau)\, d\tau , \tag{48}$$

with

$$\Lambda(t) = M(t, \infty)$$

$$\hat{K}(\tau, t) = K(\tau, \infty, t)$$

and noting that

$$\hat{v}(-\infty) = \sigma .$$

Equation (48) describes the evolution of populations of BTSs including only birth- or creation processes but neglecting annihilation. This equation has been derived and discussed in earlier papers (Keller, 1984).

Examples of the application of the generalised evolution equation (27) and special application to populations of BTSs are in preparation and will be reported in a forthcoming paper (Keller, 1987).

Monitoring Oscillating Chemical Reactions: The Rate of Heat Production and the Simultaneous Measurement of Other Physical Signals*

Th. Plesser, I. Lamprecht

Introduction

Recent comprehensive surveys on periodic time behaviour of chemical (Field, Burger, 1985; Vidal, Hanusse, 1986) and biochemical (Hess, Boiteux, 1971) systems have underlined the increasing importance of experiments in the understanding of nonlinear phenomena far from equilibrium. Different experimental techniques have been applied to the study of oscillating chemical reactions, among them spectroscopy, polarography, ion sensitive electrodes, manometry, and chromatography.

Application of calorimetry and thermometry to oscillating reactions

All oscillating reactions studied so far have been accompanied by high heat production (Busse, 1971; Körös et al., 1973; Lamprecht, Schaarschmidt, 1978; Plesser et al., 1985;

---

*dedicated to Professor Benno Hess on the occasion of his 65th birthday

Thermodynamics and Pattern Formation in Biology

Nagy-Ungvarai, Körös, 1985; Rodrigues et al., 1986). Therefore calorimetry is a favourable method for analysis of the energy balances associated with different steps in the various reaction mechanisms. Although several authors have noticed highly exothermic reactions, only a few calorimetric experiments on oscillating chemical reactions have been reported. More often, temperature fluctuations have been monitored by means of thermistors or thermoelements. Using this method temperature fluctuations between 0.01 ^{O} and 1 ^{O}C have been recorded in batch or continuous stirred tank reactor (CSTR) systems, but no calorimetric or energetic consequences were drawn from these findings. Roux and Vidal (1979) stated a 'slight variation of temperature' in their CSTR during study of the Briggs-Rauscher reaction; Wirges (1980) and Hugo and Wirges (1980) compared experimentally detected temperature oscillations in the catalytic decomposition of hydrogen peroxide by iron nitrate with some extended computer models. Several experiments have been performed on the Belousov-Zhabotinskii (BZ) reaction in batch and CSTR systems. Frank and Geiseler (1971) detected temperature oscillations up to 1 ^{O}C around a constant level, Busse (1971) measured a temperature change of only 0.03 ^{O}C from which he calculated a heat production of 25 mcal/ml. Körös et al. (1973 a, b) stated that a sudden temperature jump, i. e. a high rate of heat production, is coupled with an increase in the redox potential and a decrease in bromide ion concentration. For the uncatalysed oxidation of gallic acid Körös and Orban (1978) described distinct steps in the rate of temperature change.

There are some oscillating reactions in homogeneous systems with distinct temperature variations (Hafke, Gilles, 1968; Vidal, Noyau, 1980) where the heat produced in the system is an essential part of the oscillation mechanism. In such 'thermokinetic reactions' temperature may be considered as an autocatalytic substance exerting a feed-back inhibition to the system (Degn, 1972; Vidal, Noyau, 1980). In all other reactions studied so far temperature oscillations are not essential in the autocatalytic behaviour of the mechanism. The reactions presented here are exclusively of the nonthermokinetic type.

Only a few true calorimetric experiments have been performed on oscillatory reactions. Lamprecht and Schaarschmidt (1978), Körös et al. (1979) and Rodrigues et al. (1986)

investigated the BZ reaction and measured high heat productions. Plesser et al. (1985) and Schaarschmidt et al. (1986) determined heat production and optical density in the oscillating glycolysis of yeast extracts and Lamprecht et al. (1987) described some calorimetric set-ups for simultaneous measurements of different oscillating parameters. These examples will be described in more detail later in this chapter.

Modern microcalorimetry with its small probe volumes and with rates of heat production matching the heat flow to the thermostated heat sink offers the chance to determine enthalpy changes under isothermal conditions.

Calorimetric measurements deliver data integrated over the whole sample volume in the calorimetric vessel. Thus spatial inhomogeneities may disturbe the results by affecting the heat diffusion within the sample as well as across sample boundaries. To render clear and reproducible temporal oscillations in the rate of heat production perfect homogeneity must be achieved by intensive stirring of the reacting mixture. On the other hand, the spatial distribution of heat production in a flat layer of an excited medium is of great interest. Winfree (1985) observed a local heating up of 0.1 ^{o}C in the wavefront of a BZ system and Bewersdorff (1986) demonstrated in a Spacelab experiment the transport of gas bubbles by surface tension due to temperature gradients in the vicinity of a wavefront. Oscillating glycolytic extracts were spatially scanned for concentration differences in NADH (Boiteux, Hess, 1980; Müller et al., 1984 a, b) but should also show temperature fluctuations. Modern infrared cameras of the IR-thermography type as used for gynaecologic, oncologic, or dermatologic investigations and their spatial resolution of 1 mm and their temperature discrimination, better than 0.1 ^{o}C, could be of great advantage in such experiments.

Instrumental set-up

The examples cited above emphasise the importance of modern calorimetry for such investigations, especially when this method is combined with other techniques. If at least one other physical signal of the reaction mixture is recorded simultaneously with the rate of heat production, phase relations can be established. Furthermore the shape of the oscillations allows for a deeper insight into the time correlation of the different parts of the reaction mechanism. To this end an isothermal isoperibol twin microcalorimeter was modified to permit measurement of optical density, electrode potential, or gas pressure together with the rate of heat production (Lamprecht et al., 1987).

The quantitative interpretation of this simultaneously measured data is not straightforward, because of differences in the time resolution of the devices. The thermal inertia of modern microcalorimeters results in a relaxation time in the range of the period of the oscillations. Ion sensitive electrodes have relaxation times around one minute. Only the measurement of optical density may be considered as an instant technique.

Instrumental time constants in the time range of the mechanism under investigation require deconvolution or 'desmearing' of the measured signal via a carefully determined resolution or transfer function of the apparatus (Hemminger, Höhne, 1984).

Details of the design of the different experimental arrangements, the measurement of the resolution functions, and the problems arising from numerical deconvolution of the recorded data are described in the following two sections. The last section is devoted to applications. Results are presented for a biochemical system (the periodic degradation of sugar in the glycolytic pathway) and for a chemical system (the metal ion catalysed bromination of malonic acid (Belousov-Zhabotinskii (BZ) reaction). The examples are intended to demonstrate the utility of modern calorimetry in the study of temporal patterns in chemical and biochemical reactions and to stimulate research in this field.

Experimental arrangements

A twin microcalorimeter has been modified for simultaneous measurements of the rate of heat production and one other signal. The time constant of the instrument is in the order of two minutes, but for each experimental setup the exact relaxation curve has to be carefully determined. Both the calorimetric and the other signal are digitised and stored on a microcomputer floppy disk for further analysis.

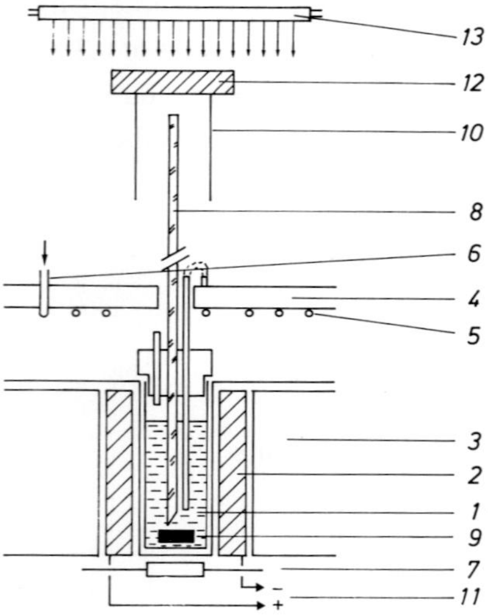

Fig. 1 Calorimetric device for simultaneous determination of the optical density and rate of heat production in a batch calorimeter.
(1) calorimetric vessel with cover and inlet/outlet tubes
(2) heat flow meter with leads (11) to the two channel recorder
(3) heat sink (thermostat)
(8) quartz rod serving as light guide
(9) photodiode
(12) optical filter
(13) light source (visible or UV).

Heat flow and optical density measurements

The device for simultaneous measurement of the rate of heat production and optical density is shown in Fig. 1. The simple optical recording system consisted of a quartz light guide and a photo diode inserted into the calorimeter vessel. The various parts of the equipment are listed in the figure caption. More details can be found in Schaarschmidt et al. (1986).

Fig. 2 displays two relaxation curves of the system recorded for a neutralisation reaction; the rapid mixing of 0.1 M HCl with an excess of 0.3 M NaOH in the presence of bromethymol blue as indicator. The difference in the relaxation behaviour is clearly visible. From the heat curve the relaxation constant of the calorimetric signal can be estimated to be about 117 seconds.

Heat flow and electrode potential measurements

The equipment for the simultaneous measurement of the rate of heat production and an electrode potential is shown in Fig. 3. A platinum electrode dips into the reaction vessel, a flexible salt-bridge connects the specimen with a calomel reference electrode (type 303, Ingold, Nürnberg) on top of the calorimeter. The various parts of the equipment are listed in the figure caption.

Fig. 4 exhibits the two relaxation curves recorded using this device. The relaxation time of the heat curve is 142 seconds, about 25 % larger than that of the optical output. This is due to the higher amount of solid material present in the reaction vessel.

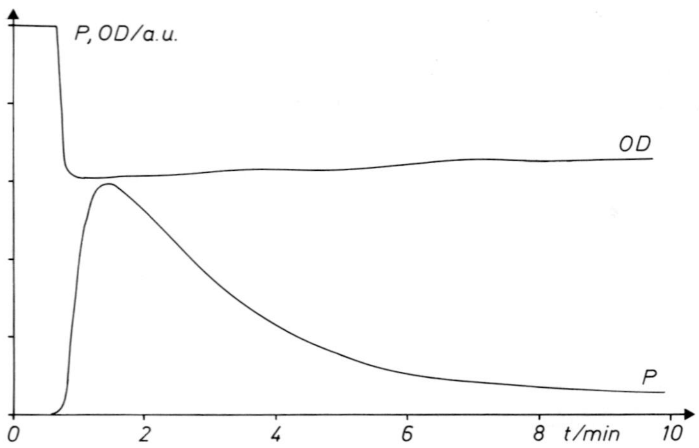

Fig. 2 Response curves measured with the apparatus shown in Fig. 1 for the neutralisation of hydrochloric acid (0.1 M) by sodium hydroxide (0.3 M) in the presence of the indicator bromethymol blue (3.95 mM). A relaxation time of 117 s is estimated for the rate of heat production from the dashed line. The optical density (solid line) follows the reaction faithfully.

Heat flow and pressure measurements

Many periodic reactions are known to result in considerable gas production (Bowers, Noyes, 1985), usually monitored by Warburg techniques or by gas sensitive electrodes

The equipment developed for simultaneous measurement of the rate of heat production and the pressure built up in a closed calorimetric vessel by the procution of gas is shown in Fig. 5. A semiconductor pressure transducer is connected to the air space above the liquid sample by means of a capillary. Its output, is fed directly to the second channel of the recorder. The time constant of the heat curve (Fig. 6) for this set-up is only 97 s, as no additional components are introduced into the liquid.

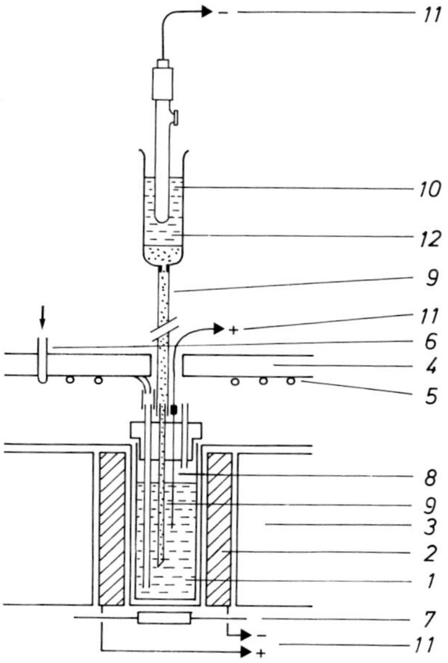

Fig. 3 Calorimetric device for the simultaneous determination of an electrode potential and the rate of heat production in a batch calorimeter.
(1) Calorimetric vessel with cover and inlet/outlet tubes
(2) heat flow meter with leads (11) to the two channel recorder
(3) heat sink (thermostat)
(8) platinum electrode
(9) salt bridge
(10) reference electrode

In oscillating reactions simultaneous measurement of heat and gas production is not as straightforward as the other techniques because liquids possess strong buffering capacities for gases. Therefore, gas production is detectable by this method only after the buffering capacity is exhausted. Preliminary calibration curves were run with sodium carbonate and sulfuric acid giving a large and nearly instantaneous pressure increase due to gas production (Fig. 6).

Deconvolution of the measured signal

A signal recorded by means of any time dependent physical instrument results from the interplay of at least two components, the 'true' signal produced by the specimen under investigation and the signal broadening inherent in the measuring device. The most significant impact of the apparatus is its inertia in instantaneously following the variation of the 'true' signal. The inertia of a device is quantitatively described by its response to a signal with pulse-like characteristic having a pulse width of order zero. The response curve r(t) is called the relaxation function, response function, or transfer function of the device.

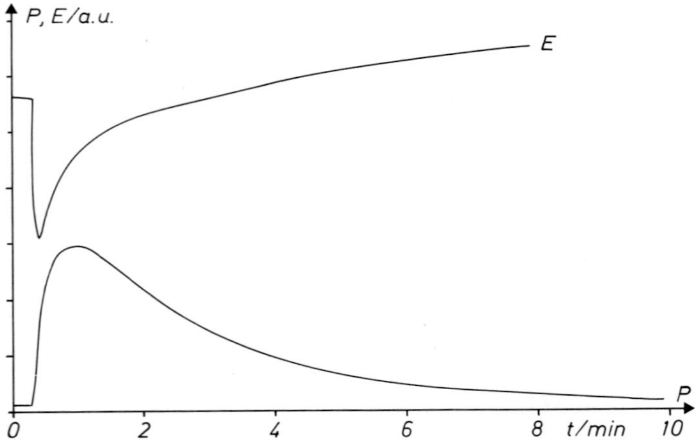

Fig. 4 Response curves measured with the apparatus shown in Fig. 3 for the neutralisation of hydrochloric acid (0.1 M). A relaxation time of 142 s is estimated for the rate of heat production from the dashed line and 40 s for the electrode signal (solid line).

In calorimetric experiments, often the broadening or 'smearing' of the signal due to the heat capacity of the calorimetric vessel and its content is taken into account by the so-called 'Tian-equation' (Calvet, Prat, 1963; Hemminger, Höhne, 1984). It is the sum of the signal and its first time derivative multiplied by the time constant of the apparatus. The applicability of the Tian equation depends on the construction of the calorimeter. The equation is exact for a Calvet calorimeter. Here we follow a more general approach.

The functional relation between the 'true' and the recorded signal is given by the following integral expression (Bracewell, 1965)

$$\varphi(t) = \int_{-\infty}^{+\infty} r(t-\tau)\, f(\tau)\, d\tau . \tag{1}$$

This formula means that the measured signal $\varphi(t)$ is a convolution of the 'true' signal f(t) with the response function r(t). It implies linear behaviour of the apparatus, i. e. the response r(t) is independent of the signal f(t).

For a pulse-like response function, r(t) can be approximated by Dirac's delta function (t), and from equation (1) is found to be

$$\varphi(t) = \int_{-\infty}^{+\infty} \delta(t-\tau)\, f(\tau)\, d\tau = f(t). \tag{2}$$

This means that the recorded signal is identical with the 'true' signal.

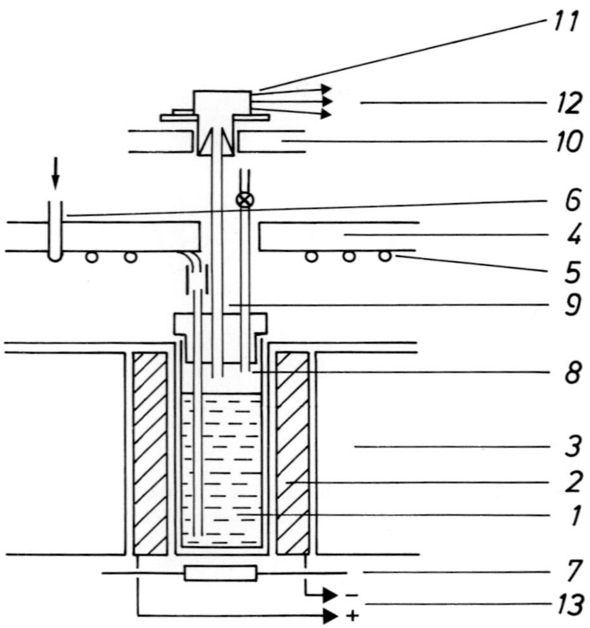

Fig. 5 Calorimetric device for the simultaneous determination of gas pressure and rate of heat production in a batch calorimeter.
(1) Calorimetric vessel with cover and inlet/outlet tubes
(2) heat flow meter with leads (12), (13)to the two channel recorder
(3) heat sink (thermostat) (4) upper calorimetric cover

Sophisticated deconvolution techniques must be applied if the response of the apparatus has a significant effect on the physical signal one is interested in. Three different, general numerical approaches have been considered in the literature:
- the 'direct' method by matrix inversion
- several iterative procedures,
- deconvolution by Fourier transforms.

We discuss only the last. A discussion of the advantages and disadvantages of the different methods may be found in (Carley, Joyner, 1979; Cesari et al., 1982 a, b; Cesari et al., 1984).

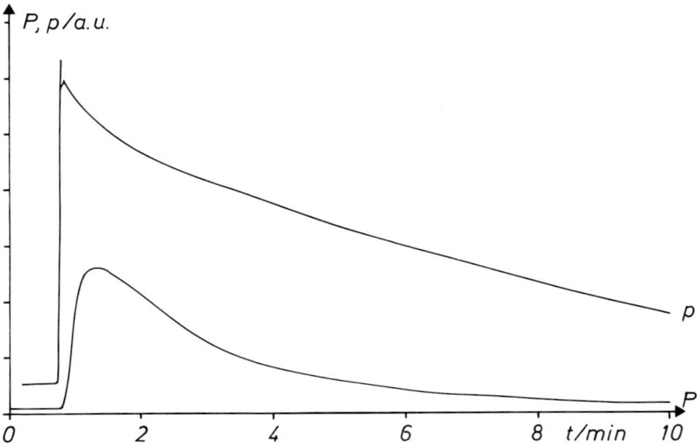

Fig. 6 Response curves measured with the device shown in Fig. 5 for the reaction of a saturated solution of sodium carbonate with sulfuric acid (1 M). A relaxation time of 97 s is estimated for the rate of heat production (dashed line).

Deconvolution by Fourier transforms is based on the convolution theorem for an equation of type (1). This theorem states that the Fourier transform of the convolution integral is given by the product of the Fourier transform of each function (Bracewell, 1965). The application of this theorem to equation (1) results in (3)

$$\mathcal{F}(\varphi(t) = \mathcal{F}(r(t)) \cdot \mathcal{F}(f(t)). \qquad (3)$$

(3) means the transition from the time domain into the frequency domain by the Fourier integral is as shown in the following example

$$\hat{\varphi}(\omega) = \mathcal{F}(\varphi(t)) = \int_{-\infty}^{+\infty} e^{i\omega t} \varphi(t)\, dt. \qquad (4)$$

The calculation of f(t) is straightforward since with

$$\mathcal{F}(f(t)) = \mathcal{F}(\varphi(t)) / \mathcal{F}(r(t)) \quad (5)$$

the inverse Fourier transform of (5) reveals the true signal

$$f(t) = \mathcal{F}^{-1}(\mathcal{F}(\varphi(t)) / \mathcal{F}(r(t))). \quad (6)$$

This very elegant solution of the deconvolution problem has some drawbacks. The most obvious is the large computation time required for the Fourier transforms. This problem can be minimised by the application of Fast Fourier Transform (FFT) techniques (Cooley, Tukey, 1965; Elliot, Rao, 1982; Harwell, 1985). The only prerequisites for the use of FFT are that the data points are equidistant in time and that their number is a power of two. More serious are the implicit properties of Fourier transforms themselves; the amplification of noise and the sensitivity with respect to cut-off errors. Noise reduction may be achieved by smoothing techniques, for example by splines (de Boor, 1978; Harwell, 1985) or polynomes (Savitzky, Golay, 1964); cut-off effects can be reduced by suitable extension of the measured functions.

Examples

In this section the deconvolution of periodic signals by FFT is demonstrated for two examples:

- periodic oxidation of glucose in a cytoplasmic medium extracted from yeast (Hess, Boiteux, 1971; Plesser et al., 1985)

- periodic bromination of malonic acid in the presence of metal ions (Belousov-Zhabotinskii reaction) (Field, Burger, 1985).

The evaluation procedure of the data proceeds in six steps:

Step 1 Smoothing of the data from experiments for the determination of the response function r(t) in the time interval $0 < t < t_r$. Method: spline fit. Algorithm: VC03AD from the Harwell Subroutine Library (Harwell, 1985)

Step 2 Smoothing of the measured signal $\varphi(t)$ in the interval $0 \leqslant t \leqslant t_{max}$. Method: spline fit. Algorithm: VC03AD from the Harwell Subroutine Library (Harwell, 1985)

Step 3 Extension of the time intervals of $\varphi(t)$ and r(t). Extension of $\varphi(t)$ for times less than zero: $\varphi(t) = \varphi(t = 0)$ in the interval $t_{min} \leqslant t < 0$ with $t_{min} \leqslant - t_r$. Extension of r(t) beyond t_r: $r(t) = 0$ for the interval $t_r \leqslant t \leqslant t_{max} - t_{min} - t_r$. The time interval for both functions is now $t_{max} - t_{min}$.

Step 4 Discretisation of $\varphi(t)$ in the range $t_{min} \leqslant t \leqslant t_{max}$: Discretisation of r(t) in the range $0 \leqslant t \leqslant t_{max} - t_{min}$. The range is divided by N intervals of the same length. N has to be a power of 2. Depending on the smoothness of the data and the range of the resolution function N = 256, 512 or 1024 is applied.

Step 5 Fourier transform of $\varphi(t)$ and r(t). Method: Fast Fourier Transform in place. Algorithm: FT01AD from the Harwell Subroutine Library (Harwell, 1985).

Step 6 The inverse Fourier Transform reveals

$$f(t) = \mathcal{F}^{-1}(\mathcal{F}(\varphi(t))/\mathcal{F}(r(t)))$$

the deconvoluted signal f(t). Method: Fast Fourier Transform in place. Algorithm: FT01AD from the Harwell Subroutine Library (Harwell, 1985).

The extension of the measured signal beyond zero minimises the cut-off errors at the beginning of the experiment. Details will be published elsewhere. The algorithm of the whole procedure is coded in FORTRAN. The set of subroutines as well as an executable code for microcomputers is available on request (Müller, Plesser, 1987).

Example 1 Periodic oxidation of glucose

The oxidation of glucose under anaerobic conditions in yeast is performed by a set of enzymes described as the glycolytic pathway. This metabolic process produces two moles of ethanol and of CO_2 per mole of glucose and a large amount of heat (Minakami, de Verdier, 1976)). There is a large endothermic step at the beginning of the pathway when the phosphorylated glucose molecule is split by the enzyme aldolase into the trioses, dihydroxyacetone phosphate and glyceraldehyde 3-phosphate (ΔH = + 42 kJ/mole glucose (Minakami, de Verdier, 1976) and a large exothermic step at the end of the pathway when acetaldehyde is transformed into ethanol and CO_2 by the enzyme alcohol dehydrogenase (ΔH = -94 kJ/mole glucose (Burton, 1974)).

The actual experiments are carried out with the disaccharide trehalose. Trehalose is split by the enzyme trehalase into two glucose molecules. This initial enzymatic step generates a constant glucose flux to the glycolytic pathway from the large trehalose pool thus maintaining the system just within the oscillatory window (Hess, Boiteux, 1969; Plesser et al., 1985).

Fig. 7 a displays the optical density and the rate of heat production recorded simultaneously with the experimental setup shown in Fig. 1. The variation in optical density reflects the concentration changes of the glycolytic metabolite nicotinamide adenine dinucleotide (NADH). From this figure one can estimate a phase shift between the two curves of about three minutes, one fourth of the 12 minute period.

Example 2 Periodic oxidation of malonic acid

The oxidation of malonic acid by bromide ions, the Belousov-Zhabotinskii reaction, is the most studied periodic chemical reaction (Field, Burger, 1985). It is a rather complex reaction mechanism with about 18 coupled reactions containing at least 21 chemical species which may oscillate for hours - even in batch systems. In its different modifications the BZ reaction exhibits periodic changes of optical density in the visible as well as in the ultraviolet range. Moreover the redox potential and the step-wise production of carbon dioxide are indicators of periodic turnover in the BZ reaction. In addition, several calorimetric and thermometric investigations have been made on this system so that it is a suitable subject for our deconvolution procedure.

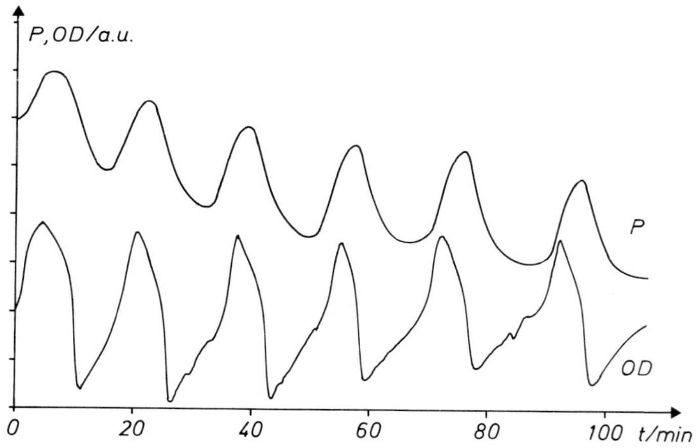

Fig. 7 a Simultaneous recording of the rate of heat production (P) and optical density (OD) for oscillating glycolysis in a cytoplasmic medium extracted from yeast. Both curves exhibit the same period and show a time lag between the maxima of about three minutes or 25 % of a period.

Fig. 7 b shows the optical signal and the heat signal deconvoluted via the dashed curve of Fig. 2. Both signals are now exactly in phase, the amplitudes are larger and the shapes of the oscillations are more alike. This leads to the suggestion that in glycolysis the redox couple NADH/NAD and the heat producing processes are highly coupled.

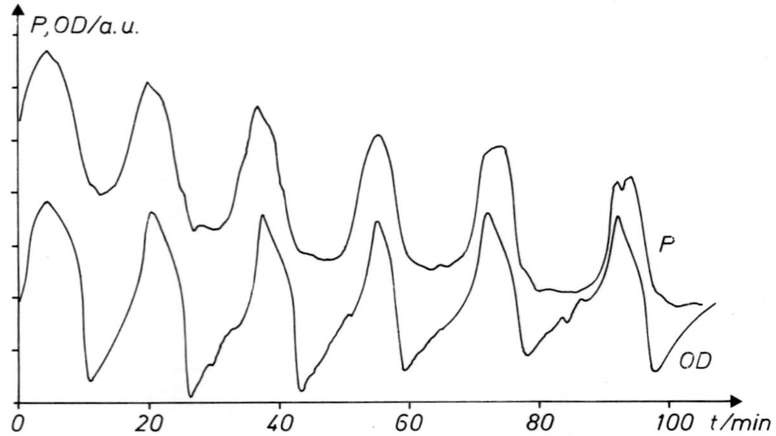

Fig. 7 b The same curves as in Fig. 7 a after the deconvolution of the calorimetric signal (P). The maxima of both curves now coincide well.

Figs. 8 a and 8 b show the simultaneous recording of heat production and electrode potential before and after deconvolution. As in the glycolytic case mentioned above, both signals are exactly in phase after the deconvolution procedure. This direct determination of the rate of heat production in correlation to the electrode potential confirms the observations of Körös et al. (1973 a, b) of sudden temperature change steps in their system.

The curves in Fig. 9 indicate that an increase in the rate of heat production (dashed lines) is coupled with an upward trend in gas pressure (solid line), i. e. higher rates of CO_2 production. A quantification of this statement is not possible, since the resolution function of pressure measurements is not yet fully understood.

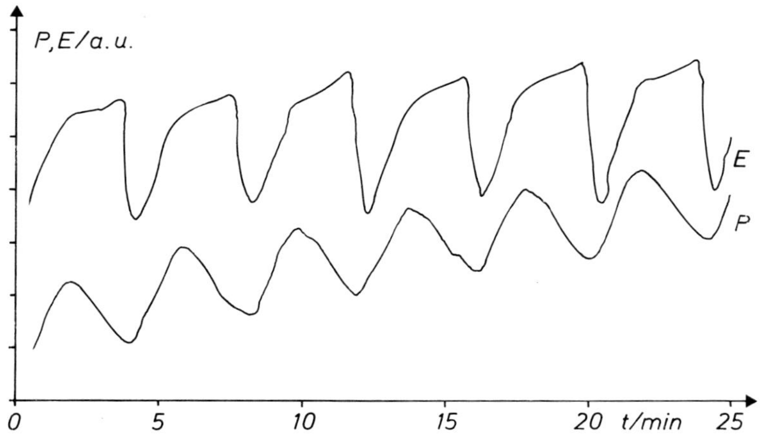

Fig. 8 a Simultaneous recording of the rate of heat production (P) and the electrode potential (E) for the Belousov-Zhabotinskii reaction.

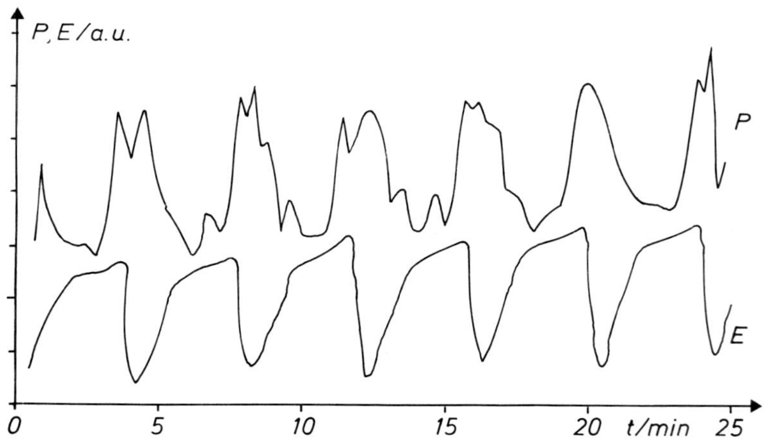

Fig. 8 b The same curves as in Fig. 8 a after the deconvolution of the calorimetric signal (P). The irregularities in this graph are artefacts due to the enhancement of experimental noise by the mathematical treatment and do not reflect changes in the rate of heat production.

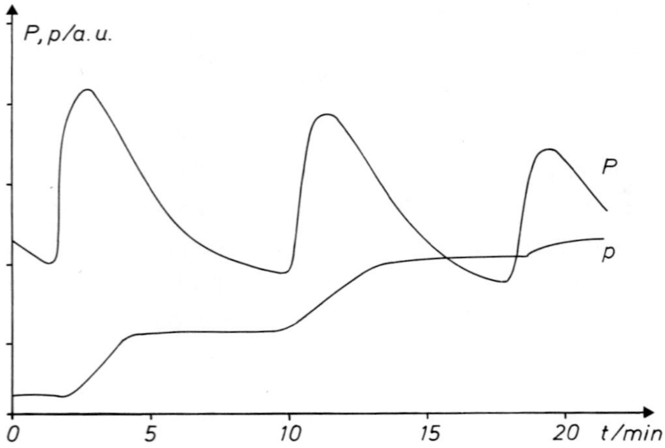

Fig. 9 Simultaneous recording of the rate of heat production (P) and the gas pressure of CO_2 (p) for the Belousov-Zhabotinskii reaction.

Summary

The literature survey and the few examples presented here emphasise the view that modern microcalorimetry is a promising tool for the study of temporal patterns in chemical and biochemical reactions - complementing other methods like the determination of optical density, electrode potential or substrate concentration. The long relaxation times of calorimeters have to be balanced by an appropriate choice of instrument and sophisticated mathematical treatment of the measured data. Moreover, the sensitivity of modern calorimeters is significantly improved so that even small specimens of oscillating systems in the range of a few hundred microliters may be investigated. As another supplementary approach, the use of tank reactors as dynamic calorimeters should be mentioned, a technique which has been successfully applied in fermentation measurements (Mou, Cooney, 1976; Marisa, von Stockar, 1986).

Since calorimeters measure only the variation of the overall rate of heat production of a reaction system, the data will only allow for a detailed mechanistic interpretation when enthalpies of individual reaction steps are known and computer models are available for calculation of the overall rate of heat production. Efforts in this directions will result in a deeper understanding of the coordination of the reaction steps in a mechanism with respect of time and energy.

Rate Limiting Steps in Oscillating Plant Glycolysis: Experimental Evidence for Control Sites Additional to Phosphofructokinase

K. Kreuzberg, A. Betz

Introduction

Phosphofructokinase (PFK) is recognised as a rate limiting step in glycolysis (Passoneau, Lowry, 1962; Bloxham, Lardy, 1973; Hofmann, 1976), but its predominant role has recently been questioned (Bosca, Corredor, 1984; Masters, 1985; Ackerman, 1985). With quantitative estimation of metabolic control at enzyme levels (e. g. Heinrich, Rapoport, 1974), the regulatory cooperation of at least two enzymes within a metabolic chain has been postulated (Westerhoff, Chen, 1984). Furthermore experimental data from oscillatory yeast cells and from yeast extracts clearly pointed to the presence of multiple regulation within the glycolytic chain (Betz et al., 1981).

In this study experimental data from oscillating yeast cells are summarised with respect to metabolic control as a consequence of cells being forced to accomodate to changing physiological conditions. The transition from growth to starvation and the onset of glycolytic activity following the addition of substrate have been followed under

Thermodynamics and Pattern Formation in Biology

aerobiosis and anaerobiosis. In addition attention will be paid to compartmental glycolysis during starch degradation in the unicellular green alga *Chlamydomonas reinhardii*. Its starch, previously accumulated in the light, is anaerobically degraded with metabolic fluctuations typically found in yeast (Kreuzberg, Martin, 1984). In oscillating algal fermentation at least three subcellular compartments, e. g. chloroplast, cytoplasm and mitochondria, are tightly cooperative (Kreuzberg et al., 1987). Here shuttle mechanisms between subcellular compartments are postulated as additonal rate limiting steps in glycolytic starch degradation by autotrophic plant cells.

Yeast cell glycolysis

Prerequisites for oscillatory control: Metabolic events in starvation.

The capability of yeast cells to generate glycolytic oscillations after feeding with hexose (glucose or fructose) firstly depends on the cellular growth phase. Thus induction of oscillations is only observable after depletion of hexose in growth medium. This transition is accompanied by induction of enzyme activities for the glyoxylic acid cycle and for gluconeogenesis (Kreuzberg, Betz, 1979). During aerobic starvation of harvested yeast cells typical oscillations are only inducible after depletion of cellular glycogen reserves (Fig. 1) and a concomitant rise in trehalose. Flux rates of anaerobic glucose consumption increases inversely with cellular glycogen and follows the time course of amplitude development in glycolytic oscillations (Fig. 1).

Another feature of oscillatory glycolysis is evident from Fig. 1 as well: that is partitioning of the glucose utilised. Only half of it is transformed to ethanol and glycerol whereas 50 % are incorporated into trehalose and glycogen (cf. Fig. 4 b, also Betz, Hinrichs, 1968).

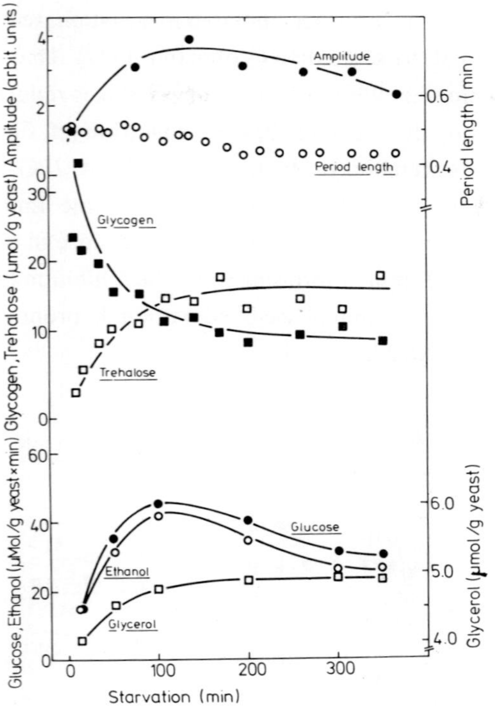

Fig. 1 Glycogen and trehalose content in S. carlsbergensis cells during aerobic starvation. The generation of glycolytic oscillations (period length and amplitude as the flux rates) in hexose consumption and ethanol and glycerol production has been monitored after addition of glucose (50 mM).

Detailed analyses of glycolytic metabolites during aerobic starvation show a characteristic pattern of metabolic events (Fig. 2). ATP rises at the expense of ADP and AMP and the sum of adenine nucleotides follows a characteristic time course. Most interesting however, with respect to control of glycolysis is the observation that 3PGA,

2PGA and PEP* are transitorily accumulated in a first phase of starvation, whereas pyruvate is depleted at the same time (Vohmann, 1977). That means that PK is the enzyme which becomes inhibited when metabolic flux is reduced in order to accomodate starvation. In a second phase of starvation (cf. Figs. 1, 2) glycogen, trehalose and adylates remain constant, whereas 3PGA, 2PGA and PEP decline to lower levels. There is no reason to postulate another rate limiting step besides PK within the glycolytic chain. This is further supported since in the reverse transition i. e. when starved cells are supplied with substrate, the inhibition of PK is actually reversed (see below). Only in this second phase of starvation do pronounced glycolytic oscillations become inducible (Fig. 1).

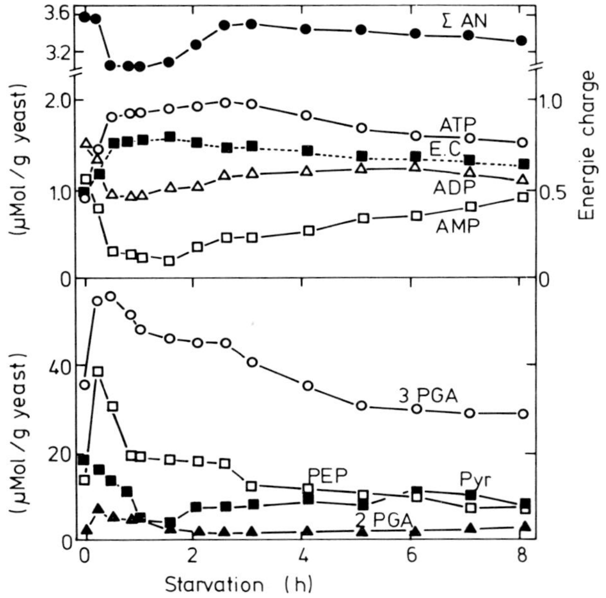

Fig. 2 Changes in concentrations of adenylates and metabolites of the lower glycolytic chain during aerobic starvation of S. carlsbergensis.

---

* Abbreviations are listed at the end of this chapter.

Transitions to aerobic and anaerobic hexose consumption

If starved yeast cells are aerobically supplied with substrate the metabolite levels change as indicated in Fig. 3. Metabolites of the upper glycolytic chain (including triosephorsphates) are accumulated as well as pyruvate, whereas 3PGA, 2PGA and PEP are dissipated (cf. Mizani et al., 1974; Betz et al., 1981). There are obviously two control points: PK is activated after feeding, whereas one of the enzymes between DHAP and 3PGA becomes rate limiting. In this case, i. e. in the transition from starvation to an activated aerobic metabolism control at PK and GAPDH (or PGK) is obviously sufficient and PFK is not involved.

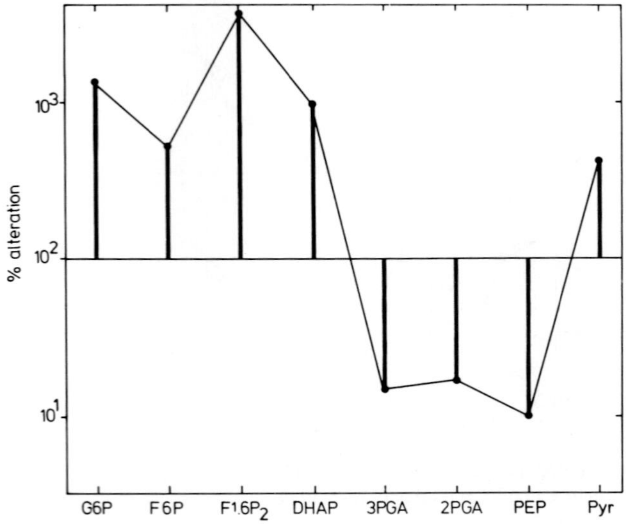

Fig. 3 Changes in glycolytic metabolites after addition of glucose (50 mM) to starved S. carlsbergensis. Alterations during aerobic glucose consumption were given with regard to starved control cells.

Feeding under anaerobiosis results in the generation of typical NADH oscillations. The amplitudes and period lengths are strongly dependent on the hexose applied, i. e. glucose or fructose (Kreuzberg, Betz, 1979). Moreover the time course of fructose consumption and the accumulation of ethanol, glycerol, glycogen and trehalose reveal distinct periodicities (Fig. 4 a, b, cf. Betz, 1966; Maitra, 1966).

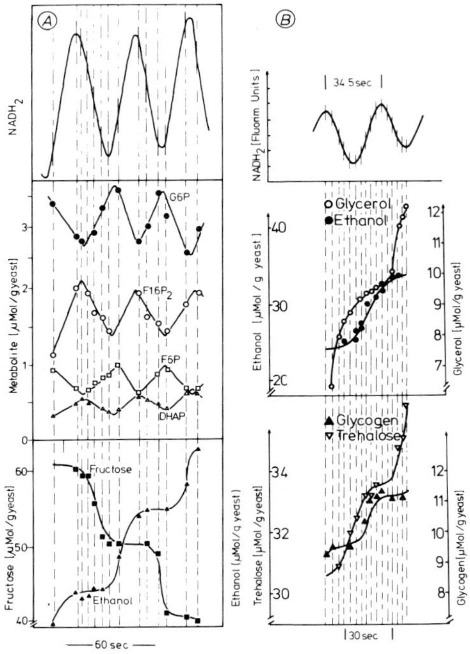

Fig. 4 a - Oscillations of NADH fluorescence, of metabolites in upper glycolysis and concomitant pulsations in fructose uptake and ethanol formation.
b - Detailed analysis of the time course of formation of ethanol and glycerol as of the synthesis of glycogen and trehalose during glycolytic oscillation.
Note: a and b are results from two different experiments, where S. carlsbergensis cells have been anaerobically supplied with 50 mM fructose.

The capacity for the generation of oscillations in yeast strongly depends on the metabolic feasibility of the synthesis of new storage compounds (cf. time course of glycogen and amplitudes in Fig. 1, also Betz, Hinrichs, 1968). Differences in period lengths and amplitudes as observed between cells supplied with glucose or fructose (Kreuzberg et al., 1977; Kreuzberg, Betz, 1979) are due to smaller differences in the metabolic fate of fructose in comparison with glucose. Fructose and its metabolite F1P interact directly with PFK in changing its regulatory properties (Betz et al., 1975; Kreuzberg, 1978, respectively).

Transition to anaerobiosis gives rise to more pronounced changes in metabolite concentration when yeast cells have been fed previously with hexose (Fig. 5) than with cells that have consumed their endogeneous glycogen reserves. The decrease in G6P and F6P points to an increase in PFK activity, whereas the significantly higher concentrations of $F1.6P_2$ might depend on additional flux limitations within the lower glycolytic chain (e. g. GAPDH, PGK). A similar but less dramatic change in metabolite concentrations is observed with yeast cells fermenting their glycogen reserves (Fig. 5).

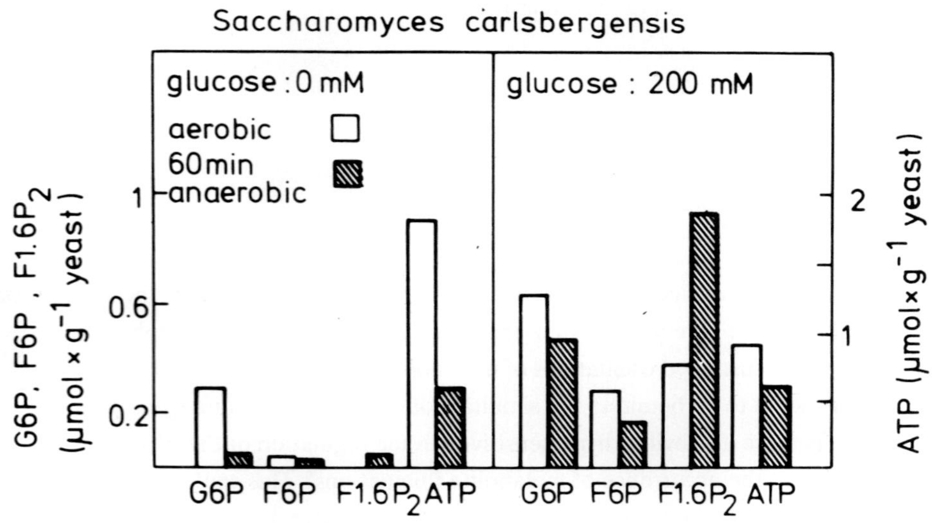

Fig. 5 Changes in metabolite concentrations after the transiton to anaerobiosis in S. carlsbergensis. Left: starved cells; right: cells supplied with glucose.

Phase relations of metabolites and enzymes

Detailed analysis of metabolite oscillations in yeast cells has revealed unique phase relations (cf. Betz, 1967; Hess, Boiteux, 1971; Giersch et al., 1975). Obviously some groups of metabolites are in phase, others deviate or are even opposite (Betz, Giersch, 1978). Adenine- and pyridine nucleotides are both present in limiting concentrations. A phase shift of 180 ^{o}, as usually observed between ATP and ADP, is indicative for pulsatory differences in the prevailance of reactions which yield ATP and others which dissipate it. If substrate and product of a reaction are in phase there is obviously no major metabolic control step in between. This is true in anaerobic glycolysing yeast cells for PGI, PGM, enolase and probably for ALD.

Phase shifts between metabolites were often taken as evidence for metabolic control steps. However, computation of flux rates at different enzymes derived from the fluctuations in metabolite concentrations revealed, that all glycolytic enzymes are in phase (Giersch et al., 1975; Betz, Giersch, 1978). Thus phase shift analysis alone (cf. Fig. 4 a) does not seem to be sufficient to identify metabolic control points. Additional disturbation experiments are indispensible.

Metabolic control points

Transitions from one metabolic state to another offer efficient tools for the elucidation of control points. It is very convenient that yeast cells regulate their glycolytic flux so slowly, that self-sustained oscillations of metabolites become observable. Glycolytic oscillations seem to be bound to the simultaneous activity of glyconeogenesis and glycolysis. Is there an obvious hypersensitivity in the regulation of the PFK/FBPase cycle leading to the occurrence of metabolite fluctuations? This could explain why yeast mutants lacking FBPase activity are unable to generate glycolytic oscillations although hexoses are fermented exactly as in wild type strains (Kreuzberg, unpublished).

It does not seem to be accidental that in oscillatory glycolysis energy generation is well balanced with energy dissipation in glycogen and trehalose accumulation. This balance seems to be favoured in the second phase of cell starvation, where glycolytic oscillations are preferably inducible, and limitations in storage metabolism and within the lower glycolytic chain (at PK) are diminished.

From the transition of starved cells to aerobic hexose metabolism two control points are obvious, one at GAPDH/PGK and one at PK. Detailed investigations during the time course of starvation revealed a major regulation of glycolytic flux at GAPDH (Kreuzberg, 1977). In the aerobic to anaerobic transition a most efficient control site can be pinpointed at PFK. However, in this transition pyruvate degradation becomes rate limiting.

Glycolytic oscillations in green algae

Periodicities induced during formate fermentation

Green algae like *Chlamydomonas reinhardii, Chlorogonium elongatum* and *Chlorella fusca* ferment their plastidic starch anaerobiosis in the dark and produce formate, acetate and ethanol (Kreuzberg, 1984). In contrast to glycolysis in yeast, fermentation of photoassimilated starch in algae is highly complex as evidenced by the interaction of different cell compartments such as the chloroplast, cytoplasm and mitochondrion (Kreuzberg et al., 1987). Nevertheless after transition to anaerobiosis, an oscillatory degradation of starch and accumulation of formate, acetate and ethanol has been observed with a mean period length of 59.6 min (Fig. 6, Kreuzberg, Martin, 1984). Clear periodicities of G6P and dramatic changes in pyruvate concentrations (Fig. 7) have been shown together with concomitant fluctuations in the levels of adenylates (cf. Kreuzberg, Martin, 1984).

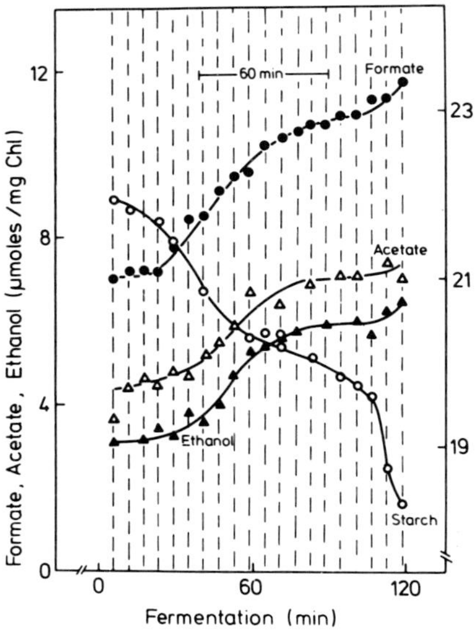

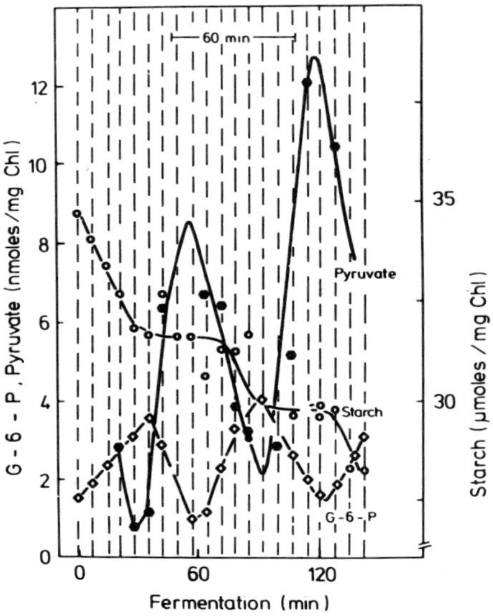

Fig. 6 Starch degradation and accumulation of formate, acetate and ethanol during anaerobic dark incubation of C. reinhardii. Starch content is given in glucose units. Mean period length of all parameters determined in the shown experiment was 64 ± 4.9 min.

Fig. 7 Oscillation in cellular content of G6P and pyruvate as pulsations in starch degradation during anaerobic incubation of C. reinhardii.

Analysis of phase relations yields a slightly different spectrum of relationships to that known to occur in yeast glycolysis. Differences of 180° between ATP and both ADP and AMP, were observed. The phase relationship of starch degradation to G6P (difference 60°) is difficult to interprete and indicates a possible regulatory step between starch phosphorolysis and chloroplastic glycolysis. Phase shifts of G6P to both, pyruvate (56°) and ATP (190°) are in accord with those found for yeast oscillations (45° and 235°, respectively; Betz, Giersch, 1978). Even the phase relation from pyruvate to ATP (134°) resembles that for yeast cells (180°).

Subcellular localisation of metabolic control points

Rapid fractionation of *C. reinhardii* by silicon oil centrifugation allows detailed analysis of metabolite levels in protoplasts as in cytoplasm and chloroplasts (Klöck, Kreuzberg, 1987 b b).

In transition to anaerobiosis in the dark analysis of whole protoplasts reveals metabolic regulation points within glycolysis at PFK, GAPDH and especially in reactions of pyruvate catabolism (Fig. 8 a). Further evidence for subcellular localisation can be drawn from the metabolite crossover in cytoplasm and chloroplasts. Thus in chloroplasts regulation can be pinpointed at PFK and GAPDH, whereas crossover at pyruvate might be still relevant, but is drastically reduced (Fig. 8 b). It should be noticed, that pyruvate in chloroplasts represents at most one percent of that in the cytoplasm. The observed control point at pyruvate catabolism in chloroplasts could therefore be due to cytoplasmic contamination. However, metabolic regulation in cytoplasmic glycolysis can be detected in GAPDH and escpecially in pyruvate catabolism, whereas that at PFK is not observable (Fig. 8 c).

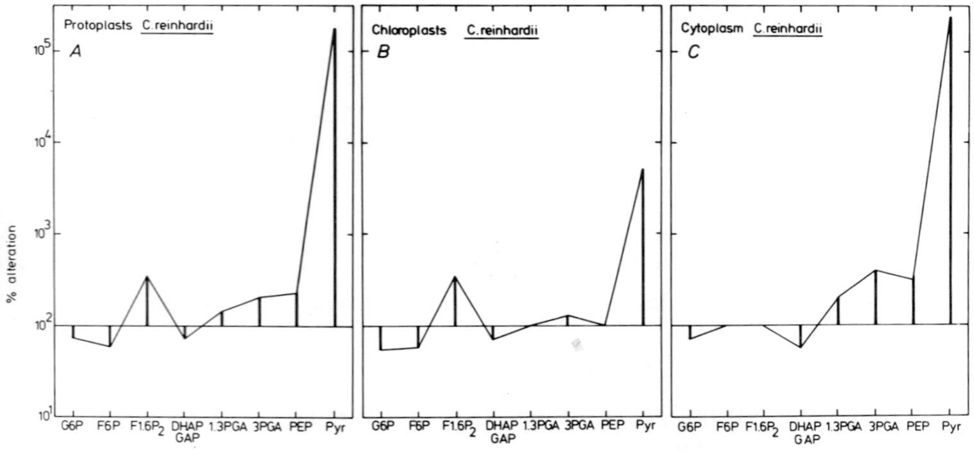

Fig. 8 Changes in glycolytic metabolites after transition of C. reinhardii; cells to anaerobiosis in the dark. Subcellular analysis has been performed by rapid separation of (a) algal protoplasts in (b) chloroplastic and (c) cytoplasmic (including mitochondria) fractions. Alterations are given with regard to aerobic control. Data were corrected for cross contamination.

In *C. reinhardii* starch is degraded exclusively in the chloroplast via an upper part of chloroplast glycolysis, which yields DHAP, Gly3P and 3PGA (Klöck, Kreuzberg, 1987 a). These glycolytic products are exported from the chloroplast via the triose phosphate shuttle (Fliege et al., 1978) and further metabolised via the lower cytoplasmic glycolytic chain to pyruvate. Therefore crossover (Fig. 8) between triose phosphates and the phosphoglyceric acids might depend on membrane translocation being the rate limiting step. Similarly the increase in pyruvate could be due to transport limitations into mitochondria in so far as it does not depend on rate limitations by mitochondrial enzymes of pyruvate catabolism. During anaerobiosis all pyruvate is thiolytically cleaved by PFL to AcCoA and formate within the mitochondria of *C. reinhardii* (Kreuzberg et al., 1987). However PFL itself underlies complex regulation by activation to a radical bearing active enzyme species at the onset of fermentation (Kreuzberg, 1985).

Recent investigation of subcellular enzyme distribution in *C. reinhardii* revealed the presence of a chloroplastic PFK only (Kreuzberg et al., 1986). This chloroplastic localisation of PFK excludes its modulation by common effectors (e. g. PEP) of lower glycolysis, in contrast to cytoplasmic PFK's (Hofman, 1976), but requires direct control by intermediates of the modified chloroplast glycolysis. PFK isolated from *C. reinhardii* chloroplasts differs in its kinetic data from yeast PFK, where it is cytoplasmic (Table 1).

This is most striking in its affinity to F6P and in its sensitivity toward different effectors. The algal enzyme is inhibited by NADH, $F1.6P_2$, 3PGA and activated by Gly3P and P_i.

These compounds are all involved in chloroplast glycolysis and no direct feed back other than via P_i from cytoplasmic glycolysis can be observed.

This alga thus shows typical glycolytic oscillations over the complete fermentation pathway, but with a plastidic upper and cytoplasmic lower glycolytic chain and multiple different compartmented regulation points.

Table 1.

Comparison of cytoplasmic (yeast) and chloroplastic (algal) PFK with regard to kinetic data and the mode of effectors

	PFK[1] cytoplasm	PFK[2] chloroplast	
Substrates	K_m	K_m	
F6P (ATP 0.1 mM)		260 µM	36 µM
F6P (ATP 0.3 mM)		480 µM	66 µM
ATP		25 µM	21 µM
F1P		2.500 µM	n.d.[3]
pH optimum		6.9	7.5
Effectors			
AMP,	50 µM	A[4]	A
NAD(P)H,	50 µM	N[5]	I[6]
$F1,6P_2$,	100 µM	A	I
$F2,6P_2$,	20 µM	A	A
3PGA,	100 µM	N	I
Gly3P,	100 µM	N	A
Pep,	100 µM	I	N
P_i, 3m	M	A	A

1 - *S. carlsbergensis* (Kreuzberg, 1978)
2 - *C. reinhardii* (Kreuzberg et al., 1986)
3 - not determined
4 - activation
5 - no influence
6 - inhibition

Conclusions

The role of PFK in plant glycolysis

PFK is generally assumed to function as main control point in glycolysis and to be responsible for glycolytic periodicities together with suitable feed back loops (Chance et al., 1964; Betz, Chance, 1965; Betz, Sel'kov, 1969; Hess, Boiteux, 1971; Hofmann, 1976). Fluctuations in yeast (cytoplasmic) PFK activity results from periodic shifts in the concentration of adenylates (ATP and AMP) or glycolytic metabolites (e. g. PEP). However, generation of oscillations by PFK does not necessarily mean that flux is controlled completely at this enzyme step, since further control sites could be detected.

In chloroplastic starch degradation in plants cytoplasmic PFK is not involved in glycolytic regulation (Fig. 8). Indications for the participation of plastidic PFK in metabolic control can only be observed under anaerobiosis, a rare physiological condition for autotrophic plants in the natural environment.

Recently $F2.6P_2$ was discovered as the most efficient effector of cytoplasmic PFK (Avigad, 1981; Bartrons et al., 1982; Hers, van Schaftingen, 1982). A 4-fold increase in $F2.6P_2$ is observed in yeast after transition to anaerobiosis (Furuya et al., 1982; Hers, van Schaftingen, 1982), but information on the time course of $F2.6P_2$ during glycolytic oscillations is still lacking. However, Schellenberger and coworkers (1985) could recently show the generation and quenching of glycolytic oscillations in a reconstituted yeast PFK/FBPase system by varying the $F2.6P_2$ concentration from 0 to 2 μM.

As shown in Table 1 even PFK from *C. reinhardii* chloroplasts, since it is also the enzyme from castor bean plastids (Miernyk, Dennis, 1982), is effectively activated in the presence of $F2.6P_2$ (K_a = 7 nM, Kreuzberg et al., 1986). Since the occurrence in plant cells is thought to be restricted to the cytoplasm (Hers, van Schaftingen, 1982), the physiological importance of this activation remains still open. But it should be noted, that detection of $F2.6P_2$ is limited to nM concentration because of methodological shortcomings today, which might not be sensitive enough to determine the concentration present in chloroplasts.

The absence of a cytoplasmic PFK in *C. reinhardii* raises the question of a possible contribution from a pyrophosphate dependent PFK (EC 2.7.1.90) in oscillating plant glycolysis. Activity of this enzyme depends on the presence of $F2.6P_2$ (Sabularse, Anderson, 1981; Kowalczyk et al., 1984) and might establish an important bypass to the irreversible ATP-PFK. Thus the adenylate dependent metabolic regulation of the latter control point could be effectively reduced. The contribution of a PP_i PFK and the influence of $F2.6P_2$ on oscillating glycolysis still has to be elucidated.

Compartmentation and subcellular transport

Fermenting algal cells offer a tool to study the influence of subcellular transport steps and to elucidate their importance in metabolic regulation. As is depicted in Fig. 8, DHAP, Gly3P and 3PGA have to be transported across the chloroplast envelope (Fig. 9). Membrane transport can thus be imagined to be the major metabolic control site. The same holds true for the translocation of pyruvate into the algal mitochondria. Details of the coordination of their subcompartmented metabolic chain have to be elucidated further.

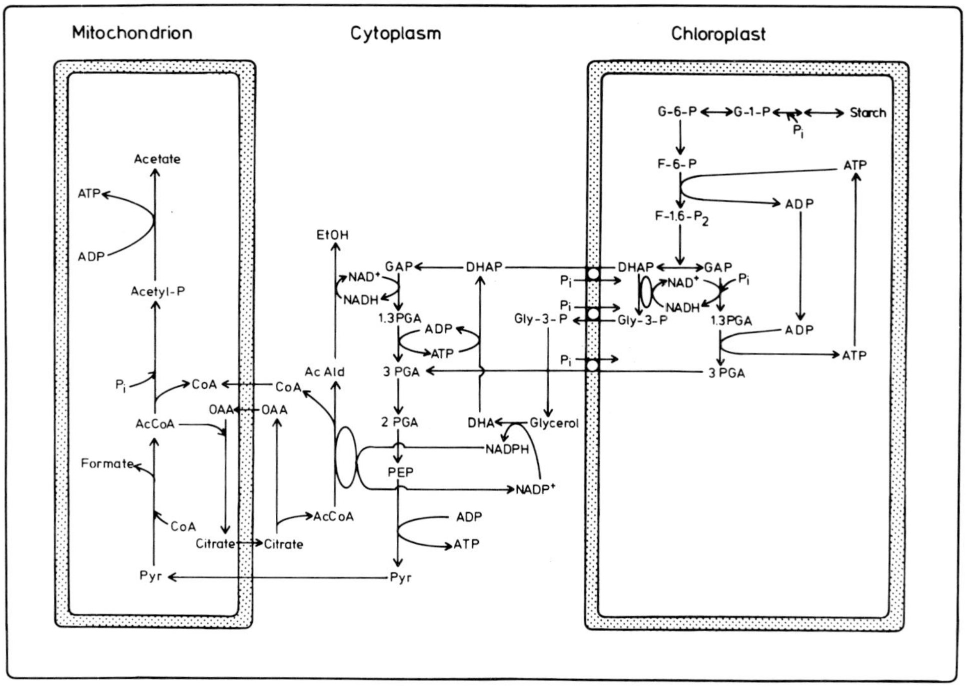

Fig. 9 Tentative scheme for the interactions of cell compartments during fermentative starch degradation in C. reinhardii.

Most competent information exists for the interaction of algal chloroplast and cytoplasm during anaerobiosis in the dark as summarised in Fig. 9. Here an autonomous ATP recycling exists between PFK (sink) and PGK (source). Reducing equivalents are balanced within the chloroplast by the stoichiometric production of Gly3P and 3PGA (Klöck, Kreuzberg, 1987 a), which delivers enough ATP for an equimolar reaction at plastidic PFK. Thus all enzymes of plastidic glycolysis are tightly coupled, but the export rate of DHAP, Gly3P and 3PGA might be regulated by the availability of their counter-ion P_i.

During anaerobiosis Gly3P is presumably dephosphorylated and subsequently oxidised to DHA, which can then, in turn, be phosphorylated to DHAP (Fig. 9). Thus DHAP, Gly3P and 3PGA are finally metabolised in the cytoplasm to pyruvate via the lower glycolytic chain. However, anaerobic pyruvate degradation to AcCoA is exclusively localised within the mitochondria of *C. reinhardii* during formate fermentation (Kreuzberg, 1984; Kreuzberg et al., 1987). Further catabolism of AcCoA can, however, be performed in mitochondria (acetate formation) as in the cytoplasm (ethanol production, Fig. 9).

Importance of pyruvate metabolism

The accumulation of pyruvate in anaerobic yeast cells and in algal cells pinpoints a rate limitation at subsequent metabolic steps. Continuous pyruvate catabolism is essential for a persisting glycolytic flux in yeast, as in plant cells, since it is indispensible for the reoxidation of NADH (or FADH). Thus pyruvate catabolism in algae (perhaps its translocation into mitochondria?) becomes rate limiting in anaerobiosis. This is obvious from its accumulation. With higher pyruvate concentrations PFL becomes activated and pyruvate is thiolytically cleaved to formate and AcCoA, and the latter delivers acetate via PTA (cf. Fig. 9). These organic acids can freely permeate the mitochondrial membrane. Their accumulation thus gives rise to a decreasing pH in the cytoplasm and results in an activation of PDC, which can be followed in *C. reinhardii* by an increase in CO_2 and ethanol production after 7 h of anaerobiosis (Kreuzberg, 1984). In both, yeast and green algae, glycolysis is regulated at 4 enzymic steps: PFK, GAPDH, PK and PDC (or PFL, respectively). In plants three subcellular compartments cooperate with membrane transport as additional control sites. Which control site is predominant depends on the metabolic condition or on which of these control sites are actually cooperating. Thus our results agree with theoretical considerations on multiple site regulation (e. g. Heinrich, Rapoport, 1974; Westerhoff, Chen, 1984).

To promote our understanding of plant glycolysis further it would be most helpful to know how far individual steps contribute quantitatively to overall control.

Acknowledgement

Work was supported by grants of the Deutsche Forschungsgemeinschaft to A. Betz and K. Kreuzberg.

Abbreviations

Enzymes

ALD	aldolase (EC 4.1.2.13)
FBPase	fructose-1.6-bisphosphatase (EC 3.1.3.11)
GAPDH	glyceraldehydephosphate dehydrogenase (EC 1.2.1.12)
PDC	pyruvate decarboxylase (EC 4.1.1.1)
PFL	pyruvate formatelyase (EC 2.3.1.54)
PFK	ATP-phosphofructokinase (EC 2.7.1.11)
PGI	phosphoglucose isomerase (EC 5.3.1.9)
PGK	phosphoglycerate kinase (EC 2.7.2.3)
PGM	phosphoglycerate mutase (EC 2.7.5.3)
PK	pyruvate kinase (EC 2.7.1.40)
PTA	phosphotransacetylase (EC 2.3.1.8)

Metabolites

Chl	chlorophyll a + b
DHA	dihydroxyacetone
DHAP	dihydroxyacetone phosphate
G6P	glucose-6-phosphate
F6P	fructose-6-phosphate
F1P	fructose-1-phosphate
$F1.6P_2$	fructose-1.6-bisphosphate
$F2.6P_2$	fructose-2.6-bisphosphate
GAP	glyceraldehyde phosphate
Gly3P,	Sn glycerol-3-phosphate
PEP	phosphoenol pyruvate
3(2)PGA	phosphoglyceric acid
Pyr	pyruvate

Dynamic Structures in the Fructose-6-Phosphate/Fructose-1,6-Bisphosphatase Cycle

W. Schellenberger, M. Kretschmer, K. Eschrich, E. Hofmann

Introduction

Metabolic pathways exhibit a great dynamic capacity due to the occurrence of nonlinear interactions. Nonlinearity results from the structure of the underlying network of chemical reactions, from allosteric interactions and from the cooperation of enzymic reactions with diffusion processes (Reich et al., 1981). Rhythmic behaviour and hysteresis are the most ubiquitous consequences of nonlinearity in biological systems (Berridge et al., 1979; Winfree, 1980). Among biological oscillations, those arising in yeast and muscle glycolysis (Hess et al., 1971; Goldbeter et al., 1976; Sel'kov, 1979) are the best understood at the molecular level. In glycolysis, the source of periodic behaviour is the phosphofructokinase/fructose-1,6-bisphosphatase cycle (Goldbeter et al., 1976; Sel'kov, 1979). This cycle was shown to be a strong candidate for switching phenomena in biochemical systems (Reich, Sel'kov, 1981; Hofmann et al., 1985). There is evidence for a regulation of phosphofructokinase and fructose-1,6-bisphosphatase by allosteric and epigenetic mechanisms (Hers et al., 1984). With respect to allosteric regulation the effects of fructose-2.6-bisphosphate and AMP are regarded as most

Thermodynamics and Pattern Formation in Biology

important. These metabolites exert reciprocal effects on the activities of the two cycle enzymes: both are potent activators of phosphofructokinase and inhibitors of fructose-1,6-bisphosphatase (van Schaftingen et al., 1981, 1982, 1984). While AMP is directly connected to cellular energy metabolism by the action of adenylate kinase, synthesis and degradation of fructose-2,6-bisphosphate is accomplished by a cyclic bypass of glycolysis catalysed by phosphofructokinase-2 and fructose-2,6-bisphosphatase. In liver, both enzyme activities reside at the same protein and are regulated by cAMP dependent phosphorylation as well as by effectors (Furuya et al., 1982; Pilkis et al., 1983; McGrane et al., 1983).

In this paper, the dynamic behaviour of reconstituted enzyme systems is investigated which operate under open conditions far from the thermodynamic equilibrium. It will be shown that in enzyme systems containing the phosphofructokinase/fructose-1,6-bisphosphatase cycle, various types of multistability and oscillations can occur. The effect of fructose-2,6-bisphosphate on dynamic structures of the phosphofructokinase/fructose-1,6-bisphosphatase cycle is discussed and stationary states of the phosophofructokinase-2/fructose-2,6-bisphosphate cycle are investigated. The experimental data are analysed by mathematical models based on the kinetic properties of the individual enzymes involved.

Materials and methods

Phosphofructokinase/fructose-1,6-bisphosphatase cycle

The experimental conditions, the purification of the enzymes, the properties of the enzymes involved and the mathematical modelling are described by Hofmann et al., 1985.

Experimental approach

The F-6-P/F-1,6-P_2[1)] cycle was investigated in a reconstituted enzyme system using a stirred flow-through reactor (Fig. 1). From two reservoirs, containing the substrates and the enzymes, respectively, the thermostatted reactor was fed continuously. The maximum activities of the enzymes, the influx concentration of the substrates and the residence time (τ) are adjustable parameters. The state of the system is defined by the metabolite concentrations in the reaction chamber. These were measured by determining the metabolite concentrations in the flux solution using enzymatic methods (Eschrich et al., 1983; Hofmann et al., 1985).

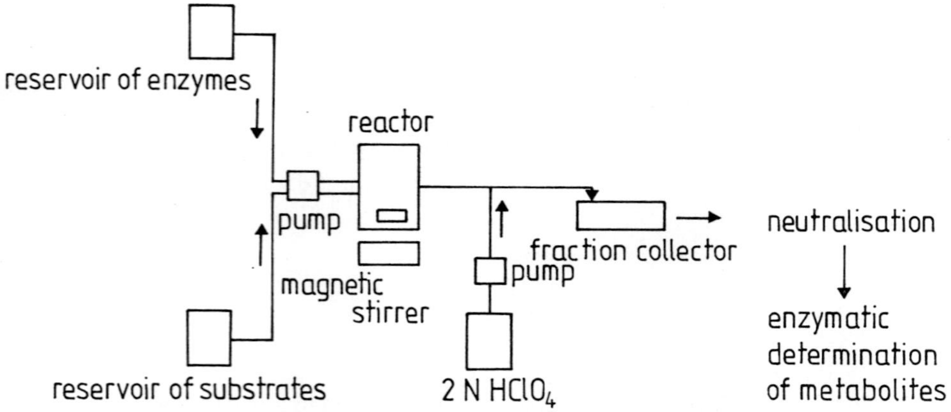

Fig. 1 Outline of the experimental device for investigation of the dynamics of the F-6-P/F-1,6-P2 cycle

---

1) Abbreviations are listed at the end of this chapter.

Mathematical modelling

The cooperation of the enzymes in the reaction chamber is governed by a set of coupled differential equations for the metabolite concentrations. These equations take into account the enzymatic conversions as well as the influx and efflux processes. The action of the enzymes is described by expressions derived from kinetic investigations of the individual enzymes (Schellenberger et al., 1981, 1985; Eschrich et al., 1983).

Phosphofructokinase-2/fructose-2,6-bisphosphatase cycle

The experimental conditions, the purification of the enzymes and the properties of PFK-2 and FBPase-2 are described by Kretschmer et al., 1985, 1987.

Experimental approach

The dynamics of the F-6-P/F-2,6-P_2 cycle was investigated in a closed reaction tube containing the PFK-2/FBPase-2 protein and creatine kinase for ADP-rephosphorylation from creatine phosphate. Creatine phosphate was found to be without effect on the activities of PFK-2 and FBPase-2 (Kretschmer et al., 1987). PFK-2/FBPase-2 was phosphorylated using a cAMP dependent protein kinase (Pilkis et al., 1984). After inactivation of the protein kinase, the fractional phosphorylation of the protein was adjusted by mixing appropriate amounts of the native and the phosphorylated PFK-2/FBPase-2. The kinetic consequences of phosphorylation of the protein could be described by variation in the maximum activities of PFK-2 and FBPase-2.

Results and discussion

The fructose-6-phosphate/fructose-1,6-bisphosphatase cycle

The reaction system

The cooperation of the enzymes in the reaction chamber is shown in Fig. 2. In addition to phosphofructokinase and fructose-1,6-bisphosphatase, the system contains pyruvate kinase for ATP-recycling from PEP, adenylate kinase and glucose-6-phosphate isomerase. Quasi-equilibria of the adenine nucleotides and the hexosemonophosphates are maintained by high concentrations of adenylate kinase and glucose-6-phosphate isomerase, respectively (Hofmann et al., 1985). The influx concentrations of the

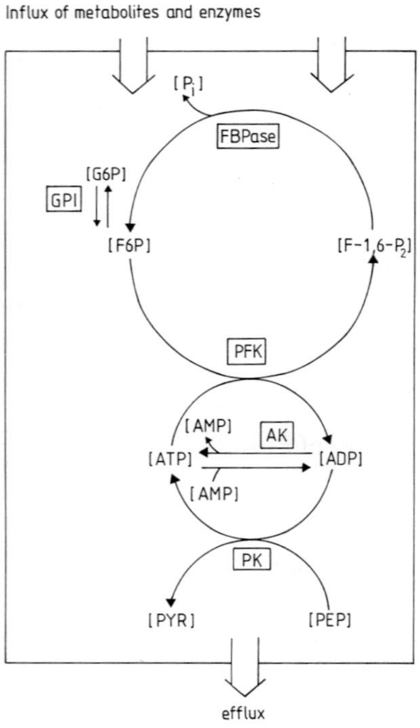

Fig. 2 Cooperation of PFK, FBPase, PK, AK and GPI in the reconstituted enzyme system

substrates ($[ATP]_{in}$, $[PEP]_{in}$, $[F\text{-}6\text{-}P]_{in}$, $[F\text{-}1,6\text{-}P_2]_{in}$, $[F\text{-}2,6\text{-}P_2]$), the rate of flow through reaction chamber (τ), and the concentrations of the enzymes (V_{PFK}, V_{PK}, V_{FBPase}) can be adjusted experimentally and are the control parameters of the mathematical model. The metabolite concentrations in the reactor result from the enzymatic conversions and the flow processes and are the variables of the mathematical model.

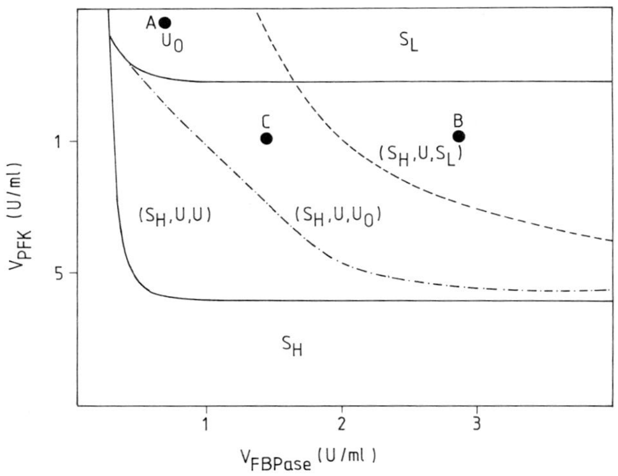

Fig. 3 Characterisation of stationary solutions of the model in respect to stability, multiplicity and typical metabolite profile.

$[ATP]_{in}$ = 3 mM; $[F\text{-}6\text{-}P]_{in}$ = 3 mM; $[PEP]_{in}$ = 9 mM; τ = 40 min; V_{PK} = 7.5 U/ml.

S_H high energy stationary states

S_L stable low energy stationary states

U_0 unstable stationary solution, related to an oscillatory motion

U unstable stationary state.

Stationary states

In Fig. 3 the stationary solutions of the mathematical model are characterised in respect to stability, multiplicity, and typical metabolite profile. At low concentrations of PFK and FBPase and at high maximum activities of both enzymes unique and stable stationary states occur. The stationary states at high maximum activities of PFK and FBPase are characterised by very low concentrations of PEP and low ATP (low energy states, S_L). At low concentrations of both enzymes unique stationary states with high ATP/ADP ratios occur (high energy states, S_H). At high V_{PFK} a parameter domain exists in which unique unstable states (U_o) occur, which are related to sustained oscillations. Fig. 4 shows the result of an experiment in this parameter domain

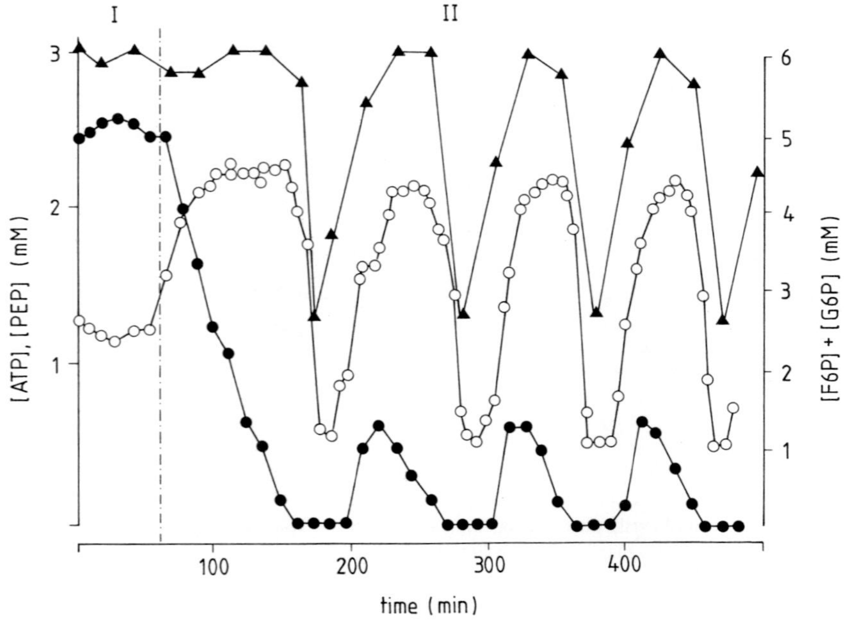

Fig. 4 Experimentally observed sustained oscillations: ▲ : [ATP] ; ● : [PEP] ; ○ : [HMP] ; Phase I: V_{FBPase} = 0; Phase II: V_{FBPase} = 0.25 U/ml. The conditions applied in phase II correspond to point A in Fig. 3.

(conditions refer to point A in Fig. 3). In phase I, in which FBPase is absent, a unique and stable stationary state is approached. Then, after 60 min, FBPase was added and a constant concentration of this enzyme is maintained in phase II. The addition of FBPase induces the emergence of sustained oscillations of the metabolite concentrations. Fig. 5

shows a computer simulation of the rhythmic changes of the metabolite concentrations in a phase space. The graph demonstrates, that the main part of the oscillatory motion proceeds in the ATP/HMP plane at very low concentrations of PEP. When [ATP] reaches its upper possible level ($[ATP]_{in}$), PEP increases and the motion continues in the HMP/PEP plane.

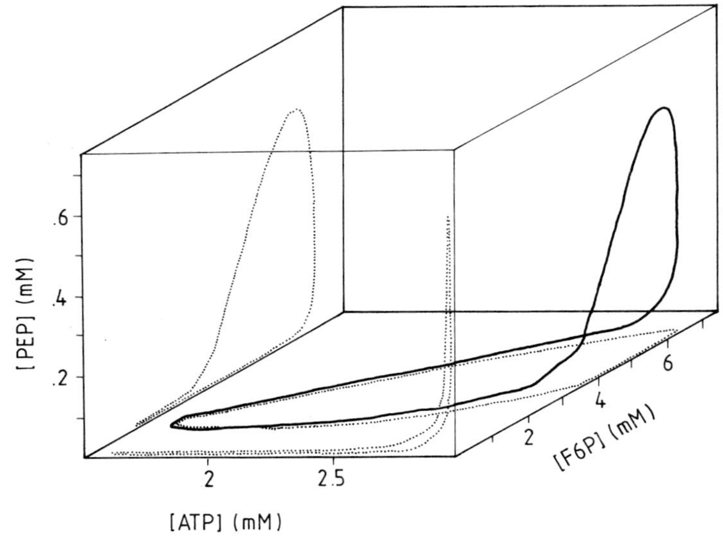

Fig. 5 Phase portrait of oscillations approaching a closed orbit. The dashed curves are projections of the curves on the phase planes. The conditions refer to Fig. 4.

Within the parameter region covered by the solid curves (Fig. 3), three alternate stationary solutions are obtained. One of the states is an unstable saddle point (U), a second one is a stable attractor characterised by high concentrations of ATP and PEP (S_H). According to the properties of the third state subdomains can be distinguished. At high maximum activities of FBPase two attractors coexist. The experiment shown in Fig. 6 a was carried out under conditions corresponding to point B in this domain. Starting with high concentrations of ATP, PEP, and HMP a high energy state (S_H) is approached in phase I. In phase II, the system was perturbed by interrupting the substrate supply. After restoration of the initial rate of substrate supply (phase III,

Fig. 6 a) a low energy stationary state (S_L) is approached. Fig. 6 b gives the respective phase plane. The points indicate the stationary solutions, the dashed curve is the nullcline for ATP, the full curves are the trajectories. All trajectories end either in the

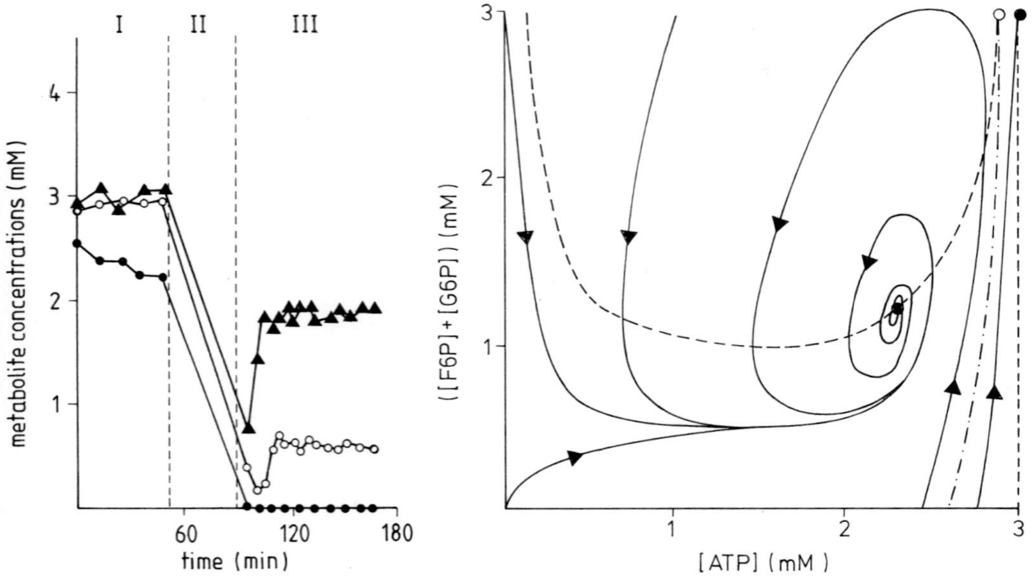

Fig. 6 Alternative stable stationary states.
a: Experimental demonstration of bistability: ▲: [ATP]; ●: [PEP]; O: [HMP].
The conditions in phases I and III fit to point B in Fig. 3. Phase II: $\tau \to \infty$.
b: Phase plane formed by [ATP] and [HMP]. ●: stable steady states;
O: unstable steady state (saddle point); – – – –: isocline to d(2 [ATP] + [ADP])/dt = 0;
–·–··: separatix; —▶— : trajectories.

stable high energy state (S_H) or in the attractor S_L. Hence two basins of attraction are defined, being separated by the separatix (dashed dotted curve). All trajectories starting left from the separatix approach the state S_L, while all curves starting right from the curve evolve to the stable high energy attractor. The perturbation of the system (phase II) gives rise to crossing of the separatix. At decreasing maximum activities of FBPase, the low energy stationary state becomes unstable. Fig. 7 a shows an experiment carried out under conditions corresponding to point C of Fig. 3. In phase I, a stable high energy stationary state is approached. In phase II the system was perturbed by a transient interruption of the substrate supply. After restoration of the initial conditions (phase III) sustained oscillations arise. In phase IV the system was perturbed by increasing the rate of flow through the reactor. As a consequence of the increased rate of substrate supply ATP, PEP, and HMP increase. In phase V the same rate of substrate is applied as in the phases I and III. The metabolites evolve to a stable high energy stationary state

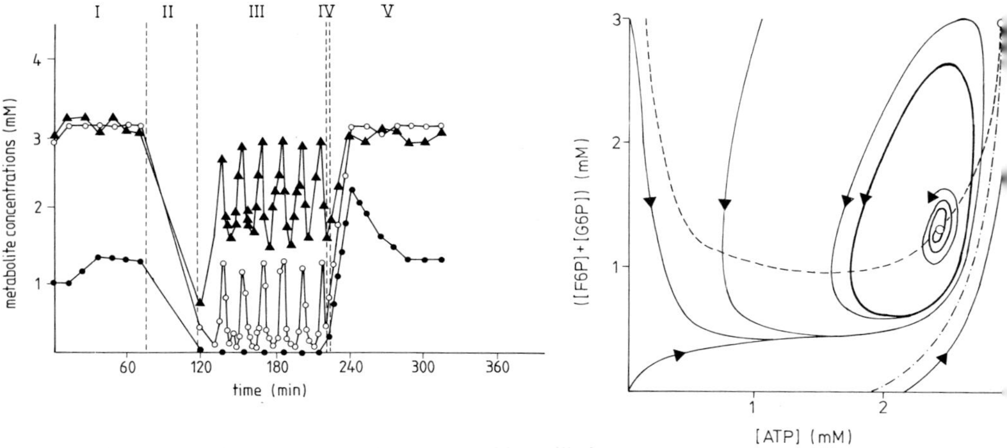

Fig. 7 Coexistence of a stable steady state and a stable oscillation.
a: Experimental demonstration: ▲ : [ATP] ; ●: [PEP] ;○: [HMP].
The conditions in I, III and V fit to point C in Fig. 3. II: $\tau \rightarrow \infty$; IV: τ = 4 min.
b: Phase plane formed by [ATP] and [HMP] ; ●: stable state (S_H); ○: unstable stationary states. The closed orbit indicates the limit cycle surrounding the unstable state U_o. All trajectories starting left from the separatix (-· – · –) approach the stable oscillation, while all trajectories starting right of this curve belong to the domain of attraction of S_H.

as in phase I. Fig. 7 b shows the respective phase plane. The unstable stationary state (U_o) is surrounded by a stable limit cycle, which is approached by all trajectories starting left from the separatix. At a further decrease of V_{FBPase} the oscillation becomes unstable. Then all trajectories belong to the domain of attraction of the stable state.

A hysteretic cycle

Fig. 8 demonstrates transitions from a unique and stable high energy state (phase I) to a unique and stable low energy state (phase II) and the reverse transition induced by variations of the maximum activity of PFK at constant V_{FBPase}. During these

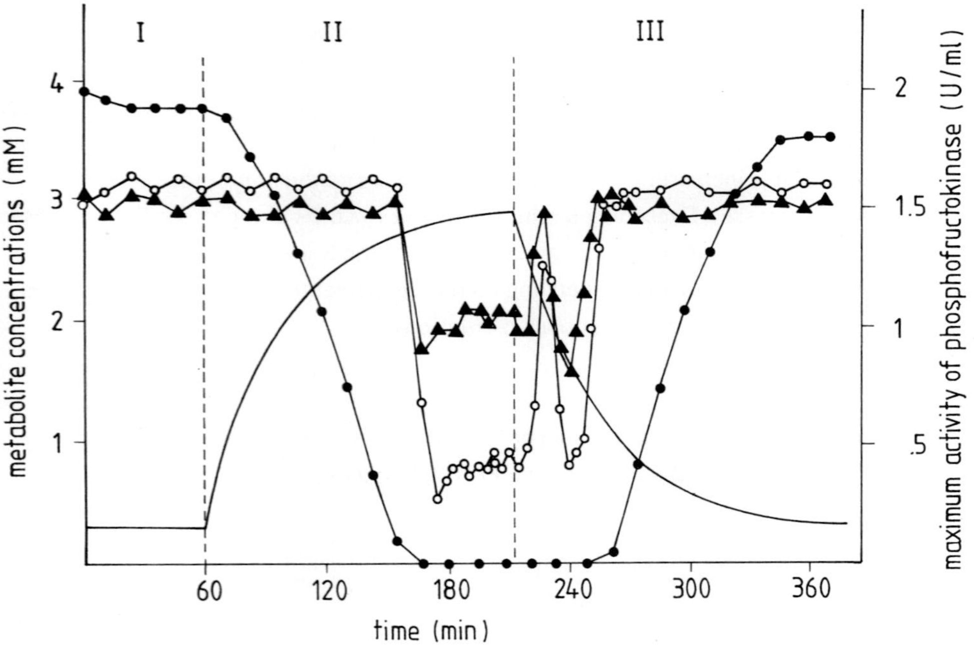

Fig. 8 Hysteretic cycle
▲:[ATP];●:[PEP];○:[HMP]; V_{FBPase} = 3.5 U/ml.
Conditions refer to Fig. 3. Transitions from an attractor S_H to S_L and vice versa induced by variations of V_{PFK}.

transitions, the domain of multiple stationary states is passed. Hence, the transitions.proceed nonsymmetrically. Starting with the high energy stationary state, a high ATP level is transiently maintained before the adenylate energy charge drops. On the other hand, the motion to the high energy stationary state induced by a decrease of the PFK concentration is related to the occurrence of transient oscillations.

The effect of fructose-2,6-bisphosphate

F-2,6-P_2 is an efficient activator of PFK. Furthermore it has been shown, that AMP and F-2,6-P_2 cooperate in the activation of the enzyme. F-2,6-P_2 increases both the AMP sensitivity of PFK and the maximum extent of activation by AMP (Nissler et al., 1983, 1984). Interactions between AMP and F-2,6-P_2 have also been reported in respect to the inhibition of FBPase (Francois et al., 1983; McGrane et al., 1983). In the experiment shown in Fig. 9 control parameters were applied fitted to the region of unique high energy steady states (Fig. 3). In phase I, F-2,6-P_2 is absent. Accordingly, a stable high energystationary state is approached. Then, F-2,6-P_2 was increased stepwise. At low concentrations, F-2,6-P_2 triggers the emergence of sustained oscillations. When its concentration is increased, the frequency enhances and the amplitude of the periodicities decreases. Exceeding a critical value of F-2,6-P_2 damped oscillations arise. The metabolites evolve to a stable stationary state of the low energy type. The experiment demonstrates, that depending on the experimental parameters, F-2,6-P_2 is capable of either to induce or to abolish oscillations.

In this type of approach, F-2,6-P_2 is a control parameter. To get deeper insight into the dynamics of the F-6-P/F-2,6-P_2 cycle, the PFK-2/FBPase-2 must be involved into the reaction network.

Stationary states of the F-6-P/F-2,6-P_2 cycle

F-2,6-P_2 is synthesised from F-6-P and ATP by 6-phosphofructo-2-kinase (PFK-2); it is hydrolysed to F-6-P and P_i by a fructose-2,6-bisphosphatase (FBPase-2). In liver, both activities are carried by the same protein and are efficiently controlled by the cAMP dependent protein kinase; phosphorylation inactivates PFK-2 and activates FBPase-2 (Pilkis et al. 1983; van Schaftingen et al., 1984; Furuya et al., 1982). The main metabolic effects of F-2,6-P_2 are activation of PFK and inhibition of FBPase. Both of these effects are synergistic with those of AMP.

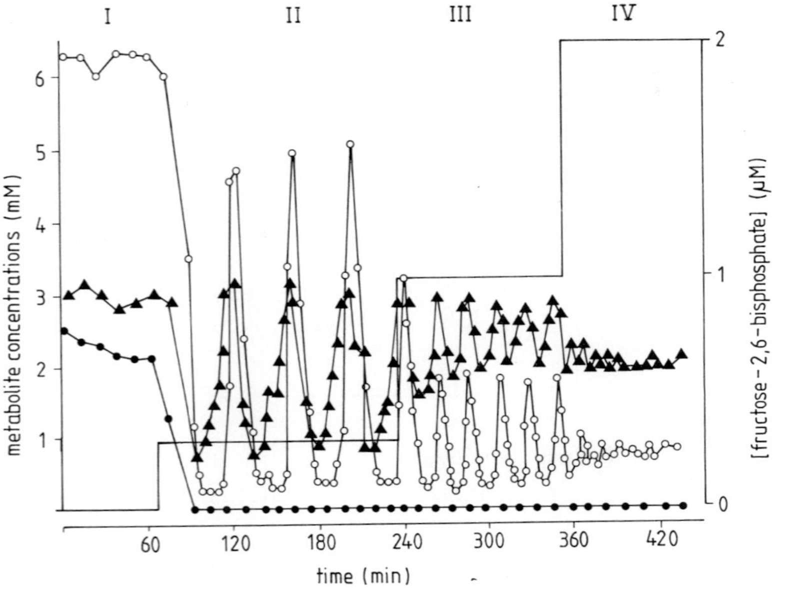

Fig. 9 Effect of F-2,6-P_2 on stationary states of the F-6-P/F-1,6-P_2 cycle.
The experimental conditons applied in phase I imply the existence of a unique attractor (S_H, Fig. 3). V_{PFK} = 0,3 U/ml; V_{FBPase} = 1.2 U/ml. The parameters fit to Fig. 3.
▲:[ATP];●:[PEP];○:[HMP];
[F-2,6-P_2] was increased stepwise as indicated.

Fig. 10 shows the reaction network of the investigated in vitro system. To achieve ATP-homeostasis, the system contains in addition to PFK-2/FBPase-2 creatine kinase and creatine phosphate for ADP-rephosphorylation.

At constant ATP, the dynamics of the F-6-P/F-2,6-P_2 cycle is governed by a differential equation for F-2,6-P_2 and a conservation equation for the total concentration of the hexose-monophosphates:

$$d[\text{F-2,6-P}_2]/dt = V_{PFK\text{-}2} - V_{FBPase\text{-}2} \quad (1)$$

$$[\text{F-6-P}]_{IN} = [\text{F-6-P}] + [\text{F-2,6-P}_2] \quad (2)$$

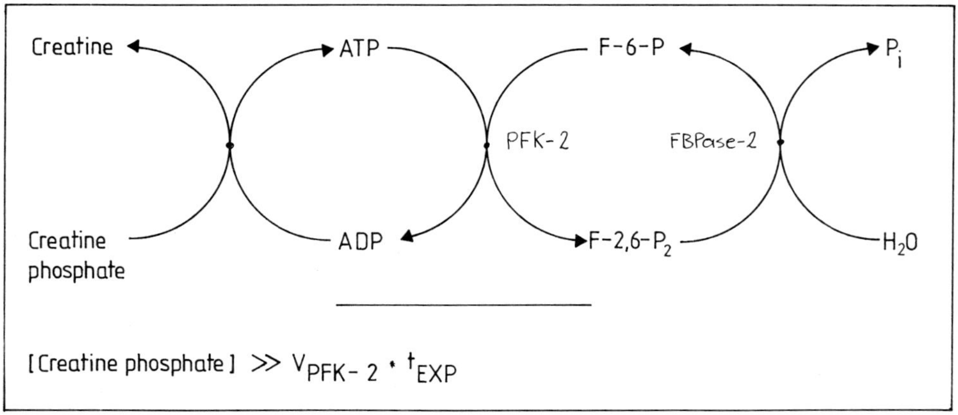

Fig. 10 Reaction scheme of the enzyme system used for investigation of the F-6-P/F-2,6-P_2 cycle. In addition to PFK-2/FBPase-2 creatine kinase for ADP rephosphorylation is involved.

By using the rate laws for PFK-2 and FBPase-2 (Kretschmer et al., 1985, 1987) an algebraic equation for the stationary concentration of F-2,6-P_2 is obtained:

$$\frac{V_{FBPase} * f(\text{F-2,6-P}_2) * ([\text{F-6-P}]_{IN} - [\text{F-2,6-P}_2] + K_{PFK\text{-}2}}{V_{PFK\text{-}2} * f(\text{ATP}) * ([\text{F-6-P}]_{IN} - [\text{F-2,6-P}_2])} \quad (3)$$

$$= 1 + ([\text{F-6-P}]_{IN} - [\text{F-2,6-P}_2])/K_{FBPase\text{-}2})$$

$f(ATP)$ and $f(F\text{-}2,6\text{-}P_2)$ are the fractional saturations of PFK-2 with ATP and of FBPase-2 with $F\text{-}2,6\text{-}P_2$, respectively:

$$f(ATP) = [ATP]/([ATP] + K_{PFK\text{-}2}) \quad (4)$$

$$f(F\text{-}2,6\text{-}P_2) = [F\text{-}2,6\text{-}P_2]/([F\text{-}2,6\text{-}P_2] + K_{FBPase\text{-}2}) \quad (5)$$

$K^{F\text{-}6\text{-}P}_{PFK\text{-}2}$, $K^{ATP}_{PFK\text{-}2}$, $K_{FBPase\text{-}2}$, $K_{FBPase\text{-}2}$ are kinetic constants for PFK-2 and FBPase-2 (Kretschmer et al., 1985, 1987). The control parameters which enter the equation for the stationary concentration of $F\text{-}2,6\text{-}P_2$ are the input concentration of F-6-P and the ratio $V_{PFK\text{-}2}/V_{FBPase\text{-}2}$. Fig. 11 a shows dependences of stationary $F\text{-}2,6\text{-}P_2$ on $[F\text{-}6\text{-}P]_{IN}$ for different proportions of native to phosphorylated PFK-2/FBPase-2. The measured stationary concentrations of $F\text{-}2,6\text{-}P_2$ were compared with the solution of (3) by using only the ratio $V_{PFK\text{-}2}/V_{FBPase\text{-}2}$ as parameter in the optimisation procedure. The obtained control curves exhibit a biphasic shape. At low $[F\text{-}6\text{-}P]_{IN}$ the stationary $F\text{-}2,6\text{-}P_2$ is very small. At high $[F\text{-}6\text{-}P]_{IN}$ a nearly linear dependence of $F\text{-}2,6\text{-}P_2$ on $[F\text{-}6\text{-}P]_{IN}$ is observed. The asymptotes of the control curves (Fig. 11 a) are obtained from (3) assuming saturation of FBPase-2 with $F\text{-}2,6\text{-}P_2$ ($f(F\text{-}2,6\text{-}P_2) = 1$):

$$[F\text{-}2,6\text{-}P] = [F\text{-}6\text{-}P]_{IN} - [F\text{-}6\text{-}P]_{IN} \quad (6)$$

The critical input concentration of F-6-P is:

$$[F\text{-}6\text{-}P]^{crit}_{IN} = \frac{K^{F\text{-}6\text{-}P}_{FBPase\text{-}2}}{2}(r-1)\left(1 \pm \sqrt{1 + 4\frac{r}{(r-1)^2}\frac{K^{F\text{-}6\text{-}P}_{PFK\text{-}2}}{K^{F\text{-}6\text{-}P}_{FBPase\text{-}2}}}\right) \quad (7)$$

with $r = V_{FBPase\text{-}2}/(V_{PFK\text{-}2} * f(ATP))$.

$[F\text{-}6\text{-}P]_{IN}$ is the intercept of the asymptotes of the control curves with the abscissa. $[F\text{-}6\text{-}P]_{IN}$ depends on the ratio of the maximum activities of PFK-2 and FBPase-2, i. e. on the phosphorylation of the PFK-2/FBPase-2 protein. cAMP-dependent phosphorylation promotes FBPase-2 and supresses the activity of PFK-2 (Pilkis et al., 1984). Hence the critical input concentration of F-6-P increases with increasing

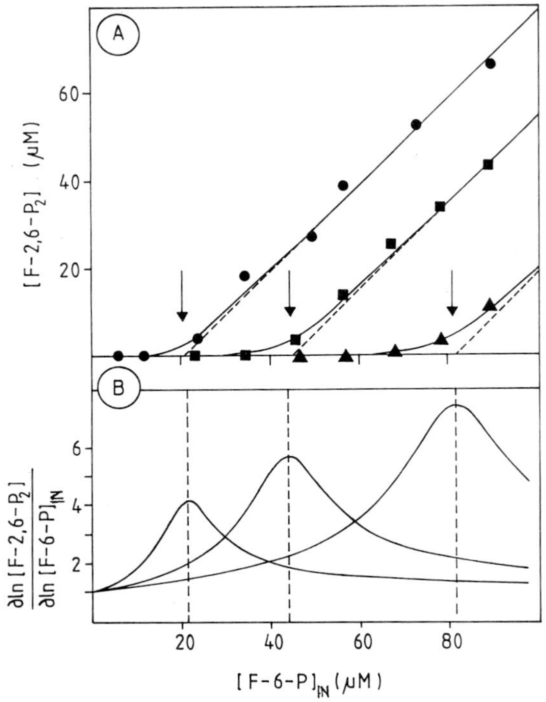

Fig. 11 Dependence of the stationary concentration of $F\text{-}2,6\text{-}P_2$ on the input concentration of F-6-P. The arrows indicate the critical input concentration of F-6-P.
a: Dependence of the stationary $[F\text{-}2,6\text{-}P_2]$ on the input concentration of F-6-P for different proportions of native to phosphorylated PFK-2/FBPase-2. The ratios $V_{PFK\text{-}2}/V_{FBPase\text{-}2}$ were estimated from the experimental data.

	●	■	▲
native enzyme	100 %	50 %	25 %
phospho enzyme	0 %	50 %	75 %
$V_{PFK\text{-}2}/V_{FBPase\text{-}2}$	3.3 + 0.2	1.3 + 0.1	0.58 + 0.4
$[F\text{-}6\text{-}P]_{IN}$ (μM)	21 + 1	44 + 2	81 + 4

b: Dependence of the concentration control coefficients on $[F\text{-}6\text{-}P]_{IN}$

phosphorylation of the PFK-2/FBPase-2 protein. In Fig. 11 b the concentration control coefficients for $[F\text{-}2,6\text{-}P_2]$ with respect to the input concentration of F-6-P are shown:

$$C^{[F\text{-}2,6\text{-}P_2]}_{[F\text{-}6\text{-}P]_{IN}} = \frac{\partial \ln([F\text{-}2,6\text{-}P_2])}{\partial \ln([F\text{-}6\text{-}P]_{IN})} \tag{8}$$

These coefficients characterise the fractional changes of stationary $[F\text{-}2,6\text{-}P_2]$ caused by small variations of $[F\text{-}6\text{-}P]_{IN}$ (Heinrich et al., 1977; Burns et al., 1985). The maxima of these control coefficients are in accordance with the critical input concentration of F-6-P. The curves approach $c_{F\text{-}6\text{-}P} = 1$ for $[F\text{-}6\text{-}P]_{IN} \rightarrow 0$ and $[F\text{-}6\text{-}P]_{IN} \rightarrow \infty$ but exhibit maximum values which are substantially greater than 1. The data show that the stationary concentration of $F\text{-}2,6\text{-}P_2$ depends both on the input of F-6-P and on the ratio $V_{PFK\text{-}2}/V_{FBPase\text{-}2}$. Interestingly, it was shown, that hepatic $F\text{-}2,6\text{-}P_2$ profiles depend not only on the phosphorylation of the PFK-2/FBPase-2, but are also causally linked to differences in F-6-P levels (Kuwajima et al., 1986).

The approach presented does not take into account metabolic interactions which arise with respect to the $F\text{-}6\text{-}P/F\text{-}2,6\text{-}P_2$ cycle. In cellular metabolism [F-6-P] is not a control parameter, but is sensitively regulated by the PFK/FBPase cycle. In addition, phosphorylation of the PFK-2/FBPase-2 is not exclusively controlled by external signals, but depends also on metabolite pattern (ATP, $F\text{-}2,6\text{-}P_2$ (Pilkis et al., 1984)). Both the interaction of the $F\text{-}6\text{-}P/F\text{-}2,6\text{-}P_2$ cycle with other metabolic structures as the $F\text{-}6\text{-}P/F\text{-}1,6\text{-}P_2$ cycle and the modulation of protein phosphorylation by metabolites as $F\text{-}2,6\text{-}P_2$ may contribute to rather complex dynamic structures with respect to the glucose metabolism.

Abbreviations

PFK	phosphofructokinase (EC 2.7.1.11)
FBPase	fructose-1,6-bisphosphatase (EC 3.1.3.11)
PK	pyruvate kinase (EC 2.7.1.40)
GPI	glucose-6-phosphate isomerase (EC 5.3.1.9)
PFK-2	6-phosphofructo-2-kinase (EC 2.7.1.105)
FBPase-2	fructose-2,6-bisphosphatase (EC 3.1.3.46)
CK	creatine kinase (EC 2.7.3.2)
F-6-P	fructose-6-phosphate
G-6-P	glucose-6-phosphate
F-1,6-P_2	fructose-1,6-bisphosphate
F-2,6-P_2	fructose-2,6-bisphosphate
PEP	phosphoenolpyruvate
HMP	hexosemonophosphate
$[X]_{IN}$	input concentration of X
$[X]$	actual concentration of X
v	enzyme activities
V	maximum activities of enzymes

Pattern Formation in Biological Systems

Morphogenesis of Behaviour and Information Compression in Biological Systems

H. Haken

Introduction

One of the most characteristic features of life is the extremely high coordination between the different parts of a biological system. In higher animals we observe such a high coordination of neurons, muscles and other tissues which then leads to a well-pronounced heartbeat, breathing, blood circulation, coordination of movements up to complex behaviour.

It is a challenging task to understand the formation of these behavioural patterns, or in other words, the morphogenesis of behaviour. One possibility which is quite well-known is based on the idea of a neuronal motorprogram. In such a case, a neuronal network similar to a computer determines the motion of the individual muscles or other tissues in order to produce a well-organised total action.

In our contribution, we wish to lend support to another paradigm which, in our opinion, plays a role at least as important as that of the motorprogram, namely the idea of self-

Thermodynamics and Pattern Formation in Biology

organisation. Over the past two decades, we have studied numerous systems in physics, chemistry and biology which can produce spatial or temporal structures without specific interference from organising centres. Instead, the activity of these systems is organised by the interaction of the individual parts of the system itself. Upon changes of external conditions which are of a rather unspecific nature, these systems can undergo dramatic transitions between different modes of their activity.

As we could show, these specific transitions between different modes of activity are not determined by specific mechanisms of the interactions between the different parts of a system but depend on general principles which can be formulated mathematically (next two sections).

Quite generally, the interdisciplinary field of synergetics (Haken, 1983 a, b) deals with systems which are composed of many subsystems. In physics such a system can be a fluid which is composed of many individual molecules, or a laser which is composed of many individual atoms. In chemistry, a system may be composed of many individual molecules which may be of the same or of different kinds. In biology, a system may be composed of neurons, muscles and perhaps other parts.

Let us assume that such a system is at rest or performs some specific activity. Now let us assume that, for instance, an energy input is increased or chemicals are added to the system at an increased rate. Then in a number of cases at critical values of these so-called control parameters the system may acquire the new state. For instance, a system which has been at rest may now start to oscillate, or a system which has been oscillating, may start a different kind of oscillation or a more complicated one. For instance, one which eventually leads to deterministic chaos.

As has been revealed by synergetics, in systems close to such transition points specific phenomena will occur jointly. One of these phenomena is the occurrence of *critical fluctuations* in which the values of observables such as the output of the laser light or the molecular concentration in a chemical reaction undergo large fluctuations. Another phenomenon consists in so-called *critical slowing down*. When a system is close to a transition point it reacts more slowly to external perturbations than in situations where the system is far away from such a critical point. These phenomena are typical for processes of self-organisation, where the system acquires its new state without specific interference from the outside, i. e. the control parameters do not anticipate the new state but rather indirectly force the system to acquire it. As is shown in synergetics, in situations where macroscopic patterns of behaviour change qualitatively, the dynamics of very many parts of a system is governed by one or a few order parameters. Incidentally, an enormous reduction of information (in the Shannon sense) takes place.

In the last four sections we present a specific example, namely involuntary transitions of human hand movements, indentify its order parameter and establish its equation. Excellent agreement with the experimental results is obtained. We believe that many more examples in biology can be found, for instance, in transitions of normal to abnormal heartbeats, breathing, blood circulation etc. which exhibit quite similar phenomena, but which have not been studied so far because of lack of an adequate theory.

A glance at the microscopic approach

In general, we may decompose a biological system in a more or less obvious way into specific elements which we shall distinguish by an index n and which are described by a set of variables which we shall abbreviate by $\underset{\sim}{e}_n$. Such variables may be concentrations of biomolecules, firing rates of neurons, elongation of muscle fibers etc. Of course, $\underset{\sim}{e}_n$ may change in the course of time. The elements are connected by information channels whose variables we shall describe by c_{mn} where m and n are the indices of the elements connected by that channel. The channels may be localised (e. g. axons and dendrites), or delocalised (e. g. blood streaming for the transport of hormones etc.). We shall model the behaviour of the system by the two sets of equations

$$\dot{\underset{\sim}{e}} = \underset{\sim}{N}_e (\underset{\sim}{e}, \underset{\sim}{c}, \alpha) + \underset{\sim}{F}_e \tag{1}$$

$$\dot{\underset{\sim}{c}} = \underset{\sim}{N}_c (\underset{\sim}{e}, \underset{\sim}{c}, \alpha) + \underset{\sim}{F}_c \tag{2}$$

where $\underset{\sim}{N}$ describes deterministic effects whereas $\underset{\sim}{F}$ incorporates external or internal noise, e. g. fluctuations in molecular concentrations, irregular electric activity etc. As it turns out $\underset{\sim}{e}$ and $\underset{\sim}{c}$ play a rather symmetric role so that we lump them together into a state vector $\underset{\sim}{q}$ according to

$$\begin{pmatrix} \underset{\sim}{e} \\ \underset{\sim}{c} \end{pmatrix} = q \tag{3}$$

In this way (1) and (2) can be written in short as

$$\dot{\underset{\sim}{q}} = \underset{\sim}{N}(\underset{\sim}{q}, \alpha) + \underset{\sim}{F}. \tag{4}$$

We may consider processes continuous in time where we make the interpretation

$$\dot{\underset{\sim}{q}} = d\underset{\sim}{q}/dt \tag{5}$$

or in discrete time steps where we use the interpretation

$$\dot{\underset{\sim}{q}} = \underset{\sim}{q}(t + \tau) - \underset{\sim}{q}(t). \tag{6}$$

Equally, in a number of cases we may consider instead of the discrete index a continuous space variable $\underset{\sim}{x}$

$$n \rightarrow \underset{\sim}{x}. \tag{7}$$

In case the fluctuating forces $\underset{\sim}{F}$ contain the variables, we must apply the Ito-Stratonovich calculus. But we shall not dwell here on these somewhat delicate questions. We mention further that instead of the Langevin type equation (4), we may equally well apply the Fokker-Planck equation for a continuous Markov process or the master equation. The general procedure is as follows.

We start from the Langevin equation

$$\dot{\underset{\sim}{q}} = \underset{\sim}{N}(\underset{\sim}{q}, \alpha) + \underset{\sim}{F} \tag{8 a}$$

which can be transformed into a Fokker-Planck equation of the type

$$\dot{P} = \widetilde{L}P \tag{8 b}$$

where $\widetilde{L}$ is a linear operator. We shall write its steady state solution as

$$P(\underset{\sim}{q}, \alpha) \tag{8 c}$$

Its information is given by

$$i = - \int P \ln P \, d^N q + i_a \tag{8 d}$$

where i_a stems from the accuracy limit imposed on our measurements of the state q. We now study the behaviour of (8 a), (8 c) and (8 d) when the control parameter α is changed. α may be fixed from the outside (e. g. administration of drugs, lower oxygen content of air, illumination etc.) or from the interior of the system, e. g. from parts other than those under consideration (hormones, nerve pulses, blood pressure etc.).

We assume that we have solved the Langevin equation without noise for a specific control parameter value

$$\dot{\underset{\sim}{q}}_o = \underset{\sim}{N}(\underset{\sim}{q}_o, \alpha_o). \tag{9}$$

This solution is called $\underset{\sim}{q}_o$. When noise is added a specific solution of the Fokker-Planck equation in the steady state case evolves. Connected with it is specific information for the control parameter value α_o. For what follows the explicit form of $\underset{\sim}{q}_o$ is not needed.

Now we change the control parameter value to a new value α and check the stability. To this end we make the hypothesis

$$\underset{\sim}{q} = \underset{\sim}{q}_o + \underset{\sim}{w}. \tag{10}$$

Under the assumption that $\underset{\sim}{w}$ is small we insert (10) into the Langevin equation (8 a) and linearise. We obtain

$$\dot{\underset{\sim}{w}} = L\,\underset{\sim}{w}. \tag{11}$$

By denoting the solution matrix of this differential equation by W we readily find that under the assumption that $\underset{\sim}{q}_o$ is time independent W can be written in the form

$$W = V e^{\Lambda t} \tag{12}$$

where V and Λ are time independent matrices. In the following we shall assume that Λ is diagonal. In order to find the exact solution of (8) we make the hypothesis

$$\underset{\sim}{q}(t) = \underset{\sim}{q}_o + V \underset{\sim}{\xi}(t). \tag{13}$$

Inserting (13) into (8) and projecting both sides of the resulting equation on the eigenvectors contained in V, we readily obtain

$$\dot{\underset{\sim}{\xi}} = \Lambda \underset{\sim}{\xi} + \underset{\sim}{N}(\underset{\sim}{\xi}, \alpha) + \underset{\sim}{F}. \tag{14 a}$$

Correspondingly, the Fokker-Planck equation (8 b) is transformed into

$$\dot{P} = L P. \tag{14 b}$$

Its steady state solution is denoted by

$$P(\underset{\sim}{\xi}, \alpha) \tag{14 c}$$

Since i is invariant against the transformation $P \to \hat{P}$, its numerical value remains unchanged, but formally it reads differently:

$$i = - \int P \ln P \, d^N \xi + i_a. \tag{14 d}$$

According to whether the real part of the eigenvalues of Λ are positive or negative, we shall distinguish between the so-called unstable and stable modes which we shall distinguish by indices u and s, respectively. According to the slaving principle of synergetics, close to an instability point we may express the slaved amplitudes by the unstable mode amplitudes

$$\xi_s = f_s(\xi_u, t). \tag{15 a}$$

According to this result, the probability distribution function in (14) can be written as a product of the conditional probabilities P_s times the distribution function P_u of the unstable mode amplitudes ξ_u.

$$P(\xi, \alpha) = \prod_s P_s(\xi_s | \xi_u) \; P_u(\xi_u). \tag{15 b}$$

It can further be shown that the information is transformed into

$$i = i_u + \Sigma \int P_u \, i_s(\xi_u) \, d^M \xi_u + i_a \tag{15 c}$$

where i_s and i_u are defined as usual by use of P_s and P_u, respectively. By means of (15 a) we may eliminate ξ_s from (14 a) and thus obtain equations for ξ_u alone

$$\dot{\xi}_u = \Lambda_u \xi_u + N_u(\xi_u, \alpha) + F_u. \tag{16}$$

The ξ_u shall be denoted as order parameters. They describe the macroscopic behaviour of the total system. In morphogenesis they determine the amplitude of the morphogenetic field, in Kelso's experiments the macroscopic motion of fingers etc.

On the right hand side of (15 c) the first part refers to the information of the order parameters whereas the second part refers to the information of the slaved modes averaged over the distribution function of the order parameters. A detailed analysis in the frame of the slaving principle shows (Haken, 1983 a, b) that while i_u depends strongly on α, the second part depends only weakly on it and represents, particularly in the transition region only some background.

Here it may suffice to indicate the main idea: In the leading approximation of the slaving principle, $P_s(\xi_s \mid \xi_u)$ can be written as a Gaussian

$$P_s(\xi_s \mid \xi_u) = N \exp[-(\xi_s - g_s(\xi_u))^2 / Q_s] \tag{17}$$

where Q_s is practically independent of α, and $g_u(0) = 0$. Now, effect in

$$i_s(\xi_u) = -\int P_s(\xi_s \mid \xi_u) \ln P_s(\xi_s \mid \xi_u) \, d\xi_s \tag{18}$$

the replacement

$$\xi_s \rightarrow \xi_s + g_s(\xi_u). \tag{19}$$

Then evidently (18) becomes independent of ξ_u, and because of the normalisation of P_u, the integral in (15 c) reduces to $i_s(0)$, which is independent of α (cf. (17) and (18)). As a consequence the information change *close to an instability point* is given by

$$\Delta i \equiv i(\alpha) - i(\alpha_o) = i_u(\alpha) - i_u(\alpha_o) \tag{20}$$

i. e. by the information change of the order parameters alone. As can be shown by an explicit study of equations (16) beyond threshold some nonvanishing order parameters occur so that we are now in the domain of self-organisation.

A specific example - One order parameter

A good deal of the interpretation of our results can be seen when we consider a single order parameter which obeys the following order parameter equation

$$\dot{\xi}_u = \lambda_u \xi_u - \beta \xi_u^3 + F_u(t). \tag{21}$$

As is well-known, the behaviour of this order parameter can be interpreted as that of a particle in a potential well which moves in an overdamped fashion (Haken, 1983 a). In the following we shall identify λ_u directly with the control parameter α which changes its sign at threshold from minus to plus. When the transition from negative values of α to positive values of α occurs, typical phenomena well-known from equilibrium phase transition occur, such as symmetry breaking, critical slowing down, soft mode and critical fluctuations. But here we are interested in the change of information.

To obtain this, we first transform (21) into a Fokker-Planck equation whose steady state solution is given by

$$f(\xi_u) = N \exp(\alpha \xi_u^2 - \frac{\beta}{4} \xi_u^4), \quad \alpha \equiv \lambda_u / 2. \tag{22}$$

Information for continuously distributed variables is infinite. It can be made finite when we prescribe a specific interval of accuracy. The variation of the information of the order parameter as a function of the control parameter has been calculated numerically. Below threshold the information has a certain value which may be put equal to zero under an appropriate choice of the accuracy interval. When we measure the system with the same accuracy above threshold, its information is now one bit. Our result means that the biological system under consideration can acquire one out of two states of macroscopic behaviour - e. g. two kinds of hand movement (see also below). The important conclusion we may draw from the above results is as follows:

We see that information compression occurs, namely instead of the very large total of information for all the individual subsystems buried in the sum over the information i_S of the modes or subsystems in (15 c), we are now concerned with the information of the order parameters alone which, in addition, can be measured macroscopically with nonvanishing values. We thus see that the biological system by its self-organisation has acquired a new macroscopic state which bears some structure, has a nonvanishing order parameter and allows us in the present case to store one bit of information. Clearly not only the information required to describe the system is reduced but, because of its macroscopic order, the system has acquired a high degree of reliability. Clearly, with more order parameters or more complicated potential fields, more information can be stored in such a system.

Quite evidently, we may now reinterpret the role of the order parameter in two ways: In our old interpretation it gave orders to the individual systems by means of the slaving principle and it described the order of the system to an external observer. In the sense of information transmission, we might call the order parameters 'informators' because they provide an informational field for the individual subsystem. The occurrence of nonvanishing order parameters (informators) is the manifestation of a new kind of information, namely *synergetic information*. It is caused by the cooperativity of the system and it is delocalised as can readily be seen by the example of the cAMP-concentration field of slime mold.

The relevant information is contained in the order parameters α which are produced by the cooperativity of the system. From the point of view of information processing our results can be interpreted as follows:

Upon the change of one or several control parameters α which may require only a few bits, the self-organising system under consideration may acquire a complicated spatial or behavioural pattern. As was shown above, this pattern is described by the superposition (13) which contains V (12). But in turn this complicated pattern is determined by one or few order parameters. The existence of few order parameters which govern the complicated pattern (e. g. in morphogenesis), or the complicated dynamics (e. g. in hand movements, gaits of horses), reflects a very high degree of coordination in the system.

One may ask why the transformation of few bits contained in the control parameters into few bits contained in the order parameters will not be a round about way of structuring in the biological system. The reason, however, is rather obvious. For instance, when we think of locomotion, very many microscopic degrees of freedom (neurons, muscles, links, tissues) are transformed into a well-coordinated macroscopic motion serving a specific purpose. Further on when we are concerned, for instance, with speech then a few bits causing, for instance, the pronunciation of a vowel are transformed into a complicated sound pattern which inflates the information in a way well-known from communication engineering in order to increase the reliability of the signal. In this way we observe an interplay between information compression and inflation depending on the specific task the biological systems fulfills.

These results shed new light on information processing in biological systems. Especially on the number of bits required to produce patterns. One may expect that by the process of evolution specific relationships between order parameters and the complicated behaviour of the subsystems were produced so that the actual number of bits required is much smaller to produce a pattern than one would expect from a genetic information.

This is indeed in accordance with previous ideas of other authors e. g. the development of morphogenetic fields. But I think here for the first time the amount of information compression in the transition region can be treated in an explicit fashion. In our opinion an important future task will be the proper identification of order parameters and control parameters in biological systems. A first step towards that goal will be done in the following sections on involuntary changes of hand movement.

Phase transitions in human hand movement

While researching voluntary oscillatory motions of the two index fingers, Kelso (1981) observed an interesting phenomenon. Under instructions to increase the frequency of out-of-phase, antisymmetrical motion (involving simultaneous flexor and extensor muscle activities), the subjet's finger movements shifted abruptly to an in-phase symmetrical mode that involved activation of homologous muscle groups. This finding was not restricted to finger movements. In later work (Kelso, 1982, 1984) that employed similar experimental manipulations, modal transitions in hand motions around the wrist were also observed: the antisymmetrical phase relationship between the hands was replaced by symmetrical phasing. Moreover, although the phase transition occurred at very different frequencies of hand motion for different subjects, it was nevertheless predictable.

The most dramatic aspect of these simple experiments is the sudden and completely involuntary change in the ordering or phasing among muscle groups that occurs at a critical, intrinsically defined frequency. In this feature, the hand movement data share a likeness to gait transitions in locomotion.

Initial development of the model: Order parameters and the potential function

To summarise, the main features of the experiments described above are:

1. the presence of only two stable phase (or 'attractor') states between the hands (which one is observed is a function of how the system is prepared, i. e. an instruction to move the hands in the out-of-phase or in phase mode);
2. the abrupt transition from one attractor state to the other at a critical cycling frequency;
3. beyond the transition, only one mode (symmetrical in-phase) is observed;
4. when cycling frequency is reduced, the system stays in the symmetrical mode, i. e. it does not return to its initially prepared state - a result that suggests coexistence of the basins of attraction for the symmetrical and antisymmetrical modes and depletion of one of them.

Taken together, these results as well as other findings in the motor control literature support the hypothesis that *phase* is a relevant macroscopic (or 'essential') parameter of certain movement patterns. For example, the internal phasing structure of activities as widely varied as chewing, locomotion, handwriting, and speech remains invariant across scalar changes in force or rate (Grillner, 1982; Kelso, 1981; Schmidt, 1982; Tuller et al., 1982). Similarly, in the experiments described above, phase is preserved constant over a wide range of frequencies, even though the magnitudes and durations of muscle activities and other kinematic variables change considerably. Only when frequency is scaled beyond a critical value does a phase shift occur.

In the present paper it seems reasonable to propose phase as an order parameter for at least two reasons. First, unlike many other possible candidates, phase is an accurate reflection of the *cooperativity* among the components of the system. Thus, we can say, in a manner consistent with synergetics, that the configuration of the subsystems (in the present context defined as the individual hand motions) specifies their phase relations, and conversely, that the phase variable specifies the spatiotemporal ordering of the subsystems. Second, it is phase that remains invariant across transformations in many motor acitivities that involve very different anatomical substrates. This highlights an important further feature of the order parameter concept, namely, that the order parameter (by hypothesis here, the relative phase) changes much more slowly than the variables describing the behaviour of the individual components (e. g. velocities of each hand motion).

Our first step in the development of the present model (Haken et al., 1985) is to provide a mathematically accurate description of the main qualitative features of the data. We therefore specify a potential function that corresponds to the layout of attractor states and how that layout is altered as a control parameter is changed. In a following section, we show how the model equations describing the potential function can be derived from the equations of motion of each hand and a (nonlinear) coupling between them.

For the sake of clarity we introduce the elongations of the finger tips x_1 and x_2 as shown in Fig. 1. In order to define the relative phase ϕ we assume that the motion of the hands is more or less harmonic, so we put

$$x_1 = r_1 \cos(\omega t + \phi_1), \tag{23}$$

$$x_2 = r_2 \cos(\omega t + \phi_2), \tag{24}$$

where ω is the basic frequency of the hand movement while the amplitudes r_1, r_2 and the phases ϕ_1, ϕ_2 are time dependent quantities whose time dependence is assumed to be much slower than that defined by the frequency ω. The relative phase is defined by

$$\phi = \phi_2 - \phi_1 . \tag{25}$$

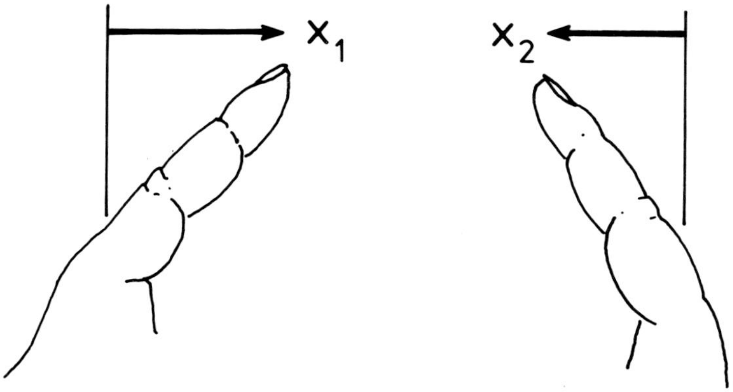

Fig. 1 Elongations of the finger tips.

In order to describe the change of phase we adopt basic ideas from synergetics. As shown in synergetics, in many cases the equations for order parameters are of the form

$$\dot{\phi} = -\partial V / \partial \phi, \tag{26}$$

where V is the so-called potential function. In our search for a model we make a few rather obvious assumptions about V. Since ϕ occurs under cosine or sine functions the properties of the physical system must not change when ϕ is replaced by $\phi + 2\pi$. Consequently, we shall postulate that the potential V is periodic:

$$V(\phi + 2\pi) = V(\phi). \tag{27}$$

We furthermore introduce the assumption that both hands play a symmetric role. In such a case the behaviour of the system must not depend on the way we label the right hand and the left hand. This means that V must remain unchanged when we exchange the indices 1 and 2 in (25). This in turn means that the potential V is symmetric:

$$V(\phi) = V(-\phi). \tag{28}$$

We assume that V obeys the conditions (27) and (28) in the simplest form which explains the above-mentioned experimental results. To this end we write V as a superposition of two cosine functions:

$$V = -a\cos\phi - b\cos 2\phi. \tag{29}$$

As is known from synergetics, the behaviour of the system obeying (26) can be easily described by indentifying ϕ with the coordinate of a particle which moves in an overdamped fashion in the potential V.

When we take the total superposition (29) but change the ratio b/a we run through a series of potential fields as shown in Fig. 2. When we initially prepare the system in a state shown by the black ball and increase the frequency, and likewise assume that b/a decreases with increasing frequency, we obtain a critical value where the ball falls to the lower minimum belonging to $\phi = 0$. This means that the hand movement made a transition from the antisymmetric ($\phi = -\pi$ state) into the symmetric state with $\phi = 0$. The hand movement stays in that state when ω is further increased. When we decrease b/a starting from high values, the system remains all the time in the $\phi = 0$ state even if ω drops below ω_o. This 'hysteresis' phenomenon is well known in many physical and biological systems.

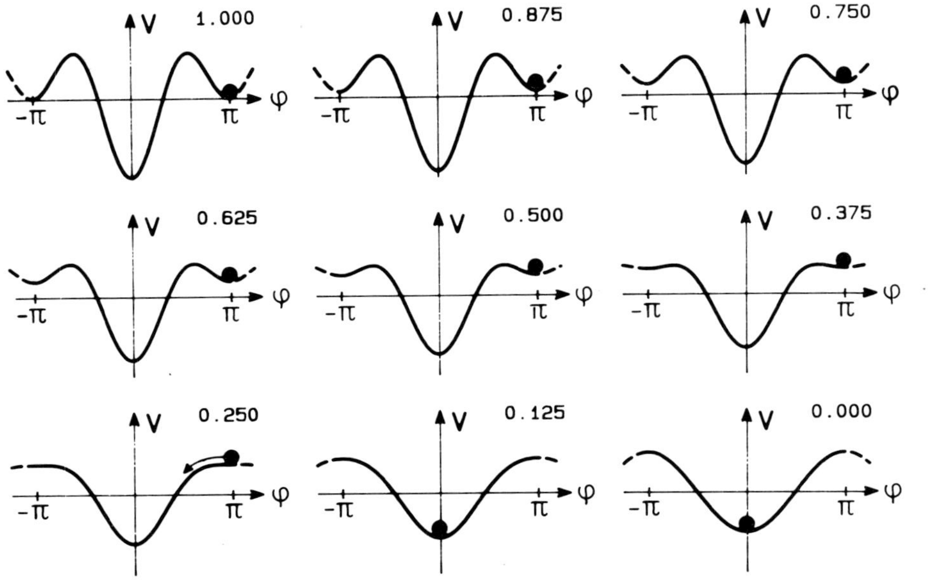

Fig. 2 The potential V for varying values of b/a. The numbers refer to the ratio b/a (after Haken et al., 1985).

Formulation at the next lower hierarchical level

The order parameter equation for ϕ can be derived from equations at the next lower hierarchical level, namely individual hand movements (Haken et al., 1985). We have studied the following type of equations

$$\ddot{x}_1 + f_1(\dot{x}_1, x_1) = K_{12}(x_1, x_2), \tag{30}$$

$$\ddot{x}_2 + f_2(\dot{x}_2, x_2) = K_{21}(x_1, x_2), \tag{31}$$

where the indices 1, 2 refer to the right and left hand, respectively, and K_{12} is the coupling term brought about via the nervous system. For the sake of simplicity, the oscillators are assumed to have identical frequencies and damping constants, respecitively, i. e. $f_1 = f_2$ and

$$K_{21}(x_1, x_2) = K_{12}(x_2, x_1). \tag{32}$$

In order to obtain (33) K_{12} must be a nonlinear function of $(x_1 - x_2)$ either via a time delay

$$K_{12} = \int e^{-\lambda(t-\tau)} [\alpha (x_1 - x_2)_\tau + \beta (x_1 - x_2)^3_\tau] \, d\tau \tag{33}$$

or by using time derivatives

$$K_{12} = (\dot{x}_1 - \dot{x}_2) [\alpha + \beta (x_1 - x_2)^2]. \tag{34}$$

These coupling terms are not unique and a huge variety of other possibilities exists, but (33, 34) are the simplest representations of the possible couplings. In order to achieve agreement with the experimental findings, we chose (34) in the following numerical simulations, but the other coupling terms show no qualitatively different behaviour.

Numerical results

In this section we present some numerical results that correspond to the analytical treatment provided above. Haken et al. (1985) solved the model of the preceding section on a digital computer using a fourth order Runge-Kutta method. To test the stability of a stationary solution small random fluctuations of finite amplitude were introduced. The resulting simulation shown in Fig. 3 compares quite favourably with the experimental data. In Fig. 3 a the displacements x_1, x_2 are plotted against time and in Fig. 3 b the corresponding phase difference between the oscillators is plotted for the same motions. As in the bimanual experiments, the coupled oscillation is prepared in the state $\phi = \pi$ and the frequency ω is increased monotonical. A transition from the out-of-phase mode to the in-phase mode is observed, when ω exceeds a critical value. However, the frequency of the oscillation changes rather quickly so that stat ionary oscillations are not reached. Thus the exact form of the curves depends strongly on the noise level and the rate of change in ω.

The steady state amplitudes for the in-phase mode and the out-of phase mode are shown in Fig. 4. The unstable branch of the out-of-phase mode is shown by dotted lines. The ω^{-1} dependence of the amplitudes is quite clear. This feature rests on our special choice of f in (30, 31) and will be different for another choice of f. As shown in Fig. 4 for ω smaller than ω_c, the in-phase mode and the out-of-phase mode are both stable. Due to the coexistence of two basins of attraction, the particular mode observed depends on the initial conditions, i. e. which coordinative state is prepared.

If one starts in the antisymmetric phase and increases ω slowly, the oscillation remains in this mode until the solution becomes unstable. At this point a jump in amplitude occurs and the only stable stationary solution revealed by the system corresponds to the in-phase mode. Such is the case when ω is increased further. On the other hand, if ω is decreased slowly the system stays within the basin of attraction of this solution even when ω drops below ω_c. As we mentioned earlier, this hysteresis phenomenon is typical for such bistable situations. To summarise, it is quite clear that the main features of the experimental data described above (Kelso, 1981) are captured by the present mathematical formulation as illustrated by these numerical results. But in

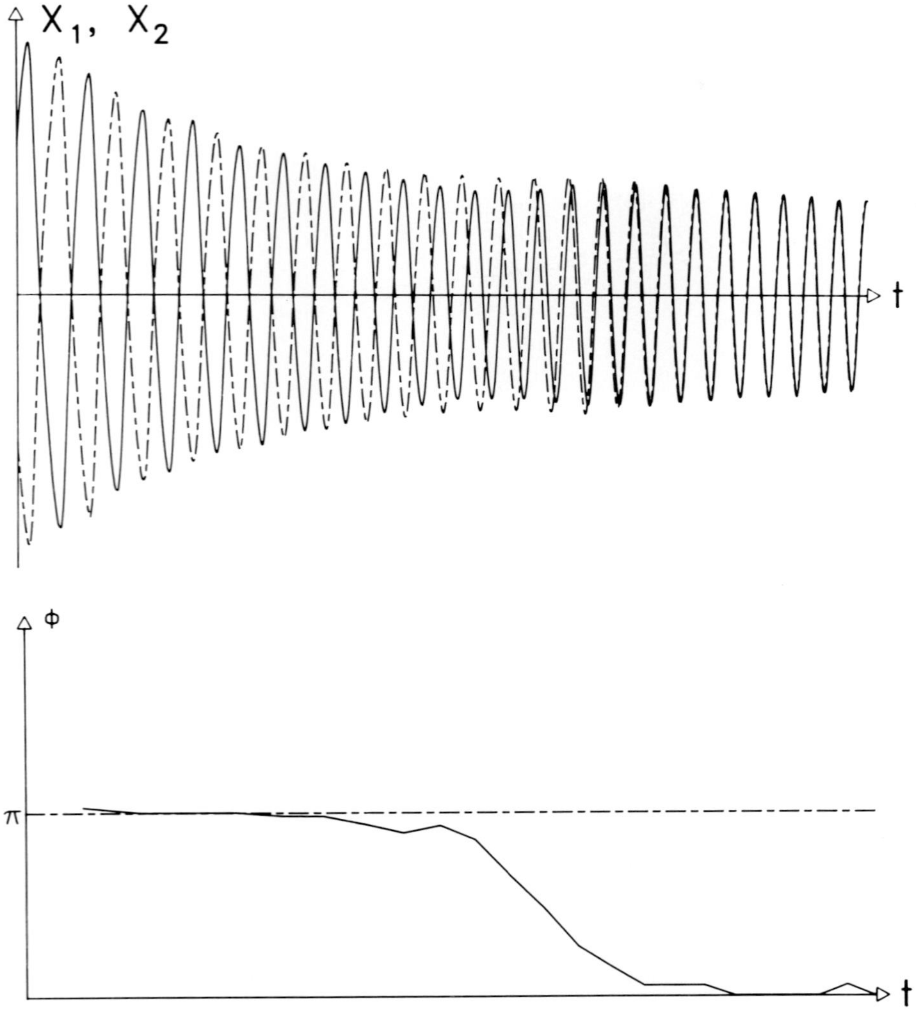

Fig. 3 In a) the calculated displacements of the finger tips and in b) the corresponding phase difference between the oscillators is plotted (after Haken et al., 1985).

addition to the main features which correspond to average values of ϕ or x_1, x_2, on the basis of synergetics we also expect a *fine structure* due to *fluctuations* close to the transition frequency.

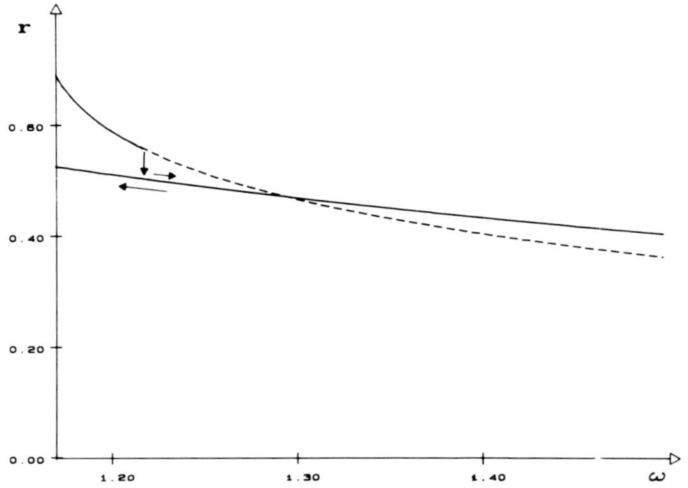

Fig. 4 The steady state amplitudes of the in-phase and the out-of-phase mode are shown as a function of ω.

Fluctuations and their significance

Looking at our above model and bearing in mind the typical critical fluctuations of synergetic systems close to their transition points, we suggested to Kelso that he should look for such fluctuations. Fig. 5 shows his experimental results (cf. e. g. Kelso, Scholz, 1985).

In this figure both the average phase and the phase fluctuations, or more precisely the root mean square of the phase fluctuations, are plotted. In the case of the transition from the antisymmetric to the symmetric hand motion large critical fluctuations indeed do occur. We have modelled this transition under the impact of fluctuations by means of adding a fluctuating force to (26), i. e. we have treated the equation (Schoener et al., 1986)

$$\dot{\phi} = - \partial V / \partial \phi + F(t). \tag{35}$$

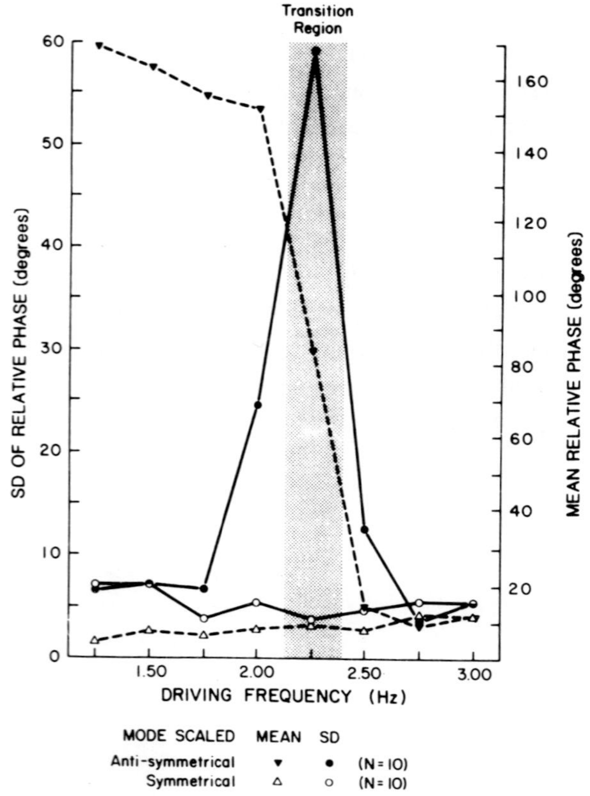

Fig. 5. The mean phase modulus $< \phi >$ (▼: AMS,△: SMS) and its SD = $\{ < \phi^2 > - < \phi >^2 \}$ (● AMS, ○ SMS) were determined in the stationary limit on each frequency plateau (i. e. for the last 3 s at each frequency). Each point on the graph represents an average from 10 runs of the experiment. AMS = antisymmetric model scaled, SMS symmetric model scaled (from Kelso, Scholz, 1985).

Using the Fokker-Planck equation, we have studied both the behaviour of the root mean square as well as correlation functions and results are found in excellent agreement with the experiment.

What is most interesting and important, is the consequence of this treatment. Namely, when we first assume that the transition between one kind of hand movement to the other kind is caused by the change of a motorprogram of the neurons, it will be very hard to understand why any fluctuation should occur at all. Indeed a motorprogram is a fixed program and no fluctuations should be expected. The way the transition occurs in the hand movement rather indicates that we are dealing here with a typical act of self-organisation. This system organises itself, that means the individual neurons and muscles act jointly as if the whole system acts as a total autonomous system.

Quite clearly, this introduces an entirely new paradigm into biology and it can be hoped that similar mechanisms and models apply to more complicated motions, where the next step will be to study the change of gaits of horses. Other highly coordinated motions may most probably be treated very much the same way, for instance, rhythmic motions like breathing and heartbeat and their coordination.

Model of Pattern Generation on Plants, Based on the Principles of Minimal Entropy Production

R. V. Jean

Introduction

This chapter deals with a reality within the core of plant morphogenesis and growth, that is, with the relative arrangement of the similar parts (leaves, scales, florets, primordia, ...) of plants, commonly called the phenomenon of phyllotaxis. These parts are conventionally represented as lattices of points in the plane, showing a constant vertical rise r, and a constant azimuthal rotation d called the divergence between consecutively born primordia. A cone, or a pineapple or a leafy stem is thus considered as a cylinder cut along one of its generators and unrolled in the plane. An equivalent descriptive model shows the primordia as arranged along two opposed families of logarithmic spirals, such as the florets in the disk-like capitulum of the sunflower. Their coordinates are then (R, D) where $D = 360\ d$, and $R = \exp(2\pi r) > 1$, where R is the ratio of the distances to the common pole of the spirals of two consecutively born primordia. To each type of phyllotactic arrangement of the primordia found in nature corresponds a series of positive integers $H(1), H(2), H(3), \ldots$, such that $H(k+2) = H(k+1) + H(k)$, $k = 1, 2, 3, \ldots$, closely related to the divergence d of the system, as underlined by the theorem described in the following section. The principal integers observed, representing the numbers of spirals in the opposed families mentioned earlier, are consecutive terms of the series 1, 2, 3, 5, 8, ..., called the Main or Fibonacci series, and the corresponding observed value of d is ϕ^{-2} where

Thermodynamics and Pattern Formation in Biology

$\phi = (5^{1/2} + 1)/2$. The theoretical understanding of this phenomenon is the classical problem of phyllotaxis, and is considered as one of the greatest mysteries of plant biology.

This chapter emphasises that

- phyllotaxis is first of all an evolutionary and functional problem
- the primordia of plants form a hierarchical or multilevel system
- the preponderance of Fibonacci phyllotaxis can be explained by an interpretative theory
- such a theory is able to include, as special cases, existing theories and models.

The first and the second point are briefly presented in the following section which underlines the physical basis of the interpretative model elaborated in the hierarchical representation. An interpretative theory interprets the phenomenon concerned, in terms of its ultimate effect. This effect is assumed here to be the minimisation of the entropy of plants. A formula for bio-entropy is thus needed which will act as a functional cost for the various phyllotactic patterns, and we will look in the set of costs for the minimal cost, taking into consideration a fundamental parameter called the rhythm of growth. This is the object of the third section where it is emphasised that all the art and difficulty of the subject is to find an appropriate functional cost. The function obtained can be compared with entropy formulae found in the literature. The well-known principle of optimal design, operationally formulated by Rosen (1967), takes here the form of the principle of minimal entropy production.

The next sections deal with the third and fourth point mentioned above, and present the main results and achievements of the model. The model is an hypothetico-deductive construct from which mathematical theorems can be derived, and predictions can be made. The main result is that the Fibonacci phyllotactic pattern is the one which minimises the entropy of plants, or maximises their energy level, provided that the rhythm of growth begins early, that is before the time $T_r = 6$ or at the time $T_r = 8$ or 13. It will be shown that some of the predictions of the model are verified, and that there is a lack of data to confirm or invalidate others. It will be seen, too, that the interpretative model generalises fundamental ideas of the diffusion theories of phyllotaxis, and basic structures of the conntact pressure theory, the current theories of phyllotaxis. These are formulated in the conventional regular lattice representations of the patterns, which can be seen to be generalised by the hierarchical representation.

About the evolutionary biological bases of the model: the hierarchical control of phyllotaxis

At the beginning of the century the botanist Church (1920) emphasised, followed by others, that we must look into the sea, at brown algae in particular, to understand the primary intentions of the remarkable phyllotactic patterns found in higher plants. The phyllotactic mechanism would be a primitive function of marine vegetation, and phyllotaxis in higher plants would be an amplification of phyto-benthic factors. One of the oldest vascular plants, *Steganotheca striata* which lived four hundred million years ago, and its descendants within the rhyniophytes, *Rhynia gwynne-vaughanii* and *Horneophyton*, respresented in Figure 1 in the usual order, show a multilevel structure of dichotomously forked branches called telome, in Zimmermann's telome theory. This theory asserts that these plants do represent the kind of primitive plants from which all others have evolved. It encompasses a consideration of all vascular plants and has permeated the thinking of many morphologists and paleontologists. These investigators advocated five fundamental elementary morphogenetic principles or tendencies, as the basis of evolution of plants, among which is the principle of overtopping illustrated in the figure. This principle means that if, at each fork, one of the two branches becomes stronger than the other, then an apparent main axis composed of one of the two forks will take place. It is held that the dichotomous branching of early land plants was derived from forked marine algae, which exhibits a mechanism of branching by apical bifurcation, and that both the stem and leaf in higher plants were derived in particular by an overtopping process. In the interpretative model the various phyllotactic patterns are considered as multilevel systems or hierarchies of partially ordered primordia, such as the one shown in Figure 2.

Fig. 1 Early vascular plants suggesting through Zimmermann's evolutionary theory and Bolle's Antriebe theory in sub-structure of a hierarchical nature in the phyllotactic patterns.

Figure 2 concerns the pattern characterized by the series, 3, 4, 7, 11, 18, 29, ..., known as the first accessory series, representing consecutive cycles. The phyllotactic fractions shown in the figure are the first few approximations of a series of approximations of the divergence $(3 + \phi^{-1})^{-1}$, point of convergence of the series, as is easily proved. This value of d is the fraction of one turn, around the stem or the tip of the shoot apex, between consecutive leaves or primordia, which, when multiplied by 360° gives $99.502...^{o}$, an angle found relatively often in phyllotaxis. Interestingly the series of cycles and that of phyllotactic fractions are linked by the following general theorem proved in the cylindrical lattice representation of phyllotaxis (Jean subm.):

cycle	phyllotactic fraction
3	1/3
4	1/4
7	2/7
11	3/11
⋮	⋮

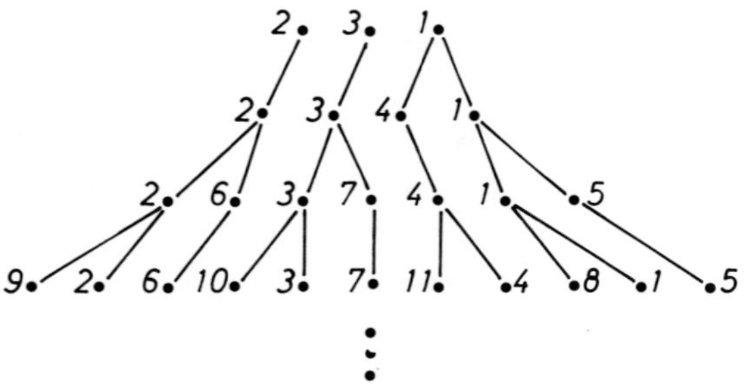

Fig. 2 A hierarchy of bifurcated and simple induction lines for a type of pattern occurring in nature, derived from a projection of the consecutive cycles of the pattern, and from Bolle's phyllotactic law which states that the induction lines from a lower cycle bifurcate for the lower leaves of the cycle, while for the upper leaves they continue their paths without bifurcating.

Theorem

Let (m, n) be a pair of families of spirals (or straight lines) going through the points in a lattice with divergence d. The following properties are equivalent:

1. the pair (m, n) is visible and opposed, meaning that there is a centre of a primordium at the intersection of any two spirals in the families of spirals winding in opposite directions with respect to the axis of the plant, the vertical of the lattice;
2. there exist unique integers μ and ν, $0 \leqslant \nu < n$, $0 \leqslant \mu < m$, such that $|m\nu - n\mu| = 1$, and d is in the closed interval whose endpoints μ/m and ν/n are smaller than 1/2.

Translated into terms of the particular case of Figure 2, the theorem gives the following correspondence: $d \in [1/4, 1/3]$ if (3, 4) is visible and opposed, $d \in [1/4, 2/7]$ if (7, 4) is visible and opposed, $d \in [3/11, 2/7]$ if (7, 11) is visible and opposed, ... To the divergence angle $d = (3 + \phi^{-1})^{-1}$ thus corresponds the series 3, 4, 7, 11, 18, ... of integers which happen to be the denominators of the principal convergents of the continued fraction of d. Specifying two consecutive integers in this series simply amounts to pointing out the obvious pair of opposed families of spirals actually displayed by the system, that is to point out the value of R or r, parameters presented earlier. That is why the series of integers corresponding to d is the fundamental structuring factor of conventional representations, and many recent mechanistic models of phyllotaxis deal only with the divergence angle. This series represents the number of points in the consecutive levels of the hierarchy in Figure 2, which has to be simply linked in some way only with simple and bifurcating induction lines in order to represent the pattern. The hierarchies show the essential mechanics and structures of the various phyllotactic patterns, recalling their ancestral origins, a point of view presented, illustrated and developed in Jean (1982, 1984, 1986 a).

Entropy in phyllotaxis

The physical meaning of the concept of entropy is much disputed; it is still considered as not very clear. According to Poincaré it is a 'prodigiously abstract concept'. Wiener indicated the need for extending the notion of physical entropy when he stated that information is negative entropy. The literature, in physics, communication theory, information theory, social sciences and life sciences, in probability theory, graph theory, and Lebesgue's integration theory, proposes many concepts and formulae of entropy and quantity of information, such as the topological structural information content of Rashevsky-Trucco, Shannon-Weaver's and Hartley-Nyquist's formulae, the chromatic information content of Mowshowitz, the entropy of measurable functions, and epsilon-entropy. Haynes et al. (1980) presented a complete list of authors who have developed their own concepts of entropy by specifying a function and applying it to a particular context. More often unfortunately the form of the mathematical function is not even specified. The word 'entropy', corresponding to an adisciplinary concept, has multiple meanings that may, however, show common denominators, and could be, for example, the idea of the entropy of partition, defined by $I = \sum p_i \log p_i$.

Schrödinger initiated the use of a concept of entropy in life sciences by developing a literal interpretation of the well-known formula $E = k\log D$, where D was a measure of disorder and k Boltzmann's constant. But it is currently recognised, that entropy in living organisms, that is bio-entropy, must be a sum of two factors, one of which is negentropy. In a hierarchy representing a phyllotactic pattern we can define parameters, denoted by $X(t)$ and $S(t)$, representing respectively the complexity and the relative frequency of bifurcation in the hierarchy up to level t. For example, in the case of Figure 2, $S(1) = 1/3$, $S(2) = 4/7$, $S(3) = 4/7$, $S(4) = 3/5$, ..., and we need a formula for $X(t)$ which will increase rapidly with t, such as $X(t) = \prod_{k=1}^{t} k^{f(k)}$, where $f(k)$, $k = 0, 1, 2, 3, \ldots$, is the number of points in the level or generation k of the hierarchy. We thus arrived at a first approximation for E, depending on the generation t, in the form

$$E = -\log (S(t)/X(t)). \qquad (1)$$

Equation (1) is linked to other formulae for entropy. For example, the entropy of a single event of probability p has been defined by Rényi by the formula $\log(1/p)$. In (1) with $p = 1/X(t)$, $\log(1/p) = \log X(t) = -\log(1/X(t))$, one of the two factors of (1), also respresenting the entropy of a set of $X(t)$ elements, and corresponding to the maximum of the more general entropy $\sum p_i \log p_i$ under minimal constraints. The formula incorporates a factor $S(t)$ of negentropy, such that $\log S(t)$ is negative when $S(t)$ is smaller than unity, as is the case here. However when we choose from among the hierarchies the one which minimises E for each consecutive generation $t = 1, 2, 3, \ldots$, unfortunately only the pattern defined by the main series 1, 2, 3, 5, 8, 13, ..., is seen to arise. We thus need a functional able to overcome such a limitation.

The bulk entropy of a hierarchy or pattern is defined by

$$E_b = -\sum_{t=1}^{\omega} \log(S(t)/X(t)) \tag{2}$$

and we look for the hierarchy which minimises E_b at the time T_r (when T_r primordia exist) at which rhythm of growth ω is induced, or shall we say preferably, is manifested as a built-in attribute revealed in time during growth. The rhythm, demanding the essential mechanism of phyllotactic pattern formation has to be represented in a formula for bio-entropy, together with the other two characteristics, namely the complexity $X(t)$ and the stability $S(t)$ arising from duplication. Church (1920) pointed out the emergence in phyllotaxis of the rhythmic sequence of duplication and nonduplication 2, 1, 2, 1, 2 - 2, 1, 2, 1, 2 - 2, ..., emphasising that if there is no rhythm there is no pattern, and the converse. The factor is translated in the model as the smallest order of the irreducible square matrices generating the hierarchy.

More precisely, such a matrix $C = (c_{ij})$ has its entries c_{ij} equal to 0 or 1, $0 < \sum_i c_{ij} < 2$ for every j, and its directed graph is strongly connected. This matrix of order n generates a language (L-system) (A, P, ω_0) where A is a set with n symbols denoted by $a_0, a_1, a_2, ..., a_n$, P is a function defined on A by $P(a_j) = (a_1{*}c_{1j})(a_2{*}c_{2j})(a_3{*}c_{3j})...(a_n{*}c_{nj})$, where for $j = 1, 2, 3, ..., n$, a_i is in $P(a_j)$ if $c_{ij} = 1$ and is not if $c_{ij} = 0$, the symbol * meaning that a_i is repeated c_{ij} times in the string; ω_0 is any string of symbols. To show how this generates a hierarchy with simple and bifurcated induction lines let us consider how the hierarchy in Figure 2 comes out of the matrix $\begin{pmatrix}1 & 1\\ 1 & 0\end{pmatrix}$ when a convenient ω_0 is chosen. The set A is made of the symbols a_1 and a_2, $P(a_1) = a_1a_2$ and $P(a_2) = a_1$ as suggested by the matrix and $\omega_0 = a_1a_2a_1a_1$. Starting with the latter string in level 0, and applying P to its elements gives level 1 of the hierarchy, that is $a_1a_2a_1a_1$, and P applied again gives level 2 having 7 symbols or points as in the hierarchy of Figure 2, and so on. The irreducibility of a growth matrix of order $\omega = 2, 3, 4, ...$, means that after every ω generations of the hierarchy the generative process is repeated, so that the rhythm produces loops. Thus a plant generates primordia, and, at T_r the existing primordia constitute the first ω levels of the hierarchy (including level 0) giving the smallest

E_b (minimal entropy production). Then the preponderant period of the rhythm is known and every further level of the hierarchy is determined. Formula (1) corresponds to a dynamic approach, while Formula (2) expresses a dynamic approach up to time T_r at which rhythm is manifested, and is static thereafter, the system having become deterministic.

Consider the series 3, 4, 7, 11, 18, ... A plant may show a conspicuous pair (7, 11) and another plant the pair (4, 7), but both patterns will display divergence corresponding to the series. The same plant may show the pair (3, 4), then the pair (7, 4) and then the pair (7, 11), a phenomenon called rising phyllotaxis. This suggests the following particular set of hierarchies. To each series of positive integers H(1), H(2), H(3), ..., where for $k = 1, 2, 3, ..., H(k + 2) = H(k + 1) + H(k)$, and $H(k) \leqslant H(k + 1) \leqslant 2H(k)$, corresponds the hierarchies denoted by (H(k), H(k + 1)), where, for a given k, H(k) and H(k + 1) respectively represents the number of points in levels 0 and 1 of the hierarchy. The second inequality means that the hierarchy is made up of simple and bifurcated induction lines only. All these hierarchies have a rhythm with period $\omega = 2$. Figure 2 shows the hierarchy (3, 4). The series S(1), S(2), S(3), ..., for the hierarchy (H(k), H(k + 1)) such that $H(k) < H(k + 1)$ (at least one bifurcated line coming from level 0) can be proved to be monotonically increasing or decreasing to $1/\phi$, depending on the parity of k. Of course there are an infinity of types of hierarchies with simple and bifurcated induction lines, but the following propositions (Jean, 1980) distinguish the above type. The value of T_r is given therein as a member of series or another, in order to reduce the symbolism used.

Proposition 1

Given $T_r = H(k + 3), k \geqslant 0$, the hierarchy

$$(H(k + 1) + [-H(k - 1)/3], \quad H(k + 2) - [-H(k - 1)/3]),$$

which has T_r points in levels 0 and 1 together, and where $[x]$ means the integral part of x, has minimal E_b among all the hierarchies, whatever their rhythms.

It can indeed be proved that the hierarchy having minimal E_b for that value of T_r is of the type

$$(H(k+1)+m, \quad H(k+2)-m)$$

where m is such so that

$$-H(k-1)/3 \leqslant m < H(k)/2.$$

As a function of m the value

$$E_b = \log[(H(k+1)+m)H(k+3)2^{H(k+3)}/(H(k)-2m)(H(k+2)-m)]$$

is increasing in the above interval, so that the minimal E_b is given by $m = [-H(k-1)/3]$. Moreover it can be proved that for $T_r = H(k+3)$, if $(H(k+1), H(k+2))$ has minimal E_b then the latter value of m is zero, so that:

Proposition 2

Given $T_r = H(k + 3)$, $k \geqslant 0$, the hierarchy $(H(k + 1), (H(k + 2))$ has minimal E_b if and only if $[- H(k - 1)/3] = 0$.

Examples: Suppose that $T_r = 10$. Then we must look for minimal E_b among the hierarchies (1,9), (2,8), (3,7), (4,6), and (5,5). The first three pairs are unacceptable given that we must have $H(k + 1) \leqslant 2H(k)$ (simple or bifurcated lines only), and the last pair produces infinite entropy (no bifurcated induction line). The hierarchy 2(2,3), belonging to the series 2(1, 2, 3, 5, 8, ...), has minimal entropy given that $H(k - 1) = 2$ for which $[-2/3] = 0$. If $T_r = 11$, the hierarchies $(H(k + 1), H(k + 2)) = (5,6)$ or (4,7) are the only ones giving acceptable series. But for the former hierarchy $H(k -1) = 4$ while for the latter $H(k -1) = 2$. It follows by Proposition 2 that (4,7) has minimal entropy at $T_r = 11$. For a given value of T_r the hierarchy producing minimal entropy is always unique.

Consequences of the interpretative approach to phyllotaxis

With formula (2) we succeeded in finding the series of normal phyllotaxis J(1, 2, 3, 5, 8, ...), $J \geqslant 1$, and $J(k, k + 1, 2k + 1, 3k + 2, ...)$, $J = 1$, or 2, $k \geqslant 3$, with the corresponding divergence $(J(k + \phi^{-1})^{-1}$ (must be multiplied by 360 to obtain degrees), and the series of anomalous phyllotaxis $2k + 1, 2k + 3, 4k + 4, 6k + 7, ...$, $k \geqslant 2$, with the divergence $(2 + (k + \phi^{-1})^{-1})^{-1}$, and only these series. The main series and the first lateral series met previously correspond respectively to $k = 2$ and $k = 3$ in the normal series with $J = 1$.

Nature shows patterns called distichous and whorled. In the former case the leaves are initiated singly at 180° ($d = 1/2$) to one another, while in the latter case they arise in groups (whorls) of $n = 2, 3, 4, \ldots$ at the same level on the stem ($d = 1/n$). The case $n = 2$ is known as decussation: the leaves aries in pairs, the leaves in a pair being on opposite sides of the stem, and the consecutive pairs making angles of 90°. It is suggested that they are limiting cases of spiral patterns: the value of d for anomalous phyllotaxis tends towards 1/2 (distichy) as k increases, and the value of d for normal phyllotaxis tends towards 0, so that the various whorls with their divergences can take place along the way, as they must have done during evolution. For the botanist Church (1920) spirality is primitive, the other systems are wholly secondary and derivative, with the transition between the former and the latter possibly lost because of evolution. To explain the transitions or generate the secondary patterns is, from this point of view, a secondary problem.

The interpretative model offers an understanding of the phenomenon of phyllotaxis in general terms which can be given particular physico-chemical meanings. For example rhythm could be the expression of the two-dimensional principle of generative resonance used by Green (1985) at the cellular level to reproduce distichy and decussation. As Goebel did, he refers to distichous phyllotaxis as pendulum-like, considering that the pattern is mimicked by a swinging pendulum. Green compares decussation to a resonating metal sheet producing potato chip-like pairs of crests and troughs. He thinks that this principle can explain, what is presented above as, phyletically remote spiral patterns.

The time T_r at which rhythm is manifested can be partly represented in Adler's contact pressure model of phyllotaxis by the time T_C at which contact pressure begins, or by the time at which Roberts' chemical contact pressure begins. In fact the former model partially produced a result stated at the end of the Introduction, that the main phyllotactic pattern is inevitable if contact pressure begins early, that is before $T_C = 6$. I suggest that rhythm expresses the way in which the primordia-inducing nutrients (morphogens) are provided in the system as a whole; it would result from the coordination of various relative growth rates $1/Y \cdot dY/dt$ of substances and sizes within the shoot apex. The factor inducing rhythm, representing one of the most fundamental problems of physiology, is not specified in the model.

Marzec (1986) developed a static approach, using an informational entropy function defined by the formula $S = - c \int_0^{2\pi} p(\Theta) \ln p(\Theta)\, d\Theta$. He treated phyllotaxis as a dissipative structure with a morphogen concentration field $p(\Theta)$ established by the leaves, depending on the constant divergence $\Delta\Theta$, and on a constant vertical rise along the stalk. The strength of each leaf as a morphogen source is decreased by a factor λ after each plastochrone (unit of time between the emergence of consecutive leaves). The entropy of morphogen distribution is considered to measure the spacing of the leaves. The processes of creation (negentropy), diffusion and destruction (entropy) of the morphogen, are modeled via a diffusion-reaction partial differential equation. The nature of the morphogen is not specified, given that its experimental identification represents one of the greatest problems of diffusion theories in phyllotaxis.

Considered as a function of $\Delta\Theta$, it is found that S is a maximum and the rates of entropy production are minimal at a $\Delta\Theta$ which asymptotically approaches, as λ approaches unity, one of the members of ϕ_G, the set of angles which do not have intermediate convergents from a certain term. This set is very large, but it includes among its elements the divergences of normal and anomalous phyllotaxis, obtained with the interpretative approach. 'In rough order of their appearance as λ increases, the most prominent members of ϕ_G are 137.5^o, 99.5^o, 78.0^o, 64.1^o, 151.1^o, 158.1^o, and 106.4^o' (the latter angle corresponding to none of those predicted by the interpretative model). This is akin to what happens in the interpretative model, when we take, from among the cases of normal phyllotaxis, only those for which $J = 1$: as T_r increases, the order of appearance of phyllotactic angles is 137.5^o ($J = 1, k = 2, T_r = 3$ or 5), 99.5^o ($J = 1, k = 3, T_r = 7$ or 11), 78.0^o ($J = 1, k = 4, T_r = 14$), 64.1^o ($J = 1, k = 5, T_r = 17$), followed by two cases of anomalous phyllotaxis 151.1^o ($k = 2, T_r = 19$) and 158.1^o ($k = 3, T_r = 25$) having between them an angle of 54.4^o ($J = 1, k = 6, T_r = 20$) - that of normal phyllotaxis.

Jean (1986 a) shows further how the interpretative approach provides a generalisation of fundamental ideas of mechanistic theories. Such an approach is needed in view of the limitations of the latter theories (see for example Jean, 1983 a and 1984). An interpretative theory, though physically grounded as expressed above, is the result of an approach aimed at a cybernetic understanding of reality, that is at a comprehension which in a way transcends the physical aspects whilst taking account of them.

Predictions and by-products

The main consequences of the interpretative approach, are that of being able to provide a more general comprehensive viewpoint, and thus more general results. Other consequences concern the predictions the model allows us to make. One of these is a classification of the various types of spiral patterns according to their frequencies of occurrence. In a recent manuscript (Jean, 1986 b) 12750 quantified observations are compiled for more than 650 species reported in the literature from the beginning of investigations into phyllotaxis. It appears from this compilation that the main series 1, 2, 3, 5, 8, ... arises in 95 % of all the cases. The bijugate series 2(1, 2, 3, 5, 8, ...), corresponding to a divergence of $\phi^{-2}/2$, is seen to arise four times more often than the first accessory series 1, 3, 4, 7, 11, ... Also the second accessory series 1, 4, 5, 9, 14, ... and the first lateral series 2, 5, 7, 12, 19, ... (anomalous series with $k = 2$) are significantly less frequently observed than the former three series. The predictions of the model in mathematical phyllotaxis, are in a good agreement with these observations in factual phyllotaxis, but many more observations are needed, to be confronted with other such predictions.

A good model would forbid certain things to happen. Accordingly, as a consequence of the interpretative model, certain types of phyllotactic patterns will not be found in nature, such as the one governed by the series 3, 8, 11, 19, 30, ... From a search in the literature, four apparently troublesome cases, with respect to the model, can be found, all of them reported more than one hundred years ago. Among them only one seems to resist to criticism, its geometrical representation being theoretically almost blameless. It concerns the series 3, 7, 10, 17, 27, ... reported for *Pothos spec.* in 1878 by Schwendener. It is neither a case of normal, nor of anomalous phyllotaxis. It seems to correspond to a single specimen in one of the seventy five species of *Pothos*. Surprisingly, to our knowledge it has not since been reported in the literature. The conclusion is that an indepth analysis of supposedly aberrant cases should be methodologically performed, and the recurrence of the patterns should be examined.

The well-known L-systems are embedded in the development of the interpretative model, via the irreducible matrices mentioned earlier, used to generate the hierarchies. This resulted in a by-product in the theory of growth functions of L-systems, which were then bridged with Perron-Frobenius spectral theory. Among the results are conditions for the convergence of the ratios $f(t + 1)/f(t)$, C^t/r^t and $f(t)/r^t$, where $f(t)$ defined as previously in relations with $X(t)$, is the growth functions of the L-system generated by the irreducible matrix C with spectral radius r. These conditions are relevant in the theory of Leslie's matrices. A table giving the value of $\lim_t f(t + m)/f(t + n) = r^{n-m}$ for various values of m and n has been built by working on the equation $\lambda^{m+n-1} - \lambda^{m-1} - \lambda^{n-1} = 0$ whose largest root in absolute value is precisely r. This table is used in the study of branching structures, cell populations with lineage control, and filamentous organisms such as the algae *Anabaena catenula* and *Chaetomorpha linum* for which $m = 4, n = 5$, and $m = 3, n = 5$, respectively (see Jean, 1981).

An allometry formula $Y = aX^b$, expressing differential growth, comes out as an indirect result of the global approach. It corresponds to the constant ratio of two relative growth rates within the apex. It is of the type $\ln R = 2\pi p(\gamma)(m + n)^{-2}$ ($Y = \ln R$, $X = m + n$, $b = -2$), where R is the plastochrone ratio defined in the introduction, (m, n) is the conspicuous pair of families of spirals with angle of intersection γ, and $p(\gamma) = (\phi^3/5^{1/2}2)(5^{1/2}\cot\gamma + (5\cot^2\gamma + 4)^{1/2})$ (Jean, 1983 b). This means that for fixed γ all the points $(\ln R, m + n)$, for any two consecutive terms m and n of the normal or anomalous phyllotactic series, are on a straight line with slope -2 in the log-log grid. This model allows us to assess with ease and precision phyllotactic patterns in specimens (Jean, in preparation).

Conclusion

This chapter is a presentation of a methodological paradigm in which mathematically formulated system concepts are seen to be useful to illuminate the central problem of phyllotaxis in plant biology. 'This approach has served to inject some essential new ideas into the study of phyllotaxis, predicated as it is on the view that reductionistic, ontogenetic approaches alone are not adequate' (Foreword in Jean, 1984). However, the treatment given here is not optimal. It is only preliminary and illustrative of the interest in the interpretative approach. More performance parameters can be defined, so as to produce a more dynamically oriented model (permitting for example changes of rhythm to occur in order to provide for the possible transitions between the patterns), the super model long awaited for, that will decidedly drive the subject of phyllotaxis off the beaten tracks of the usual mechanistic theories. These are considered by many as too simple and not logical enough. For example, Church (1920) speaks of the futility of attempting to reach a final solution by mere observation of the ontogeny of the primordia at the shoot apex, and according to Wardlaw's eighth principle of plant organisation the phyllotactic patterns would not come only from the genes but also from the environment. Above the traditional distinctions between developmental and evolutionary biology, such an interpretative model would welcome any mechanistic attempts as a partial particular case of it, expressing one of its systemic and holistic aspects.

However inspiring the mechanistic models may be, testing of their premises remains an open question. This makes the links to plant physiology and biochemistry rather weak, and lowers their heuristic value. The physiological evidence is incomplete, given that plants live in an environment in which various factors interfere, making phyllotaxis a functional problem. It is dealt with as such in the interpretative model. I do not say that this model is better as far as the testing of the premises is concerned. But these premises are more general, more in line with a body of ideas in biochemistry according to which the study of biology should start with the principles of thermodynamics, which constitute according to many the most fundamental point of view from which to study all biological phenomena.

Structures in Models of Morphogenesis

B. S. Kerner, V. I. Krinskii, V. V. Osipov

Introduction

In recent time, interest in the supercellular level of life organisation has been increasing. At this level of interest, individual cells are considered as the elementary 'bricks', of which all living organisms are built. A great deal of attention is paid to the problem of self-organisation - the generation and evolution of spatial structures under the influence of environmental parameters, i. e. the problems of morphogenesis, ontogenesis, embryogenesis etc. (Belentsev, 1983; Ebeling, 1976; Haken, 1978; Polak, Mikhailov, 1983; Romanovskii et al., 1984; Nicolis, Prigogine, 1977; Vasil'yev et al., 1979; Volkenstein, 1984).

Modelling of these processes is based on the idea that the diffusion and chemical reactions of certain agents (morphogens) initiate the formation of a pre-pattern - a morphogenetic field, which then affects the cells (Nicolis, Prigogine, 1977; Haken, 1978; Belintsev, 1983; Volkenstein, 1984; Romanovskii et al., 1984). Since Turing's classic work (Turing, 1952), morphogenetic processes have been described, as a rule, in terms of the macroscopic equations of chemical kinetics:

$$\tau_i \partial X_i / \partial t = \sum_{j=1}^{N} \vec{\nabla}(L_{ij} \vec{\nabla} X_j) - Q_i(X_1, X_2, \ldots, X_N, A, B, C), \quad (1)$$

Thermodynamics and Pattern Formation in Biology

where X_i are the morphogen concentrations and A, B, C, ... the chemical reaction constants. Gierer and Meinhardt in their experiments with Hydras (Gierer, 1981; Gierer, Meinhardt, 1972; Meinhardt, Gierer, 1974, 1980) have shown that the morphogenetic structure formation can be described using two agents, an activator and an inhibitor. The activator forms a positive and the inhibitor a negative feedback. The positive feedback is usually related to the auto- or cross-catalytic process of activator production; the negative feedback implies that the inhibitor is the controlling agent supressing activator production (Belintsev, 1983; Gierer, Meinhardt, 1972; Gierer, 1981; Haken, 1978; Meinhardt, Gierer, 1974, 1980). Thus, in the simplest (but rather general) case, the model of morphogenesis is reduced to two equations of type (1)[1] describing the distribution of concentrations of the activator Θ and the inhibitor η

$$\tau_\Theta \frac{\partial \Theta}{\partial t} = \ell^2 \Delta \Theta - q(\Theta, \eta, A), \quad \tau_\eta \frac{\partial \eta}{\partial t} = L^2 \Delta \eta - Q(\eta, \Theta, A), \tag{2}$$

where τ_Θ, τ_η and l, L are the characteristic times and variation lengths of Θ and η, respectively. For the class of type (2) equations, Turing (1952) formulated the conditions for spontaneous layering, i. e., instability of the homogeneous state with respect to aperiodically growing fluctuations with the dominating wave number $k = k_0 \simeq (l\,L)^{-1/2}$ (next section). It seemed natural to assume (Belintsev, 1983; Haken, 1978; Nicolis, Prigogine, 1977) that near the point of loss of stability ($A = A_c$) there must arise a quasi-harmonic structure of period $2\pi k_0^{-1}$. Such structures of small amplitude were central to numerous studies (Belintsev, 1983; Haken, 1978; Nicolis, Prigogine, 1977; Vasil'yev et al., 1979).[2]

---

1) More complicated (multiparametrical) models of morphogenesis are also easily reduced to a system of two equations of type (2) (Romanovskii et al., 1984).

2) This assumption forms the basis for Haken's technique of structure formation, in which the mode of period $2\pi k_0^{-1}$ is considered as basic and as parameter of the kinetic phase transition determining the structure evolution.

The development of self-organisation was associated with the fact that the structure of period $2\pi k_0^{-1}$, following change in one of the system parameters (A), becomes, in turn, unstable and gives way to a more complicated structure (Prigogine's principle, 'from dissipative structures to order through fluctuations' (Nicolis, Prigogine, 1977)).

However, very often the pattern of self-organisation turns out to be different, as follows from a series of studies by Kerner and Osipov (Kerner, Osipov, 1978, 1979, 1980, 1981, 1982 a, b, c, 1983 a, b, 1985 a, b, c). In these studies the authors have developed methods for the construction and stability analysis of large-amplitude structures being formed in a broad class of physical, chemical and biological systems, whose properties are described by the equations of type (2). As shown (Kerner, Osipov, 1978, 1979, 1980, 1981, 1982) the layering of the homogeneous state (point $A = A_c$) gives rise not to a quasi-harmonic small-amplitude structure, but to the sudden formation of a large-amplitude structure with a period not necessarily equal to $2\pi k_0^{-1}$. In this case the inhibitor $\eta(\vec{r})$ varies smoothly in space, while the activator exhibits a rapidly varying distribution - the change of $\Theta(\vec{r})$ in some size regions of the order of l is sharp (Fig. 1). The increase in A may result in a dynamic, not fluctuational, rearrangement of the structure (Kerner, Osipov, 1982 a, 1985 a), i. e. that coupled not with the loss of the stability of the initial state, but with its disappearance (see below).

This scheme of self-organisation (pattern formation and evolution) is realised at $L >> l$. We emphasise that the condition $L >> l$ is one of the conditions of the Turing instability. It implies local autocatalysis and long range inhibition and is considered the most important morphogenesis-determining condition (Belintsev, 1983; Gierer, Meinhardt, 1972; Gierer, 1981; Haken, 1978; Meinhardt, Gierer, 1974, 1980; Nicolis, Prigogine, 1977).

At $L >> l$, structures exist also in the stability region of the homogeneous state of the system, i. e. at $A < A_c$. They can be excited in the homogeneous system by a short-term inhomogeneous perturbation of final amplitude. The simplest structure excitable in a stable medium is an autosoliton (AS), a solitary stationary state (Kerner, Osipov, 1978, 1981, 1982 c, 1985 a, b, c) (Fig. 2). Such a localised state can go over on the periphery into a stable homogeneous state of the system. The shape of AS is independent of the mode of the initial short-term local perturbation and is determined only by the system parameters. The system can generate two, three or more AS, spaced any distance apart. There is some minimum value of the distance between two AS (Kerner, Osipov, 1978, 1979) below which they become unstable against the 'pumpover' phenomenon (Kerner, Osipov, 1978) and one of them vanishes. These properties of AS can explain the results obtained in experiments with Hydras (Belintsev, 1983; Gierer, Meinhardt, 1972; Gierer, 1981; Haken, 1978; Meinhardt, Gierer, 1974, 1980) (see below).

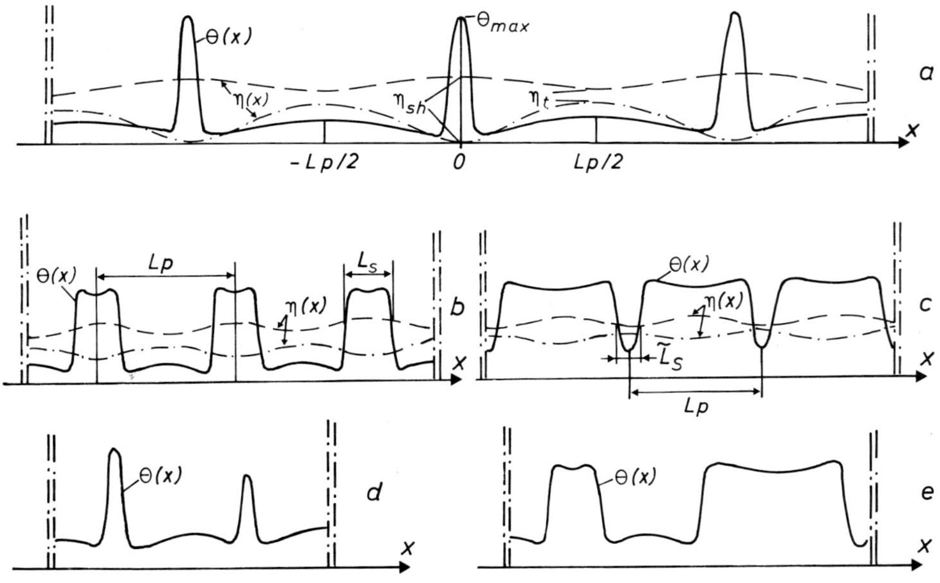

Fig. 1 Major types of one-dimensional large-amplitude structures (strata). a - narrow periodic strata; b - broad periodic strata; c - strata with decreased activator value; d, e - fragments of the activator distribution in a stochastically inhomogeneous structure. Solid lines in Fig. 1 a - 1 c show the distribution of the activator Θ (x), dashed lines the distribution of the inhibitor η(x) in V- or И -models, and dashed- and dotted lines the distribution of the inhibitor η (x) in Λ - or N-models.

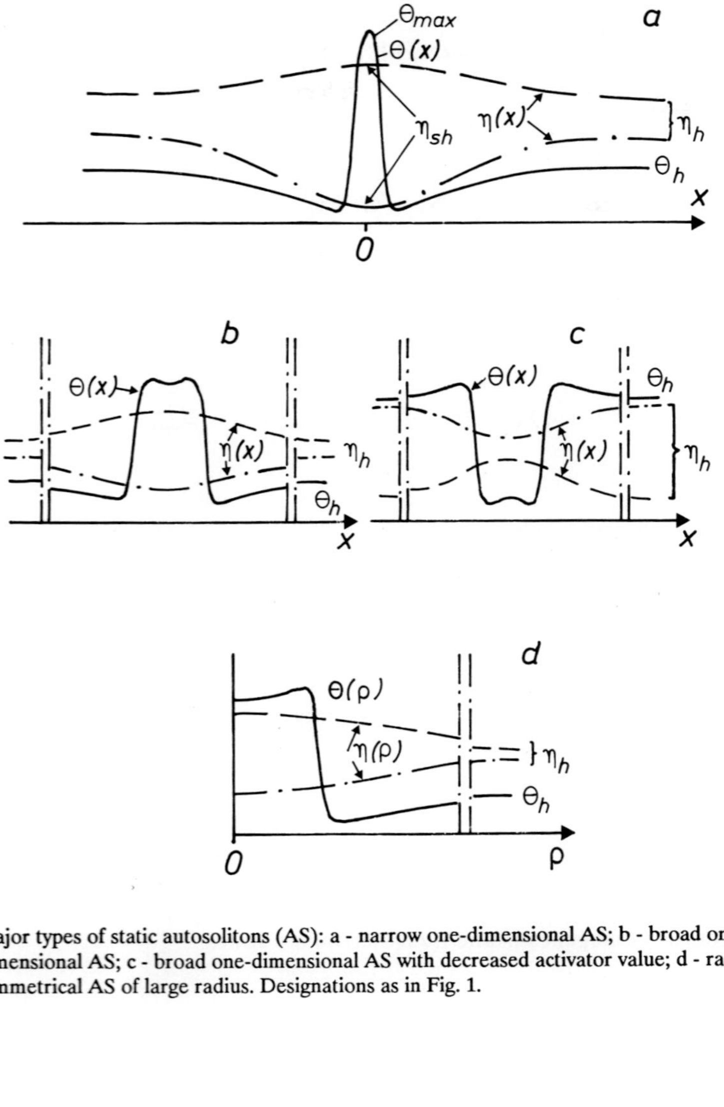

Fig. 2 Major types of static autosolitons (AS): a - narrow one-dimensional AS; b - broad one-dimensional AS; c - broad one-dimensional AS with decreased activator value; d - radially symmetrical AS of large radius. Designations as in Fig. 1.

In the present paper the results of the general theory (Kerner, Osipov, 1978, 1979, 1980, 1981, 1982 a, b, c, a, b, 1985 a, b, c) are described in relation to structure analysis in various biophysical models (Kerner, Osipov, 1980, 1982 c, 1983 a). Most attention is paid to the Gierer-Meinhardt (GM) model of morphogenesis, for which (Belintsev, 1983; Gierer, Meinhardt, 1972; Gierer, 1981; Haken, 1978; Meinharrdt, Gierer, 1974, 1980)

$$q = \Theta - A\Theta^2/\eta - B, \quad \Theta = \eta - C\Theta^2, \tag{3}$$

and the model of the Brussels school ('Brusselator') (Nicolis, Prigogine, 1977)

$$q = (A+1)\Theta - B - \Theta^2\eta, \quad Q = \Theta^2\eta - A\Theta. \tag{4}$$

Structures in models with cubic and more complicated types of nonlinearity are also considered below.

Before describing the results of the nonlinear theory of structures, we will discuss the structure stability near the point of loss of stability of the homogeneous state, $A = A_c$.

Instability of the homogeneous and quasi-harmonic states

For the homogeneous distribution in the GM model

$$\Theta \equiv \Theta_h = B + A/C, \quad \eta \equiv \eta_h = C\left(\frac{A}{C} + B\right)^2. \tag{5}$$

Then, according to (3),

$$Q'_\eta \equiv \partial Q/\partial \eta = 1, \quad q'_\Theta \equiv \partial q/\partial \Theta = -(A-CB)(A+CB)^{-1}, \tag{6}$$

i. e. at $A > A_0 = CB$, $q'_\Theta < 0$. The requirements that over a certain range of parameters $q'_\Theta < 0$, and $Q'_\eta > 0$, must be fulfilled for any model, inasmuch as they are a mathematical reflection of the fact that the activator locks the positive and the inhibitor the negative feedback.

Linearising equations (2) near the homogeneous state with respect to fluctuations $\delta\Theta$, $\delta\eta \sim \exp(-\gamma t + i\vec{k}\vec{\eta})$, it is easy to show that its stability is broken (Re $\gamma < 0$) at a certain $A = A_c$, when one of the two conditions is satisfied

$$\tau_\eta (q'_\Theta + k^2 \ell^2) + \tau_\Theta (Q'_\eta + k^2 L^2) < 0, \tag{7}$$

$$k^4 \ell^2 L^2 + k^2 \ell^2 Q'_\eta + k^2 L^2 q'_\Theta + Q'_\eta q'_\Theta - q'_\eta Q'_\Theta < 0. \tag{8}$$

The first condition is more readily satisfied for $k = 0$ at $\alpha << 1$, and the second at $\varepsilon << 1$ (where $\alpha = \tau_\Theta/\tau_\eta$; $\varepsilon = \ell/L$) with respect to fluctuations with the wave number

$$k = k_0 = (\ell L)^{-1/2} (Q'_\eta q'_\Theta - q'_\eta Q'_\Theta)^{1/4}, \tag{9}$$

when

$$q'_\Theta < -[(\ell/L)^2 Q'_\eta + 2(\ell/L)(Q'_\eta q'_\Theta - q'_\eta Q'_\Theta)^{1/2}]. \tag{10}$$

When condition (7) is satisfied, stability is lost with respect to homogeneous fluctuations (with $k = 0$ and frequency $\omega = \omega_0 \simeq (\tau_\eta \tau_\Theta)^{-1/2}$. It is, however, known from the qualitative theory of differential equations (Andronov et al., 1981) that the homogeneous fluctuations that arise at $\alpha << 1$ are not quasi-harmonic ones, but they are the successive combinations of the segments of the slowly- and fast-varying distributions with characteristic times, τ_η and τ_Θ, respectively. In a sense, an analogous situation occurs also in the Turing instability, i. e. when condition (10) is satisfied and the homogeneous state is unstable against fluctuations with $\omega = 0$ and $k = k_0 \simeq (\ell/L)^{-\frac{1}{2}}$. Structures arising spontaneously at $\varepsilon \ll 1$ (Fig. 1) are also qualitatively different from quasi-harmonic structures (Kerner, Osipov, 1978, 1979, 1980).

By (3), the condition for spontaneous generation of homogeneous fluctuations (7) in the GM model is satisfied for $\omega = (\tau_\Theta \tau_\eta)^{-1/2}$ at $\alpha = \tau_\Theta / \tau_\eta < 1$, when

$$A > A_0 (1 + \alpha)(1 - \alpha)^{-1}. \tag{11}$$

The condition of layering (10) is satisfied for fluctuations with $k = (l L)^{-1/2}$ at $\varepsilon < (\sqrt{2} - 1) \simeq 0.4$, when

$$\frac{2A}{A + A_0} > (1 + \varepsilon)^2 \text{ , i. e. at } A > A_c = \frac{(1+\varepsilon)^2 A_0}{(1 - 2\varepsilon - \varepsilon^2)} . \tag{12}$$

Thus, the GM model demonstrates that spontaneous formation of structures occurs at $\alpha > 1$ and $\varepsilon << 1$, and the smaller the value $\varepsilon = l/L$ the smaller A_c at which stratification takes place. From (12) it follows that at $\varepsilon \to (\sqrt{2} - 1)$, the value $A_c \to \infty$, and at $\varepsilon \to 0$, $A_c \to A_0 = AB$, i. e. A_c tends to the value at which the derivative q'_Θ reverses sign. It is readily seen that these results remain qualitatively true also for other models. For instance, for the Brusselator (4)

$$\Theta_h = B, \quad \eta_h = A/B, \quad Q'_\eta = B^2, \quad q'_\Theta = 1 - A, \quad A_0 = 1, \tag{13}$$

$$k_0 = B(\ell L)^{-1/2}, \qquad A_c = (1 + \varepsilon B)^2. \tag{14}$$

At the point of layering ($A = A_c$), along with the homogeneous distribution ($\eta = \eta_h$, $\Theta = \Theta_h$), there arise inhomogeneous quasi-harmonic structures of the type $\eta = \eta_h \pm \Delta\eta \cdot \cos(k_x x)$, where $\Delta\eta \sim |A - A_c|^{1/2}$ (Nicolis, Prigogine, 1977; Haken, 1978; Belintsev, 1983). When such small-amplitude structures are realised at $A > A_c$, we speak of supercritical branching and at $A < A_c$ of subcritical branching (Belintsev, 1983; Haken, 1978; Nicolis, Prigogine, 1977). Using general methods (Gershuni, Juhovitskii, 1978; Joseph, 1976; Nitzan, Ortoleva, 1981), it can be shown that at $\varepsilon = 1/L << 1$ only subcritical branching is realised. This implies that in models of morphogenesis, where, as a rule, $L >> 1$, supercritical branching is not realised. As to subcritical branching, the small-amplitude quasi-harmonic structures arising in this case are unstable (Joseph, 1976).

In other words, in defiance of the existing views (Belintsev, 1983; Haken, 1978; Nicolis, Prigogine, 1977), small-amplitude quasi-harmonic structures do not arise even near the point of layering when the condition $L >> 1$, which is important for models of morphogenesis (Belintsev, 1983; Gierer, Meinhardt, 1972; Gierer, 1981; Haken, 1978; Meinhardt, Gierer, 1974, 1980) is satisfied. In the Brusselator and GM model at $A = A_c$, there appear suddenly large-amplitude structures of the type shown in Fig. 1 a (Kerner, Osipov, 1978, 1982 c, 1983 a).

Structures in small-size systems (Λ - and V-models)

Based on the results of previous studies (Kerner, Osipov, 1978, 1981, 1983 a), we consider first the shape and properties of structures of size $\mathcal{L}_{x,y,z} \ll L$. From equations (2) it follows in this case that inhibitor distribution in the system is practically homogeneous, i. e. $\eta \neq f(\vec{r})$. Because of this, the distribution $\Theta\,(\vec{r})$ is found from the first of the two equations (2) at η = const.

In the one-dimensional stationary case this equation takes the form:

$$\ell^2 d^2\Theta/dx^2 + dU_\Theta/d\Theta = 0, \quad U_\Theta = -\int^{\Theta} q(\Theta,\eta,A)\,d\Theta, \quad \eta = \eta_{sh} = \text{const}. \tag{15}$$

The homogeneous distribution of the inhibitor η, in turn, depends on the mode of the activator distribution $\Theta\,(\vec{r})$ and the value of A. It is found from the second equation of (2) averaged over the system volume (Kerner, Osipov, 1979)

$$\tau_\eta \partial\eta/\partial t = -\langle Q(\eta,\Theta,A)\rangle\,, \tag{16}$$

under periodical or neutral conditions on its boundary S:

$$\vec{\nabla}\eta\,|_S = \vec{\nabla}\Theta\,|_S = 0. \tag{17}$$

In (16) and henceforth the symbol <...> implies averaging of a function over the system volume.

Equation (15) coincides in form with the equation for the conservation of motion of a particle with coordinate Θ and time x in the potential U_Θ. The shape of U_Θ depends on the value $\eta = \eta_{sh}$, which, in turn, depends functionally on the solution of Θ (x). By (15), extrema of the potential U_Θ are determined by the points where $\partial U_\Theta / \partial \Theta = - q(\Theta, \eta) = 0$, i. e. by the points of intersection of the curve $\eta(\Theta)$, found from the equation

$$q(\Theta, \eta, A) = 0 , \tag{18}$$

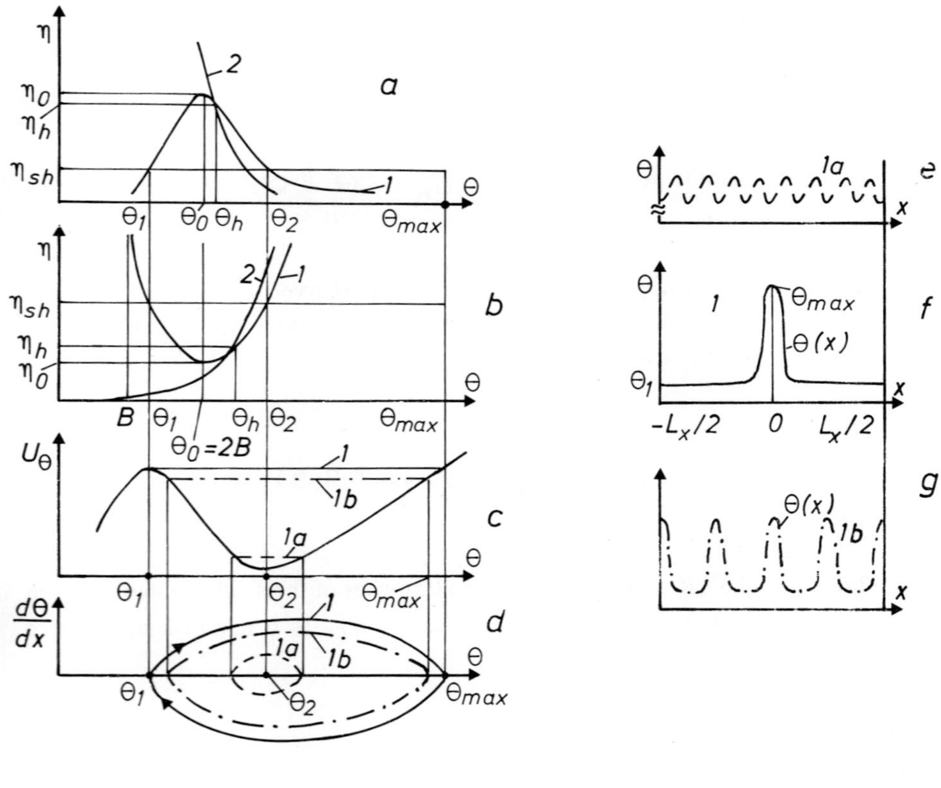

Fig. 3 see text

with the line $\eta = \eta_{sh}$. In what follows, we shall call the curve $\eta(\Theta)$ the local relation (LR) curve. Thus, by the form of LR (Fig. 3 a and b) it is possible to reconstruct the shape of the potential U_Θ (Fig. 3 c) at various values of $\eta = \eta_{sh}$ (Kerner, Osipov, 1977, 1978).

When constructing the LR curve, we note that, by (18), the derivative

$$d\eta/d\Theta = -q'_\Theta / q'_\eta . \tag{19}$$

For the Brusselator (4)

$$q'_\Theta = 1 + A - 2\Theta\eta \qquad q'_\eta = -\Theta^2 < 0, \tag{20}$$

i. e. by (18) - (20), the LR curve has an Λ-shaped form (curve 1 in Fig. 3 a). Its extremum, where the derivative $d\eta/d\Theta = 0$, i. e. $q'_\Theta = 0$ (19), corresponds to the values of $\Theta = \Theta_0$ and $\eta = \eta_0$, where

$$\Theta_0 = 2B(1+A)^{-1} \qquad \eta_0 = (1+A)^2 (4B)^{-1} . \tag{21}$$

Fig. 3 a shows that the LR curve consists of two branches of the single-valued function $\Theta(\eta)$: branch I, where $\Theta \leqslant \Theta_0$ and the derivative $q'_\Theta \geqslant 0$, and branch II, where $\Theta \geqslant \Theta_0$ and the derivative $q'_\Theta \leqslant 0$.

At the extrema of the potential U_Θ (15) the value

$$\partial^2 U_\Theta / \partial \Theta^2 = - q'_\Theta . \tag{22}$$

Therefore, the point $\Theta = \Theta_2$ of intersection of the line $\eta = \eta_{sh}$ branch II (Fig. 3 a), where $q'_\Theta < 0$, corresponds to the minimum of the potential U_Θ and the point $\Theta = \Theta_1$ on branch I to its maximum. As a result, the potential U_Θ takes the form of a potential well at $\eta_{sh} < \eta_0$ (Fig. 3 c). This means that equation (15) can have periodic solutions O(x) with a characteristic length of the order of 1 (Fig. 3 e), corresponding to the closed trajectories of the particle moving in the potential U_Θ (Andronov et al., 1981) (dashed lines 1 a in Fig. 3 c, 3 d). However, such oscillating distributions $\Theta(x)$ at $\eta =$ const are unstable (Kerner, Osipov, 1978).

In the GM model the form of the potential U_Θ (Fig. 3 c) is the same. By (3) we have for this model

$$q'_\Theta = 1 - 2A\Theta\eta^{-1}, \qquad q'_\eta = A\Theta^2\eta^{-2} > 0 . \tag{23}$$

It follows from (18), (19), (23) that the LR curve in the GM model has a V-shaped form (curve 1 in Fig. 3 b), and its extremum, where $d\eta/d\Theta = 0$, i. e. $q'_\Theta = 0$, corresponds to

$$\Theta = \Theta_0 = 2B, \qquad \eta = \eta_0 = 4BA . \tag{24}$$

The LR curve is formed by two branches of the single-valued function $\Theta(\eta)$: branch I, where $\Theta \leqslant \Theta_0$ and $q'_\Theta \geqslant 0$, and branch II, where $\Theta \geqslant \Theta_0$ and $q'_\Theta \leqslant 0$. It is seen from Fig. 3 b that also in this case the potential U_Θ takes the form of a potential well (Fig. 3 c) in the range of the inhibitor values $\eta_{sh} > \eta_0$, i. e. the possible distributions of the activator $\Theta(x)$ in the GM model are qualitatively similar to those in the Brusselator (Fig. 3 e, 3 f, 3 g).

Of particular form (Fig. 3 f) are the solutions of $\Theta(x)$ corresponding to the separatrix - the trajectories of a particle passing through a saddle point (Andronov et al., 1981) (singular point of equation (15)), i. e., through the maximum of the potential U_Θ ($\Theta = \Theta_1$, Fig. 3 c). Near the singularities $\Theta = \Theta_1$ and Θ_2 (Fig. 3 a, 3 b) the solutions of $\Theta(x)$ take the form

$$\Theta - \Theta_i \sim \exp(\lambda_i x), \qquad \lambda_i = \pm [q'_\Theta(\Theta_i)]^{1/2}. \tag{25}$$

In this case the point $\Theta = \Theta_1$ where $q'_\Theta > 0$, is a 'saddle' point and $\Theta = \Theta_2$, where $q'_\Theta < 0$, is a 'centre' (Andronov et al., 1981). From (25) it is seen that the distribution of $O(x)$ becomes increasingly smoother when the saddle point $\Theta = \Theta_1$ is approached and at $x \to \infty$ it goes over asymptotically into the line $\Theta = \Theta_1$. According to (15), $O(x)$ varies with a characteristic length of l. Therefore, in a system of size of $\mathcal{L}_x >> l (\mathcal{L}_x << L)$ the trajectory that coincides virtually with separatrix 1 in Fig. 3 c describes the activator distribution as a single narrow layer, a stratum (Fig. 3 f), and its neighbouring trajectories (1 b in Fig. 3 c, 3 d) as several such strata (Fig. 3 g). The distributions of $\Theta(x)$ in the form of several strata are unstable at $\eta(x)$ = const. The only stable distribution $\Theta(x)$ in a small-size system (more exactly, at $L \to \infty$) is that in the form of a single stratum (Fig. 3 f)[3] (Kerner, Osipov, 1977, 1978).

---

3) Strictly speaking the single stratum $\Theta(\dot{x})$ (Fig. 3 f) is stable, to the approximation $L = \infty$, only at cyclic boundary conditions, when the neutral conditions (17) are satisfied, the stratum is unstable (Kerner, Osipov, 1977, 1978) against its shift from one of the boundaries. The only potentially stable distribution (Kerner, Osipov, 1981) is that in the form of a single half-stratum located at one of the boundaries of the system (distribution $\Theta(x)$ at $x \geqslant 0$ or $x \leqslant 0$ in Fig. 3 f). The results of studies of strata evolution under change in A hold also in this case (Kerner, Osipov, 1978, 1983 a).

Strata parameters and evolution in the Brusselator model

It follows from (15) and (4) that at $\mathcal{L}_x >> 1$ the activator distribution in the form of a single stratum (Fig. 3 f) can be written (accurate to an exponentially small value (Kerner, Osipov, 1978) as

$$\Theta(x) = \Theta_1 + (\Theta_{max} - \Theta_1)\,ch^{-2}(\xi x/l), \tag{26}$$

where

$$\begin{aligned} &\Theta_1 = (1+A)[1-(1-\eta/\eta_0)^{1/2}]/2\eta, \quad \Theta_{max} = (1+A)[1+2(1-\eta/\eta_0)^{1/2}]/2\eta \\ &\xi^2 = (1+A)(1-\eta/\eta_0)^{1/2}/4, \qquad \eta_0 = (1+A)^2(4B)^{-1}. \end{aligned} \tag{27}$$

By substituting (27) in (16), we find $\eta = \eta_{sh}$. For the stationary case at $\eta << \eta_0$, which is valid at $L_x > 10\,l$, we arrive at (Kerner, Osipov, 1983 a)

$$\eta = \eta_{sh} = A(1+A)^2[1 \pm (1-24l/\mathcal{L}_x A^2(1+A)^{1/2})^{1/2}]/2B. \tag{28}$$

Expression (28) determines the dependence of η_{sh} on A, the bifurcation characteristic (BC) of the model (curve II, Fig. 4 a). From (28) it follows that the value of $A = A_b$, at which $d\eta_{sh}/dA = \infty$ (point 2 in Fig. 4 a), satisfies the equation (Kerner, Osipov, 1983 a):

$$A_b^2(1+A_b)^{1/2} = 24l/\mathcal{L}_x. \tag{29}$$

Linearising equations (15) to (17) and (4) with respect to the fluctuations $\delta\Theta$, $\delta\eta \sim \exp(-\gamma t)\cos k_x x$, where $k_x = \pi n/\mathcal{L}_x$ ($n = 0,1,\ldots$), we show with ease that there arises layering in the homogeneous state of the system relative to fluctuations with $k_x = k_o = \pi/\mathcal{L}_x$ at $A = A_c = 1 + (\pi l/\mathcal{L}_x)^2$. At the same time, as follows from (29), $A_b < A_c \simeq 1$ at $L_x > 24\,l$. Consequently, when $L >> \mathcal{L}_x > 24\,l$ and $A = A_c$, this layering results in a sudden formation (jump $0 \rightarrow 1$, Fig. 4 a) of a narrow stratum (Fig. 3 f). Thus, the characteristic scale of variation of the stable activator distribution at $A = A_c$ is of the order of l (Fig. 3 f), i. e. it is much less than the period $\pi k_o^{-1} = \mathcal{L}_x$ of the critical fluctuation. From (27), (28) it follows that at the point of layering $A = A_c \simeq 1$ (Kerner, Osipov, 1983 a):

$$\Theta_{max} - \Theta_1 = \Theta_h \mathcal{L}_x / 4\sqrt{2}\,l\,, \quad \xi = 2^{-1/2}$$
$$\Theta_1 = \Theta_h/2 = B/2\,, \quad \eta_{sh} = 12\sqrt{2}\,l/\mathcal{L}_x\,\Theta_h\,. \tag{30}$$

The amplitude of the stratum that appears increases with increasing A (Kerner, Osipov, 1983 a)

$$\Theta_{max} - \Theta_1 = \Theta_h A \mathcal{L}_x (4l)^{-1}(1+A)^{-1/2}, \quad \Theta_1 = \Theta_h(1+A)^{-1}, \quad \eta_{sh} = 6(1+A)^{3/2} l/\Theta_h A \mathcal{L}_x\,, \tag{31}$$

and it drops when A decreases. It is seen from (28) that in the range of values $A_c > A > A_b$, where the homogeneous state of the system is stable, there exist two inhomogeneous solutions $\Theta(x)$ in the form of a stratum (Fig. 3 f). Only the one which has the larger amplitude, is stable (Kerner, Osipov, 1977, 1978, 1982). It corresponds, according to (27), to the lesser value of η_{sh} (solid portion of curve II in Fig. 4 a). In accordance with the general result described by Kerner and Osipov (1978, 1981), we can assert that with decreasing A the stratum will abruptly vanish (jump $2 \rightarrow 3$ in Fig. 4 a) at $A = A_b$, where the derivative $d\eta_{sh}/dA = \infty$.

The value of $A = A_b$ determines the minimum value of the bifurcation parameter. At this point a short-time local perturbation can excite a large-amplitude stratum in the system (Fig. 3 f). From (29) it follows that at $L_x > 24\,l$ we have $A_b = (24\,l/L_x)^{1/2}$. It is clear from (27) and (28) that $A = A_b$

$$\Theta_{max} - \Theta_1 = \Theta_h (3\mathcal{L}_x/8l)^{1/2}, \qquad \xi = 1/2, \tag{32}$$
$$\eta_{sh} = (6l/\mathcal{L}_x)^{1/2}\sigma^{-1}, \qquad \Theta_1 = \Theta_h\,[1 + (24l/\mathcal{L}_x)^{1/2}]^{-1}.$$

Strata parameters and evolution in the Gierer-Meinhardt model

For the GM model, the distribution $\Theta(x)$ in a stratum (Fig. 3 f), according to (15) and (3), is also given by expression (26) (Kerner, Osipov, 1982 c), in which

$$\Theta_{max} - \Theta_1 = 6\eta\xi^2 A^{-1}, \quad \xi^2 = (4\eta)^{-1}(\eta^2 - 4AB\eta)^{1/2}, \quad \Theta_1 = \eta(2A)^{-1} - 2A^{-1}\eta\xi^2. \tag{33}$$

Substituting (33) into (16), at $L_x >> l$, or more exactly at $\eta >> \eta_0 = 4\,AB$, we get

$$\eta = \eta_{sh} = A^2\mathcal{L}_x/12Cl \cdot (1 \pm (1 - A_b^2/A^2)^{1/2}), \tag{34}$$

where

$$A_b = (24l/\mathcal{L}_x)^{1/2} A_0. \tag{35}$$

The BC derived from (34) is shown in Fig. 4 b.[4)] In the range of values $A_b < A < A_o = CB$ there are two solutions $\Theta(x)$ in the form of a stratum (Fig. 3 f). Also in this case, only a larger-amplitude stratum is stable which corresponds to the greater value of the inhibitor η_{sh} (solid portion of curve II, Fig 4 b).

The scheme of evolution of a stratum in the GM model is similar to that considered in a former section for the Brusselator. In a system of size of $L_x > 24\,l$, at $A = A_c \approx \approx A_o = CB$ there appears by jump (jump $0 \to 1$, Fig. 4 b) a large-amplitude stratum, for which, according to (33) and (34), we have

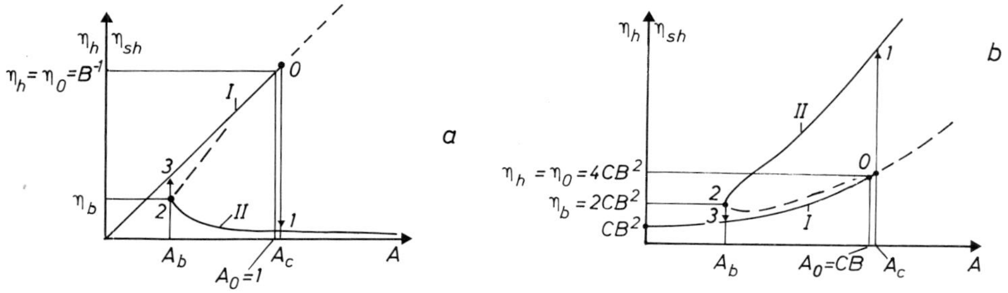

Fig. 4. Illustration of stratum evolution in the Brusselator (a) and GM model (b). Curve I corresponds to the homogeneous state of the system, curve II to the inhomogeneous state in the form of the stratum shown in Fig. 3 f. Dashed portions of the curves correspond to unstable states.

$$\Theta_{max} - \Theta_1 = B\mathcal{L}_x/4l, \quad \Theta_1 = B(1 + 6l/\mathcal{L}_x), \quad \eta_{sh} = B^2 C\mathcal{L}_x/6l, \quad \xi = 1/2. \tag{36}$$

The stratum amplitude Θ_{max} grows with increasing A:

---

4) Note that the form of BC for the GM model coincides qualitatively with that in the model of a heated electron-hole plasma in semiconductors (Kerner, Osipov, 1978, 1979).

$$\Theta_{max} - \Theta_1 = A\mathcal{L}_x/4\ell C, \quad \Theta_1 = B(1 + A_c 6\ell/\mathcal{L}_x A), \quad \eta_{sh} = A^2\mathcal{L}_x/6\ell C, \quad \xi = 1/2. \tag{37}$$

With decreasing A, the value of Θ_{max} decreases and at the point $A = A_b$, where $d\eta_{sh}/dA = \infty$, the stratum vanishes jumpwise (jump 2 → 3, Fig. 4 b). From (33) to (35) it follows that at $A = A_b$

$$\Theta_{max} - \Theta_1 = B(3\mathcal{L}_x/8\ell)^{1/2}, \quad \Theta_1 = B(1 + (6\ell/\mathcal{L}_x)^{1/2}), \quad \eta_{sh} = \eta_\ell = 2CB^2, \quad \xi = 1/2. \tag{38}$$

The conclusion of the qualitative coincidence of the forms of stable strata and their evolution patterns for the Brusselator and for the GM model is of a rather general character (Kerner, Osipov, 1980, 1981). The results of this section are applicable to many systems with a Λ - and V-shaped LR[5)], for which the conditions (Kerner, Osipov, 1980)

$$Q'_\eta > 0, \qquad q'_\Theta \Theta'_\eta - q'_\eta Q_\Theta > 0 \tag{39}$$

are satisfied. The second condition means that a smooth increase of the activator in the system is damped by a smooth change of the inhibitor. For determinacy, in the next sections we will mainly deal with the Brusselator and GM model, though the results to be presented are general for many models with a Λ- and V-shaped LR, which will be called, for short, the Λ- and V-models (systems) (Kerner, Osipov, 1985 a, c).

---

5) In the systems, for which the function $Q(\Theta, \eta, A)$ for η = const is bounded at $\Theta \to \infty$, the size of a stratum can be as large as $\sim \mathcal{L}_x/2$, i. e. it exceeds considerably 1. In this case Θ_{max} is a sharply increasing function of $\mathcal{L}_x$. This is also the case when $\mathcal{L}_x \sim L$ (Dubitsky et al., 1986 a, b).

Structures in large-size systems (Λ- and V-models)

The distribution of $\Theta(x)$ (Fig. 3 f) obtained in the last section at $\eta = \eta_{sh}$ = const corresponds to the approximation $1 << \mathcal{L}_x << L$, or more precisely, $L \to \infty$ (Kerner, Osipov, 1981). It is clear that at a certain finite L and $\mathcal{L}_x >> L$ the values of the inhibitor η and the activator Θ outside the stratum become $\eta = \eta_h$ and $\Theta = \Theta_h$, which corresponds to the homogeneous state of the system. At $A > A_c$, this state is stable (see above) and the value of η changes smoothly from $\eta = \eta_{sh}$ within the stratum to $\eta = \eta_h$ at a distance $\sim L$ away from it. Fig. 4 a shows that for the Brusselator $\eta_{sh} < \eta_h$, therefore, the value of η outside the stratum will grow smoothly from $\eta = \eta_{sh}$ to $\eta = \eta_h$ (dashed-and-dotted line, Fig. 2 a). For the GM model, as seen from Fig. 4 b, $\eta_{sh} > \eta_h$, hence the value of η outside the stratum decreases (dashed line, Fig. 2 a). The distributions of the activator $\Theta(x)$ (Fig. 2 a) in the Λ - and V-models (Fig. 3 a, 3 b) turn out to be qualitatively similar, inasmuch that in both of them $\Theta_1 < \Theta_h$ (Kerner, Osipov, 1978, 1982 c). Thus, in a system of size of $\mathcal{L}_x >> L$ the stratum considered in the last section (Fig.3 f) is transformed into a solitary state, an autosoliton (AS) (Fig. 2 a).

At $A > A_c$, the homogeneous state of the system is unstable and, therefore, the quantities η and Θ outside the stratum cannot reach values η_h and Θ_h. However, there may arise a structure of period $\mathcal{L}_p \sim L$ (Fig. 1 a) in which the distributions $\Theta(x)$ within the strata (regions of large values of the activator) are close to $O(x)$ at $L = \infty$ (Fig. 3 f) and outside the strata $\eta(x)$ and $\Theta(x)$ vary smoothly with a length $\sim L$, as in the case of AS (Fig. 2 a) (Kerner, Osipov, 1978, 1981). The only qualitative difference between Λ - and V-models is that the variations of the inhibitor and activator outside the strata are in phase in Λ -models and in counterphase in V-models (dashed curves η (x), Fig. 1 a, 2 a) (Kerner, Osipov, 1978, 1980, 1982 c).

The above conclusions derived from the form of large-amplitude structures in the Λ - and V-models can be substantiated (next section) by the qualitative procedure of constructing the solutions $\Theta(x)$, $\eta(x)$ developed in references (Kerner, Osipov, 1978, 1980).

Periodic strata. Maximum strata period

One dimensional static structures are described, according to (2), by the equation system

$$\ell^2 d^2\Theta/dx^2 + dU_\Theta/d\Theta = \sigma, \qquad U_\Theta = -\int^\Theta q(\Theta, \eta(\Theta), A)\, d\Theta, \tag{40}$$

$$L^2 d^2\eta/dx^2 + dU_\eta/d\eta = \sigma, \qquad U_\eta = -\int^\eta Q(\eta, \Theta(\eta), A)\, d\eta. \tag{41}$$

To construct the potentials U_η, U_Θ and the solutions $\eta(x)$, $\Theta(x)$, an iteration procedure was developed (Kerner, Osipov, 1980) which uses as the zeroth approximation the forms of U_η, U_Θ, $\eta(x)$ and $\Theta(x)$ in the regions of slowly-varying and rapidly-varying distributions.

The smooth distributions correspond to the solutions of system (40), (41) at $l = 0$, when the relation between the activator Θ and the inhibitor η is local and is found from equation (18). For this reason the dependence $\eta(\Theta)$ derived from (18) was called the local relation, LR. The slowly varying distribution of $\eta(x)$ is described by the equation (Kerner, Osipov, 1978)

$$L^2 d^2\eta/dx^2 + dU_\eta/d\eta = 0, \qquad U_\eta = -\int^\eta Q(\eta, \Theta(\eta), A)\, d\eta, \tag{42}$$

in which the dependence $\Theta(\eta)$ is given by (18). From this it follows that the characteristic length of slowly varying distributions is L.

The abrupt distributions of $\Theta(x)$ correspond to the solutions of system (40), (41) at $L = \infty$, i. e. satisfy equation (15) (Kerner, Osipov, 1978, 1980). These were considered in detail above (Fig. 3), where it was noted that the only stable solution for the class of rapidly varying distributions is that in the form of a single stratum (Fig. 3 f).

Equation (42) for slowly varying distributions coincides, formally, with the equation for conservative motion in the potential U_η of a classical 'particle' with coordinate and time x. Since the function $\Theta(\eta)$ corresponding to LR (Fig. 3 a, 3 b, 5 a) is not single-valued, the potential U_η (42) has two independent branches I and II corresponding to $\Theta \leqslant \Theta_0$ and $\Theta \geqslant \Theta_0$, respectively (Fig. 5 b). The extremum of the potential U_η corresponds, according to (42) and (18), to the values of $\Theta = \Theta_h$ and $\eta = \eta_h$ for the homogeneous state of the system. In other words, the extremum of U_η is on that branch on which is located the point of intersection of the LR and the state equation (SE) curves, i. e. the dependence $\eta(\Theta)$ given by

$$Q(\eta, \Theta, A) = 0 \tag{43}$$

The derivative $dU_\eta/d\eta$ is

$$dU_\eta/d\eta = -Q(\eta, \Theta, A), \tag{44}$$

and can reverse sign only on the SE curve, near which the change of the quantity Q along the ordinate axis $dQ = Q_\eta d\eta$. This allows reconstruction of the form of U_η for different types of LR (Kerner, Osipov, 1978, 1980, 1981, 1982 c). For determinacy, we show in Fig. 5 the construction of the potential U_η and the solutions $\Theta(x)$, $\eta(x)$ for the case of Brusselator (4) (Kerner, Osipov, 1978). In this case, considering (39) and (43), on the right of the SE curve, where $d\eta > 0$, we have $Q > 0$ and therefore $dU_\eta/d\eta < 0$; correspondingly, on the left $Q < 0$ and $dU_\eta/d\eta > 0$. At the extremal point of the LR curve η_0 (24), corresponding to the boundary of its branches I and II of LR, the derivatives $dU_\eta/d\eta$ of branches I and II of the potential U_η coincide, but the potential U_η may have arbitrary values. From the above considerations, we can reconstruct the form of branches I and II of the potential U_η (Fig. 5 b).

In the conservative case, a particle moves at constant energy and its rate $(d\eta/dx)$ becomes zero only when it is reflected from the wall of the potential U_η. A solution of $\eta(x)$ and $\Theta(x)$ corresponding to a smooth distribution everywhere exists when the values of η_h and Θ_h for the homogeneous state of the system are on branch II of the potential U_η. Indeed, the type of the potential U_η extremum is determined by the sign of

$$d^2U_\eta/d\eta^2 = -(q'_\Theta Q'_\eta - q'_\eta Q'_\Theta)/q'_\Theta . \tag{45}$$

From (39) and (45) it follows that the quantities $d^2U_\eta/d\eta^2$ and q'_Θ have opposite signs. On branch II of LR, $q'_\Theta < 0$, therefore the points $\eta = \eta_h$ and $\Theta = \Theta_h$ correspond to the minimum of U_η. In other words, at $A > A_c$, when the values of η_h, Θ_h correspond to the unstable homogeneous state ($\Theta_h > \Theta_0$) branch II of the potential U_η takes the form of a potential well (Fig. 5 b). The particle trajectory in such a well describes the periodic distribution of the activator $\Theta(x)$ and the inhibitor $\eta(x)$ with a period $\sim$ L. However, all such smooth distributions are unstable (Kerner, Osipov, 1978). This is because in the region of size less than L the values of Θ and η vary little and are close, respectively, to Θ_h and η_h. At the same time, the homogeneous state corresponding to the range $\Theta_h > \Theta_0$ is unstable to increasing fluctuations of period $\mathcal{L}_0 = 2\pi k_0^{-1} << L$ (see above). Hence it follows that in stable slowly varying distributions the activator value must be less than Θ_0. Fig. 5 a shows that the range of values $\Theta < \Theta_0$ corresponds to branch I of the LR curve. Therefore, stable slowly varying distributions can be described only by trajectories of a particle moving in branch I of the potential U_η (Fig. 5 b). Since branch I has the form of a potential 'hump' (Fig. 5 b), the motion of the particle cannot be finite (bounded). Consequently, appropriate solutions $\Theta(x)$ and $\eta(x)$ do not satisfy the boundary conditions (17). At the same time, such smooth distributions in combination with the abrupt distribution of $\Theta(x)$ (Fig. 3 f) produce a stable structure in the form of a periodic succession of narrow strata (Kerner, Osipov, 1978, 1981, 1982 a).

Let us construct an elementary fragment of such a periodic structure (Fig. 5 e). If on the boundaries $x = 0$ and $\mathcal{L}_p/2$ of this fragment

$$\partial\eta/\partial x\,\big|_{0,\mathcal{L}_p/2} = \partial\Theta/\partial x\,\big|_{0,\mathcal{L}_p/2} = 0\,, \qquad (46)$$

then its mirror reflection will lead to a structure in the form of strata of period $\mathcal{L}_p$ (Fig. 1 a).

Let the rapidly varying distribution in the form of a stratum (Fig. 3 f) be in the neighbourhood of $x = 0$, i. e., $\Theta(0) = \Theta_{max}$, $\eta(0) = \eta_{sh}$ (Fig. 5 e). In the region of slowly varying distributions, $d\eta/dx$ becomes zero only when the particle is reflected from a wall of the potential U_η (Fig. 5 c). Thus, to satisfy condition (46) at $x = \mathcal{L}_p/2$ the particle must move from the coordinate $\eta(0) = \eta_{sh}$, corresponding to the region of rapidly varying distributions, towards the wall of the potential U_η (Fig. 5 c). In this case $\eta(x)$ increases and therefore $\eta(\mathcal{L}_p/2) \equiv \eta_t > \eta_{sh}$ and $\Theta(\mathcal{L}_p/2) \equiv \Theta_t > \Theta_1$ at $x = \mathcal{L}_p/2$. The choice of the distribution of $\Theta(x)$ in the form of a single stratum (Fig. 3 f) is related to the fact that it is only for this distribution among the many other rapidly varying distributions of $\Theta(x)$ (Fig. 3 e to 3 g) that the activator value on the periphery of the stratum $\Theta = \Theta_1$ lies on branch I of the LR curve. Hence only this solution $\Theta(x)$ (Fig. 3 f) can go over near points $\Theta = \Theta_1$ and $\eta = \eta_{sh}$ (Fig. 5 a) into the smooth distribution corresponding to the particle trajectory in branch I of the potential U_η. By (40), (41), the values $\eta = \eta_{sh}$ and $\Theta = \Theta_1$ must satisfy the integral conditions (Kerner, Osipov, 1980)

$$\int_{\eta_t}^{\eta_{sh}} Q(\eta,\Theta(\eta),A)\,d\eta = \int_0^{\mathcal{L}_p/2} Q(\Theta(x),\eta(x),A)\,dx = 0\,, \qquad (47)$$

$$\int_{\Theta_t}^{\Theta_{max}} q(\eta(\Theta),\Theta,A)\,d\Theta = \int_0^{\mathcal{L}_p/2} q(\Theta(x),\eta(x),A)\,dx = 0\,, \qquad (48)$$

where $\Theta(\eta)$ and $\eta(\Theta)$ are the true dependencies of Θ on η within the stratum (solid curve 3 in Fig. 5 a). Condition (47) is analogous to the condition $< Q(\Theta(x, \eta_{sh}, A) > = 0$ used above for determination of the dependencies of η_{sh} and Θ_{max} on $\mathcal{L}_x$ and A. It follows that at $\mathcal{L}_p/2 < L$ or, more precisely, when η_{sh} is not very different from $\eta_t = \eta(\mathcal{L}_p/2)$ this value Θ_{max} in the stratum (Fig. 5 e) can be estimated from (31) (or (36) for the GM model) by substituting $\mathcal{L}_p/2$ for $\mathcal{L}_x$.

In constructing the true potential U_η in equation (41), we note from (47) that at $\mathcal{L}_p >> 1$ we have (Kerner, Osipov, 1980),

$$< Q >_{sh} \simeq - \varepsilon^{-1} \frac{\mathcal{L}_p}{2L} < Q >_{sm} , \tag{49}$$

where the symbols $<...>_{sm}$ and $<...>_{sk}$ mean averaging of the function over the regions of rapidly and slowly varying distributions, respectively. According to (49), at $\eta \simeq \eta_{sh}$ and $\mathcal{L}_p \sim L$ (Fig. 5 e) the value $|Q| \sim \varepsilon^{-1} >> 1$. Since the derivative $dU_\eta/d\eta = - Q$, the true potential U_η in (41) near $\eta = \eta_{sh}$ goes sharply upward from branch I of the potential U_η (42) for the smooth distributions, producing a steep wall, i. e. it is transformed into a potential well (Fig. 5 c). The distribution of $\eta(x)$ (Fig. 5 e) corresponds to finite motion of a particle in such a well (trajectory 1 in Fig. 5 c).

In constructing the true potential U_Θ in equation (40), we note that at $x = \mathcal{L}_p/2$ the derivative $d\Theta/dx = 0$ and $\Theta(x)$ varies with a characteristic length $\sim L$. As a result, in the region of smooth distributions near $x = \mathcal{L}/2$ we have $l^2\, d^2\Theta/dx^2 \sim - \varepsilon^2$, i. e. by (40), $q \simeq - \varepsilon^2$. In the region uniting smooth and abrupt distributions the derivative $d\Theta/dx$ must again pass through zero (Fig. 5 e), i. e., at a certain point in the smooth distribution $q = 0$. Since $dU_\Theta/d\Theta = - q$, the potential U_Θ, in which the motion of a particle describes the smooth variation of $\Theta(x)$, has the form of a shallow (of the order of ε^2) potential well. This well is united with the deep well U_Θ, corresponding to an abrupt distribution of $\Theta(x)$, only near the saddle point $\Theta = \Theta_1$, where $dU_\Theta/d\Theta \to 0$ (Fig. 3 c, 5 f) (Kerner, Osipov, 1980).

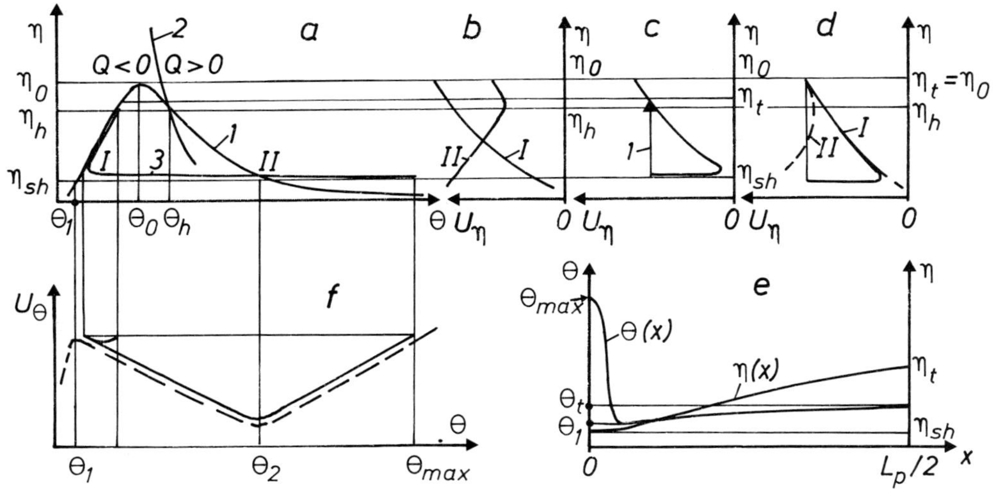

Fig. 5. See text

The distance between the strata $\mathcal{L}_p$ (Fig. 1 a) is bounded from above (Kerner, Osipov, 1978). Indeed, when we increase the length of the segment $\mathcal{L}_p/2$ (Fig. 5 e), the values η_t and Θ_t at $x = \mathcal{L}_p/2$ will, evidently, tend to the homogeneous values η_h and Θ_h (Fig. 5 a). However, the values η_h and Θ_h at $\Theta_h > \Theta_o$, i. e. at $A > A_o$ (see second section), correspond to branch II of the potential U_η and determine the position of its minimum (Fig. 5 b). The smooth transition from branch I of the potential U_η, where the point $\eta = \eta_t$ is located, to the minimum of branch II is impossible, as evidenced by Fig. 5 d. Therefore, at a given A ($A > A_o$) there is a certain value $\mathcal{L}_p = \mathcal{L}_{max}$ above which the solution in the form of strata of period $\mathcal{L}_p > \mathcal{L}_{max}$ cannot be constructed. The maximum possible period $\mathcal{L}_{max}$ of the strata at a given A is determined by the attainment condition involving the quantities η_t and Θ_t, respectively, the values η_o and Θ_o, (accurate to $\varepsilon << 1$) (Kerner, Osipov, 1978). Hence it follows that the value $\mathcal{L}_{max}$ can be estimated from the solution of the equation for smooth distributions (42) with the boundary conditions $\eta(0) = \eta_{sh}$ and

$\eta(\mathcal{L}_{max}/2) = \eta_o$. Considering (4), (21) and (31) for the case of Brusselator, we get (Kerner, Osipov, 1983)

$$\mathcal{L}_{max} \simeq 2^{1/2} L(1+A)^{3/2} B^{-1} (A-1)^{-1/2}. \qquad (50)$$

According to (50), with increasing $A > A_o = 1$, $\mathcal{L}_{max}$ decreases and at $A = 2$ reaches its minimum value

$$\mathcal{L}^{(m)}_{max} \approx 3\sqrt{6}\, L B^{-1} \qquad (51)$$

whereupon it increases practically in direct proportion to A (see Fig. 3 of (Kerner, Osipov, 1983)). These results are helpful in the study of strata evolution (self-organisation) (Fig. 1 a) under change in the parameter A.

In this study, it is convenient to use bifurcation characteristics (BC) - the dependence of the inhibitor value at the centre of the strata, $\eta = \eta_{sh}$, on the parameter A. As noted above, the dependence $\eta_{sh}(A)$ (Fig. 6 a) at $\mathcal{L}_p < L$ is close to that considered in the last section for the case $L = \infty$ (Fig. 4 a), if in (31) we put $\mathcal{L}_x = \mathcal{L}_p$. According to (31), the greater the value of $\mathcal{L}_x = \mathcal{L}_p$ the smaller is η_{sh} (Fig. 6 A) (Kerner, Osipov, 1978).

Let in a system of size $\mathcal{L}_x >> L$, at the point $A = A_c \approx A_o = 1$, there appear spontaneously N strata of period $\mathcal{L}_p^{(N)} = \mathcal{L}_x/N > \mathcal{L}_{max}$ (jump 0 → 1, Fig. 6 a) as a result of layering of the homogeneous state (strata of small period $\mathcal{L}_p = 2\pi k_o^{-1} \sim (l\,L)^{1/2}$ at $\varepsilon = l/L << 1$ are unstable; see below). As A increases the value of $\eta = \eta_t$ between the strata (Fig. 1 a) grows until at a certain $A = A_d^{(N)} > A_c$ it becomes $\eta_t \approx \eta_o$ (equation (21)). In this case $\mathcal{L}_{max}(A_d^{(N)}) = \mathcal{L}_p^{(N)}$. Further increase in A leads to increasingly rapid growth of Θ between the strata, i. e. to a local break resulting in the appearance in this region of one more stratum (Kerner,

Osipov, 1978, 1982 a). This local break is a result of loss of the solution in the form of strata with a given period $\mathcal{L}_p^{(N)}$ at $A > A_d^{(N)}$ rather than to loss of the stability of the initial state. Therefore, the new structure results from dynamic rather than fluctuational rearrangement of the initial structure (Kerner, Osipov, 1982 a). In the dynamic rearrangement the time of formation of a new structure is determined not by the mean square value and the increment in growing critical fluctuations but by times τ_θ and τ_η and the increment of A over $A_d^{(N)}$. Thus, when A increases the number of strata will grow in a jump-like fashion due to the effect of local break between the strata. Their period $\mathcal{L}_p$ will, accordingly, lessen (most probably, by a factor of 2) (jumps 2 → 3, 4 → 5, 6 → 7, 8 → 9, 10 → 11, Fig. 6 a) (Kerner, Osipov, 1978). Such period jumping in the process of self-organisation without fluctuations in the Brusselator model (equations (4)) has a limit: at $\mathcal{L}_p < \mathcal{L}^{(m)}_{max}$ an increase in A does not lead to a dynamic rearrangement of the structure.

In the GM model (equations (3)), whose BC is shown in Fig. 6 b[6)] self-organisation without fluctuations can also take place when A increases. From equation (42) for smooth distributions in the GM model, $\mathcal{L}_{max}$ can be estimated as

$$\mathcal{L}_{max} = 2L \ln [A^2 L A_0^{-1} \ell^{-1} (A - A_0)^{-1}]. \tag{52}$$

It follows from (52) that this is of the order of magnitude $\mathcal{L}_{max} \sim L$ (Romanovskii et al., 1984), i. e. it coincides with the value found for the Brusselator (Kerner, Osipov, 1983 a). According to (50) and (52), the dependencies $\mathcal{L}_{max}(A)$ in the two models also are qualitatively coincident: in the GM model at $A = 2A_0 = 2BC$, $\mathcal{L}_{max}$ reaches a minimum, and at $A > 2A_0$ $\mathcal{L}_{max}$ increases with A. As a result, at A exceeding A_c this model generates a qualitatively similar pattern of self-organisation as that described above for the Brusselator (Kerner, Osipov, 1978). In contrast to the latter, however (Fig. 6 a), the jumpwise decrease in the strata period (jumps 2 → 3, 4 → 5,

---

6) Note that the BC for the GM model (Fig. 6 b) coincides qualitatively with the BC for the model of a heated electron-hole plasma in semiconductors (see Fig. 9 a in Kerner, Osipov, 1978).

$6 \rightarrow 7, 8 \rightarrow 9, 10 \rightarrow 11$ in Fig. 6 b) is followed here by an increase in η_{sh} at the centre of a stratum.

As seen from (50) and (52), $\mathcal{L}_{max} \rightarrow \infty$ at $A \rightarrow A_0$. This is a reflection of the fact that at $A < A_0$ the system is capable of generating an AS in the form of a single stratum (see below).

Periodic strata stability. Minimum strata period

The number of strata in a system of size $\mathcal{L}_x >> L$ is bounded from above by a certain value N_{max}. This is associated with the fact that strata of period $\mathcal{L}_p < \mathcal{L}_x/N_{max} = \mathcal{L}_{min}$ are unstable (Kerner, Osipov, 1978). Critical activator fluctuations are localised in the strata walls, i. e. in the region of rapid variation of $\Theta(x)$ (Fig. 1 a). Their growth is damped by corresponding variation of the inhibitor (Kerner, Osipov, 1978). The most critical fluctuation is that with a period $2\mathcal{L}_p$ which describes the decrease in amplitude of one stratum connected with corresponding increase in a neighbouring stratum - the 'pumpover' effect (Kerner, Osipov, 1978). This critical fluctuation of Θ is coupled with the variation in η of the same period $2\mathcal{L}_p$. Such a short-wave variation of the inhibitor produces large diffusion fluxes, which considerably weaken its damping effect on the growth of the dangerous activator fluctuation. As a result, the number of strata in the system decreases jumpwise, and therefore the distances between the strata grow. In the simplest case, the instability criterion of a pumpover type has the form (Kerner, Osipov, 1978

$$\gamma = \lambda + P < \sigma, \tag{53}$$

where γ is the increment of damping of the dangerous activator fluctuation $\delta\Theta = \delta\Theta(x)\exp(-\gamma t)$; $\lambda < 0$ is the increment of the growth of this fluctuation in the absence of the damping effect of the inhibitor, i. e. $\gamma = \lambda$ at $\delta\eta = 0$; the coefficient $P > 0$ describes the damping effect. To estimate P (Kerner, Osipov, 1978, 1981, 1983 a) we can write

$$P = a_0 \mu_0 \mu_{N/2}^{-1} = a_0 \mu_0 (\mu_0 + \pi^2 \mathcal{L}_p^{-2} L^2)^{-1}, \tag{54}$$

where the coefficient $a_0 \simeq |\langle q'_\eta \rangle_{sh} \langle Q'_\Theta \rangle_{sh}| \mu_0^{-1} (1/L_0)$; μ_0 and $\mu_{N/2}$ are the increments of damping of the homogeneous and for period $2\mathcal{L}_p$ perturbations of the inhibitor, respecitvely. The value $\mathcal{L}_p = \mathcal{L}_{min}$, corresponding to the boundarry of strata stability (Re $\gamma = 0$) can be estimated from (53) (Kerner, Osipov, 1978, 1981). Such estimation for the Brusselator was performed by Kerner and Osipov (1983 a), who showed that

$$\lambda \simeq -(5/4)(1+A), \quad \mu_0 \approx \langle \Theta^2 \rangle = B^2[1+2A+A^2\mathcal{L}_p(1+A)^{1/2}/6\ell](1+A)^{-2}, \tag{55}$$

$$a_0 \approx [(5/4)(1+A)+1](15\pi^2/32\sqrt{2})^2. \tag{56}$$

Using (53) to (55), we get (Kerner, Osipov, 1983)

$$\mathcal{L}_{min} \approx L(\ell/L)^{1/3}(15\pi^2/2)^{1/3}(1+A)^{5/6}(AB)^{-2/3}. \tag{57}$$

It follows from (57) that $\mathcal{L}_{min} > 2\pi k_0^{-1} \sim (l/L)^{1/2}$ at $L >> 1$. Therefore, at the point of layering there must arise jumpwise (jump $0 \to 1$, Fig. 6 a) large-amplitude strata with a period $\mathcal{L}_p > 2\pi k_0^{-1}$ (Kerner, Osipov, 1978, 1981, 1985 a). This conclusion also applies to the GM model because by (53), $\mathcal{L}_{min} \sim L$ (Polezhaev, 1984; Romanovskii et al., 1984).

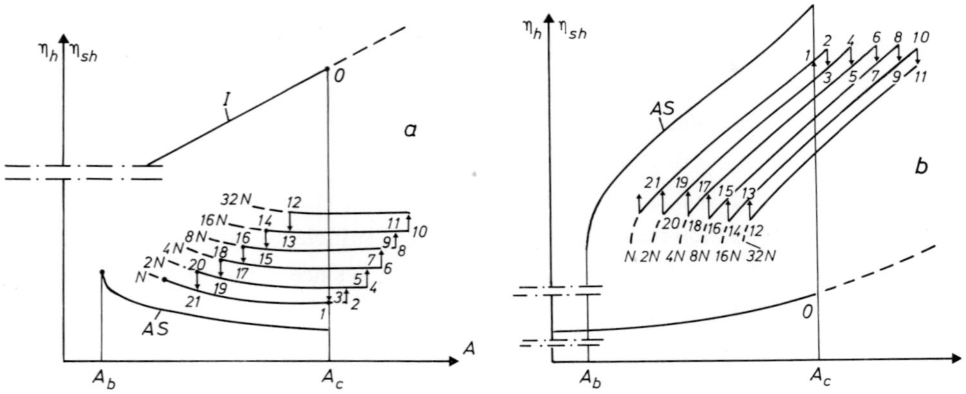

Fig. 6 Illustration of the evolution of periodic strata in the Brusselator (a) and GM model (b). Curve I corresponds to the homogeneous state; curves AS to the autosoliton shown in Fig. 2 a; numbered curves (N, 2N, ..., 32N) to the periodic strata shown in Fig. 1 a.

A structure is stable if its period $\mathcal{L}_p$ is in the range from $\mathcal{L}_{min}$ to $\mathcal{L}_{max}$ (Kerner, Osipov, 1978). From this it also follows that at a given A there are many structures for which $\mathcal{L}_{min} < \mathcal{L}_p < \mathcal{L}_{max}$ (Fig. 6) (Kerner, Osipov, 1978). The evolution of structures (self-organisation) is determined by the dependencies of $\mathcal{L}_{min}$ and $\mathcal{L}_{max}$ on A. It is clear from (57) that $\mathcal{L}_{min}$ increases with decreasing A ($A < A_c$). As a result, at a certain $A = A_b^{(N)}$ the value $\mathcal{L}_{min}(A_b^{(N)})$ becomes equal to $\mathcal{L}_p$, and therefore the strata lose their stability due to the pumpover effect. The critical activator fluctuation, as was noted, has a double period, and its growth, most probably, will lead to an abrupt decrease in the number to one half of strata (jump 12 → 13, Fig. 6 a, 6 b) (Kerner, Osipov, 1978). With further increase in A, resulting in a succession of such period-doubling bifurcations, the strata will decrease in number jumpwise (jumps 14 → 15, 16 → 17, 18 → 19, Fig. 6 a, 6 b), which, eventually, lead to a single stratum, AS (Fig. 2 a). We recall that a different pattern of structure evolution is realised when A increases. It was emphasised above that self-organisation in such a case can be associated not with the growth of fluctuations but with the dynamic rearrangement of the structure (jumps 2 → 3,....., 10 →11, Fig. 6). Such a dynamic rearrangement can be replaced by the fluctuation rearrangement when A is sufficiently large. This can be realised, in the case of the Brusselator, at $A > 4A_0 = 4$, since in this range the value $\mathcal{L}_{min}$ according to (57), increases as $A^{1/6}$.

Structures in two and three dimensions

Strata (Fig. 1 a) in the considered two- and three-dimensional Λ - and V-models of morphogenesis are unstable to the growth of a fluctuation localised in a stratum and tending to break the strata along the y (and z) axis into smaller regions (Kerner, Osipov, 1980). Radially symmetrical layers of radius $\rho >> l$, as well as any other drops whose size greatly exceeds l in at least one direction are also unstable to break up. The only stable structures are drops of size of the order of l spaced at a distance $\mathcal{L}_p$ from one another, such that $\mathcal{L}_{max} > \mathcal{L}_p > \mathcal{L}_{min} \sim L$. The variation of Θ and η at the section of the drops is close to a distribution in the form of a hot narrow stratum (Fig. 1 a). From the form of the critical fluctuation it follows (Kerner, Osipov, 1980) that a structure consisting of radially symmetrical drops (Fig. 7 a) will most probably arise on the breakup of radially symmetrical layers (Kerner, Osipov, 1982). This conclusion explains the results of a computer experiment with the GM model (Gierer, Meinhardt, 1982; Haken, 1978). Stable over the largest range of variation in A is that configuration of drops in which the drops are equally spaced from one another. The separation being the maximum possible for the given number of drops (Fig. 7 b) (Kerner, Osipov, 1982 c).

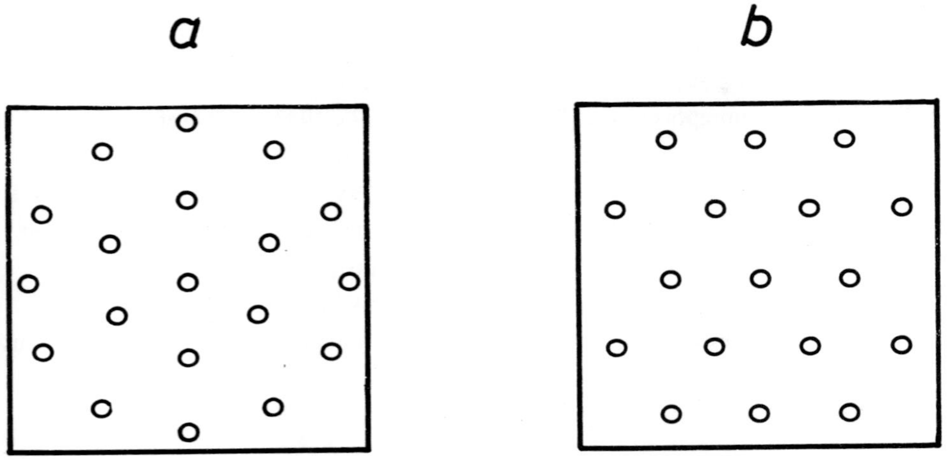

Fig. 7 Possible distributions of AS (symbolised by the circles) in a two-dimensional system.

Complex structures and turbulence

In the Λ - and V-models of morphogenesis considered, in addition to those simplest symmetrical periodic strata (Fig. 1 a), there can appear stable structures in the form of nonsymmetrical periodic strata whose periods contain several nonsymmetrical strata along with states nonperiodic at all, namely, stochastic inhomogeneous structures (Fig. 1 d). The possibility of stochastic structures was predicted by the qualitative theory (Kerner, Osipov, 1978, 1980) and the basis described by Kerner et al. (1984 b, c).

In two- and three-dimensional systems, there can exist randomly located drops $\Theta(\vec{r})$ of size ~ 1 with the distances between them lying in the interval $(\mathcal{L}_{min}, \mathcal{L}_{max})$.

The conditions determining the values of $\mathcal{L}_{min}$ and $\mathcal{L}_{max}$, as noted above, correspond to effects that are quite different in their nature: local break between strata and instability of a pumpover type (Kerner, Osipov, 1978, 1982 a). Therefore, in a certain range of A (at $A > A_0$) the condition $\mathcal{L}_{max} < \mathcal{L}_{min}$ can be satisfied when all the static structures are unstable (Kerner, Osipov, 1982 b). At $\alpha = \tau_\Theta / \tau_\eta >> 1$ stationary states in the form of homogeneous oscillations, pulsating structures (Kerner, Osipov, 1982 b) and autowave processes (Kerner, Osipov, 1985 d; Krinskii, Zhabotinskii, 1981; Polak, Mikhailov, 1983; Romanovskii et al., 1984; Vasil'yev et al., 1979) do not exist. At the same time, at $A > A_c$ the homogeneous state of such systems is unstable. Only states of the form of randomly appearing and disappearing strata or drops are realisable, a kind of turbulence (Kerner, Osipov, 1983 a).

The effect of a small inhomogeneity

The pattern of self-organisation (structure evolution) described above holds true for the ideally homogeneous systems. In real systems, however, the evolution of strata is to a large extent determined by the boundary conditions and by the presence of small inhomogeneities. Close to the bifurcation values of the parameter A, the inhomogeneities greatly affect the probability of the formation of one of the structures possible at a given A. For instance, at A close to $A_b^{(N)}$ or $A_d^{(N)}$ (Fig. 6) the number of strata may decrease or, respectively, increase not by a factor of 2, but by one, two, or more strata (Kerner, Osipov, 1982 a, 1985 a). In this case nonsymmetrical and nonperiodical located strata may arise. A small local inhomogeneity can show up as a priming (embryo) for the sudden spontaneous formation of a number of strata or a single stratum, an autosoliton (AS), in a stable system at values of A close to A_c. When the amplitude of a small inhomogeneity increases ($\Delta = \Delta\Theta/\Theta_h$), the critical value A_{cr} for the spontaneous formation of an AS can become appreciably smaller than A_c ($\Delta A = A_c - A_{cr} \backsim \Delta^{2/3}$ at $\Delta << 1$). The form of AS appearing in this case is practically independent of the form of the inhomogeneity that provokes its excitation.

Autosolitons

An AS, a single localised state (Fig. 2 a), can be excited by a short-term local perturbation of finite amplitude in a homogeneous stable system, i. e. at $A < A_0$. In experiments with Hydras the role of such perturbations can be played by the transplantation of the cells of a fragment into an adult Hydra (Belintsev, 1983; Gierer, Meinhardt, 1972; Haken, 1978) and the resulting formation of the Hydra's head can be viewed as an AS. These experiments have shown that the secondary head in a Hydra develops only in the situation where an embryo is located sufficiently far away from the primary head. In accordance with the general theory (Kerner, Osipov, 1978), this can be explained by the fact that at a distance smaller than $\mathcal{L}_{min} \sim L$ two AS (heads) turn out to be unstable to the pumpover effect. It follows from (37), (38) that the amplitude (O_{max}) of AS increases with the size of the system $\mathcal{L}_x$ (at $\mathcal{L}_x \lessdot L$). It is exactly this correlation that was noted in experiments with Hydras, where organs develop in proportion to the size of the head (Belintsev, 1983).

An AS can be excited in a stable homogeneous system of arbitrarily large size ($\mathcal{L}_x >> L$). The construction of the form of AS (Fig. 2 a, Fig. 8) is analogous to that described above for periodic strata (Fig. 5). The prinicipal difference is that in a stable system at $A < A_0$ the extremum of the potential U_η (42) corresponds to $\Theta_h < \Theta_0$, i. e. according to (44) it is on branch I (Fig. 8 a). Since in the case of a stable homogeneous state $q'_\Theta > 0$, this extremum, by (45), (39), is a maximum (Fig. 8 b). Therefore, $\Theta = \Theta_h$ and $\eta = \eta_h$ (Fig. 8 b) correspond to the saddle point of (42). The trajectory of motion of a particle in the potential U_η (42) closed on this point corresponds to the solution $\Theta(x)$ and $\eta(x)$, which at $x \rightarrow \pm \infty$ becomes Θ_h and η_h. The AS is symmetrical relative to the point $x = 0$ (Fig. 2 a). By the same argument as above uniting the abrupt distribution of $\Theta(x)$ (Fig. 3 f) with the smooth distribution corresponding to the trajectory $\eta(x)$ of a particle in the potential U_η (42) closed at the saddle point $\eta = \eta_h$, we arrive at a distribution of $\Theta(x)$ and $\eta(x)$ as shown in Fig. 2 a (dashed-and-dotted line $\eta(x)$ is for the Brusselator (4), dashed line for the GM model (3)). The distribution of $\eta(x)$ corresponds to the uppermost trajectory of the particle closing on the saddle point (maximum point) of the true potential U_η in (41) (solid curve, Fig. 8 c), which can be reconstructed, for periodic strata (Fig. 5 c) by uniting branch I of the potential U_η for the smooth distributions with the abrupt distribution of $\Theta(x)$ (Fig. 3 f) near $\eta = \eta_{sh}$. The region of AS localisation is $\sim L$ (Fig. 2 a). Therefore, the basic parameters of AS, Θ_{max} and η_{sh} (Fig. 2 a), for the Brusselator (4) and the GM model (3) can be estimated, respectively, from (27), (28) and (33), (34) by putting $\mathcal{L}_x = L$.

The AS exists in the range $A_b \leqslant A < A_c$; the critical value $A = A_b$ corresponds to that point on the BC (curve AC in Fig. 6 a, 6 b) where the derivative $d\eta_{sh}/dA = \infty$ (Kerner, Osipov, 1978, 1980, 1981). The value A_b for (3) and (4) can be estimated from (29) and (35) by assuming $\mathcal{L} \approx L$. From these formulae it follows, for both models, that at $L >> 1$

$$A_b \sim (\ell/L)^{1/2} A_0 \approx (\ell/L)^{1/2} A_c .$$

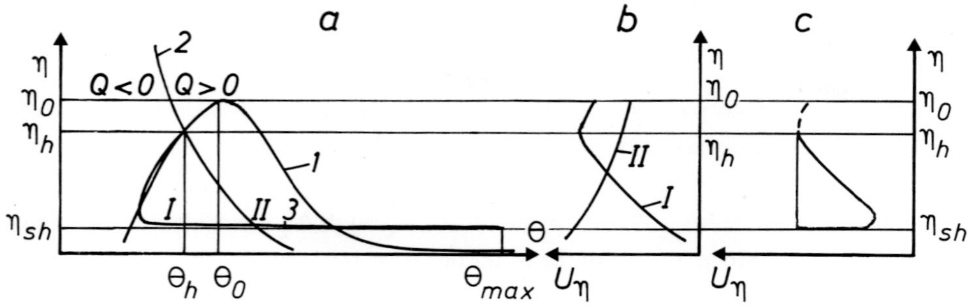

Fig. 8 See text.

It is clear from here that at $\varepsilon = 1/L << 1$ the value A_b can be several orders of magnitude less than $A = A_c$ at which the Turing instability is realised. We emphasise that, when applied to real physical systems, this conclusion implies that an AS in the form of a highly nonequilibrium region can be excited in systems not too far from thermodynamic equilibrium (Kerner, Osipov, 1985 d, 1987). At $A < A_b$, the AS vanishes in a jumpwise fashion and at $A > A_c$ it is transformed into a periodic or more complicated structure filling the entire system. These conclusions are also valid for a radially symmetrical AS which can be excited in a two- or three-dimensional system (Kerner, Osipov, 1985 b, c).

Structures in N- and И -models

For some models of morphogenesis (Chernavskii, Ruijgrok, 1978, 1982; Romanovskii et al., 1984), the LR curve has an N- or И -shaped form (curves 1 in Fig. 9 a, 9 b). In such models the homogeneous state of the system, according to (2), turns out to be unstable in a limited range of A: $[A_c, A'_c]$. The quantities A_c and A'_c coincide; this coincidence is accurate to a value of the order of $\varepsilon = 1/L << 1$, with A_0 and A'_0, respectively, at which the derivative $q'_\theta = 0$.

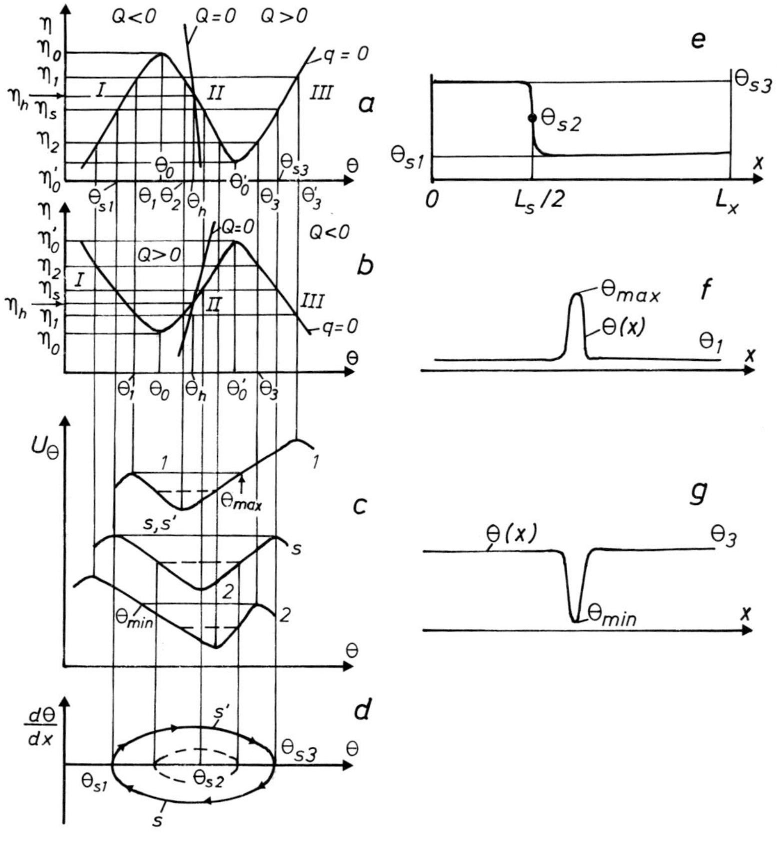

Fig. 9 See text

Small-size systems

Consider first the form and properties of structures in systems whose size $\mathcal{L}_{x,y,z} << L$ (Kerner, Osipov, 1981). In this case one-dimensional structures are described by (15) to (17). Let us reconstruct the form of the potential U_{Θ} in (15) for N- or $\mathcal{N}$-models (Fig. 9 a, 9 b). We note that here the LR plot has two extremal points: Θ_0, η_0 and Θ'_0, η'_0, at which, by (19), the derivatives $d\eta/d\Theta = 0$ and $q_{\Theta} = 0$. This plot consists of three branches of the single-valued function $\Theta(\eta)$: branch I, where $\Theta \leqslant \Theta_0$ and $q'_{\Theta} \geqslant 0$; branch II, where $\Theta_0 \leqslant \Theta \leqslant \Theta_0$ and $q'_{\Theta} \leqslant 0$; and branch III, where, $\Theta \geqslant \Theta'_0$ and $q'_{\Theta} \geqslant 0$ (Fig. 9 a, 9 b). According to (22), the point of intersection of the line $\eta = \eta_{sh}$ with branch II corresponds to the minimum of the potential U_{Θ}, and those with branches I and III to its maxima. This conclusion applies to models with an N-and $\mathcal{N}$-shaped LR (Fig. 9 a, 9 b).

Thus, in the (η'_0, η_0) range of η_{sh} values the potential U_{Θ} has the form of a potential well (Fig. 9 c). Corresponding to these values of $\eta = \eta_{sh}$ are two saddle points of (15), that correlate with the maxima of U_{Θ} (Fig. 9). The values of U_{Θ} at the saddle points $\Theta = \Theta_{s1}$ and Θ_{s3} coincide at a certain value of $\eta_{sh} = \eta_s$, which satisfies, according to (15), the equations

$$\Delta U_{\Theta} = \int_{\Theta_{s1}}^{\Theta_{s3}} q(\Theta, \eta_s, A)\, d\Theta = 0, \quad q(\Theta_{si}, \eta_s, A) = 0, \qquad i = 1, 2, 3 \tag{58}$$

When the saddle points Θ_{s1} (or Θ_{s3}) are approached, the distribution of $\Theta(x)$, by (25), becomes increasingly smoother and at $x \to \infty$ changes asymptotically into the line Θ_{s1} (or Θ_{s3}). As a result, the trajectory of the particle, which practically coincides with the separatrix S in Fig. 9 d describes the distribution of the activator in the form of a broad stratum of size $\mathcal{L}_s/2 >> 1$ at the boundary of the system (Fig. 9 e). It is seen from Fig. 9 a that in the case of N-shaped LR curve at $\eta = \eta_{sh}$ in the range $\eta_0 > \eta_{sh} > \eta_s$ the distribution of $\Theta(x)$, corresponding to the uppermost trajectory of the particle in the potential U_{Θ} (curve 1, Fig. 9 c) has the form of a narrow stratum (Fig. 9 c). In the range $\eta_s > \eta_{sh} > \eta'_0$ the distribution of $\Theta(x)$ (curve 2, Fig. 9 c) has the form of a narrow stratum of decreased activator value (Fig. 9 g). Analogously, in models of morphogenesis with N -shaped LR curves (Fig. 9 b) a narrow stratum (Fig. 9 f) is realised at $\eta_o < \eta_{sh} < \eta_s$ and a narrow stratum of decreased activator value (Fig. 9 g) at $\eta'_o > \eta_{sh} > \eta_s$.

From (16) there follows an approximate expression for determining the width of a stratum in the stationary case ($\mathcal{L}_s/2$) (Kerner, Osipov, 1981):

$$(\mathcal{L}_s/2)Q(\eta_s, \Theta_{s3}, A) = -(\mathcal{L}_x - \mathcal{L}_s/2)Q(\eta_s, \Theta_{s1}, A). \tag{59}$$

This condition is satisfied because the quantity Q has opposite signs on the right and on the left from the plot of the state equation (43) (Fig. 9 a, 9 b). Hence it follows that, if $\Theta_h \leqslant \Theta_{s1}$ (or $\Theta_h \geqslant \Theta_{s3}$), condition (59) is a priori not satisfied, i. e. it is violated at certain $A > A_s$ (or $A < A'_s$). The values of A_s and A'_s are found from equations (58) in which we put $\eta_s = \eta_h$, and $\Theta_{s1} = \Theta_h$ or $\Theta_{s3} = \Theta_h$, respectively (Fig. 10). When A is close to A_s (or A'_s), equation (16) is satisfied only by a narrow stratum (Fig. 9 f, or 9 g) (Kerner, Osipov, 1981). According to (59), at a given A the size of a broad stratum $\mathcal{L}_s/2$ (Fig. 9 e) increases with $\mathcal{L}_x$. For that reason, the larger $\mathcal{L}_x$ the larger is the range of variation of A for broad strata and the smaller is the range for narrow strata (Kerner, Osipov, 1981). The true region of existence of broad and narrow strata is determined by their stability (Kerner, Osipov, 1978, 1979, 1980, 1981). The critical width of a stratum can be estimated from (53), where (Kerner, Osipov, 1979, 1981)

$$\begin{aligned} &\lambda_o \sim -\exp(-\min(\mathcal{L}_x/2, \mathcal{L}_x - \mathcal{L}_s/2)/\ell), \\ &P = a_o \approx |\langle q'_h \rangle_{sh} \langle Q'_\Theta \rangle_{sh}| \, \ell \mu_o^{-1} (\ell/\mathcal{L}_x) \sim \ell/\mathcal{L}_x. \end{aligned} \tag{60}$$

From (53) and (60) we obtain estimates for the critical width

$$\mathcal{L}_s(A_b) = \mathcal{L}_b \sim l \cdot \ln(\mathcal{L}_*/l); \quad 2\mathcal{L}_* - \mathcal{L}_s(A'_b) = \tilde{\mathcal{L}}_b \approx \mathcal{L}_b \tag{61}$$

where A_b and A'_b are values of A, at which $d\eta_{sh}/dA = \infty$ (Kerner, Osipov, 1979, 1981).

Thus, at the point $A = A_c$ there appears in a jumpwise fashion (jump $0 \rightarrow 1$, Fig. 10) a broad stratum (Fig. 9 e) as a result of loss of stability of the homogeneous state. With decreasing A, according to (59), the width of the stratum $\mathcal{L}_s/2$ decreases and at $A = A_b < A_o$ the stratum vanishes jumpwise (jump $2 \rightarrow 3$, Fig. 10) (Kerner, Osipov, 1981). In contrast, when A increases (from $A = A_c$) the stratum becomes wider and at $A = A'_b > A'_o$ it vanishes jumpwise (jump $2' \rightarrow 3'$, Fig. 10) (Kerner, Osipov, 1981).

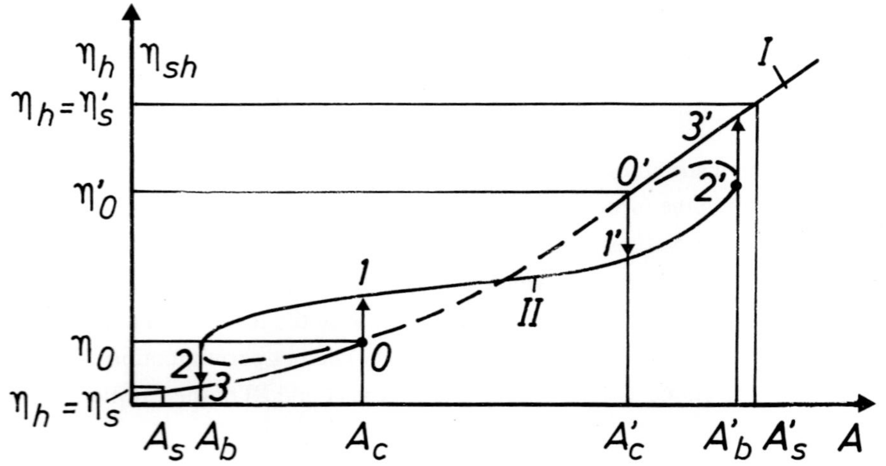

Fig. 10 Illustration of stratum evolution. Curve I corresponds to the homogeneous state, curve II to the stratum shown in Fig. 9 e.

Periodic strata form and evolution in large-size systems

The form of strata in N- and $\mathcal{N}$ -models is described by equations (40), (41). The potential U_η in equation (42) for smooth distributions has three independent branches I, II and III corresponding to $\Theta \leqslant \Theta_0$, $\Theta_0 \leqslant \Theta < \Theta'_0$ and $\Theta \geqslant \Theta'_0$, respectively. As an example, Fig. 11 shows the construction of the potential U_η for a $\mathcal{N}$ -model, (by (39), on the right of the plot of the state equation (43), where $d\eta < 0$, the value $Q < 0$ and, therefore, according to (44), $dU_\eta/d\eta > 0$; and on the left $Q > 0$ and $dU_\eta/d\eta < 0$). At the extremal points of the LR plot η_0 and η'_0 (Fig. 11 a), corresponding to the boundaries of the branches of U_η, the derivatives $dU_\eta/d\eta$ of the appropriate branches coincide, but the potential U_η may have arbitrary values (Fig. 11 b). Corresponding to the smooth distributions $\Theta(x)$, $\eta(x)$ are only motions of the particle in those branches of U_η where $q'_\Theta > 0$ (Kerner, Osipov, 1978, 1979, 1980, 1981). This condition is satisfied by branches I and III of the potential U_η. In spite of the fact that these branches intersect (Fig. 1 b), finite motion of a particle with a coordinate η in such a 'potential well' is impossible because corresponding to the point of intersection of branches I and III are different values $\Theta < \Theta_0$ and $\Theta > \Theta'_0$.

Among the abrupt distributions $\Theta(x)$ there is a unique distribution at $\eta_{sh} = \eta_s$ which satisfies (58) in the form of a half-oscillation of $\Theta(x)$, whose extremal points Θ_{s1} and Θ_{s3} correspond, respectively, to branches I and III of the LR plot (Fig. 11 a). These points are, at the same time, saddle points of the potential U_Θ for the abrupt distribution (dashed curve, Fig. 11 d). In accordance with the results of the last section, all other potentials constructed at $\eta \neq \eta_s$ (Fig. 9 c) do not permit motion on those trajectories that are closed simultaneously on branches I and III of the LR plot.

Therefore, only the rapidly varying distribution $\Theta(x)$ at $\eta = \eta_s$ (Fig. 9 a) can go over near the point Θ_{s1} into the smooth distribution, corresponding to motion of a particle in branch I of the potential U_η, and near Θ_{s3} into the smooth distribution corresponding to branch III of U_η (Fig. 11).

Let, at the point x = 0, the values $\eta = \eta_m$ and $\Theta = \Theta_m$ correspond to branch III of the LR plot (Fig. 11 a, 11 e). With increasing x, the value η (x) decreases and $\Theta(x)$ increases tending to $\eta \simeq \eta_s$ and $\Theta_{max} \approx \Theta_{s3}$ in accordance with the dependence $\Theta(\eta)$ for branch III. In the range $x = \mathcal{L}_s/2$, the value of Θ falls abruptly along the length ~ 1 to about Θ_{s1}, while η changes only by a value of the order of $\varepsilon = 1/L << 1$ (the region of the abrupt change of the activator Θ is called the wall of a stratum). With further increase in x, η continues to decrease and Θ to grow in accordance with the dependence $\Theta(\eta)$ for branch I of the LR curve (Fig. 11 a). From the theory of singular perturbations (Kerner, Osipov, 1985 a) it follows that, (accurate to $\varepsilon << 1$), $\Theta_{max} = \Theta_{s3}$, $\Theta_{min} = \Theta_{s1}$, $\eta(\mathcal{L}_s/2) = \eta_s$, i. e. these values are found from (58) and the values $\mathcal{L}_s$, η_m, Θ_m, η_t, Θ_t from the relations

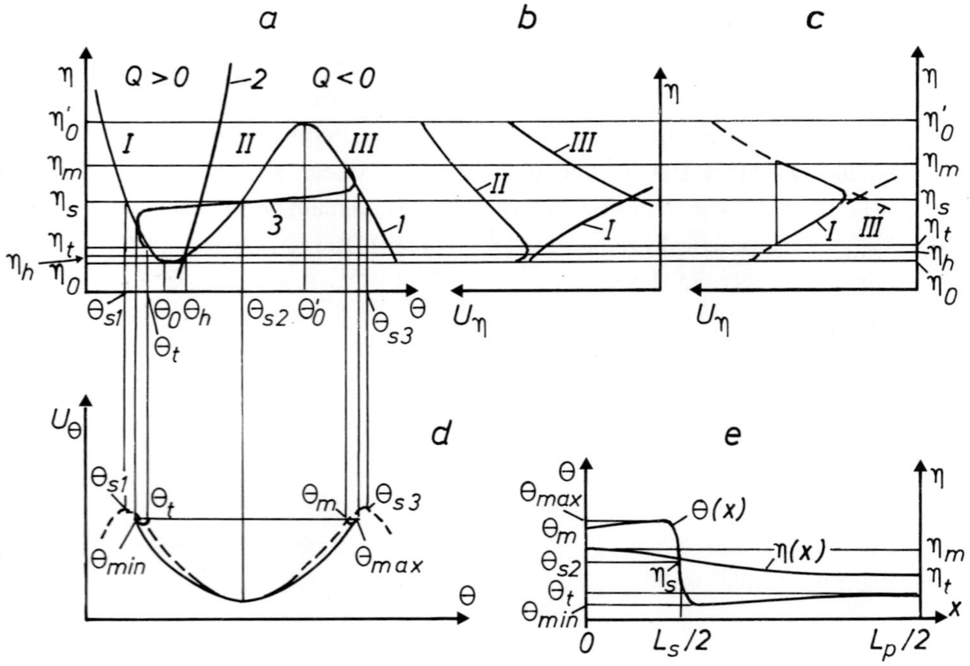

Fig. 11 See text.

$$\int_{\eta_s}^{\eta_t} Q_I \, d\eta + \int_{\eta_m}^{\eta_s} Q_{III} \, d\eta = 0, \quad \mathcal{L}_s = \sqrt{2} L \int_{\eta_m}^{\eta_s} \Big[\int_{\eta_m}^{\eta} Q_{III} \, d\eta \Big]^{-1/2} d\eta, \tag{62}$$

$$\mathcal{L}_p - \mathcal{L}_s = \sqrt{2} L \int_{\eta_s}^{\eta_t} \Big[\int_{\eta_t}^{\eta} Q_I \, d\eta \Big]^{-1/2} d\eta, \quad q(\eta_m, \Theta_m, A) = q(\eta_t, \Theta_t, A) = 0, \tag{63}$$

where $Q_j \equiv Q(\eta, O_j(\eta), A)$, the function $Q_j(\eta)$ is determined by (18), and the subscript $j = I, III$ indicates the portion of the LR curve where the function $\Theta(\eta)$ is single-valued (Fig. 9 a, 9 b).

Based on the concept of the true function $\eta(\Theta)$ (or $\Theta(\eta)$) (curve 3, Fig. 11 a), it is possible to construct (Kerner, Osipov, 1980) the form of the true potentials U_Θ (40) and U_η (41) (solid curves in Fig. 11 c, 11 d, respectively), in which conservative motion of a classical particle with coordinates O and corresponds to the distribution of $\Theta(x)$ and $\eta(x)$ in a stratum (Fig. 11 e). Periodic strata (Fig. 1 b) in N- and И -systems lose their stability due to the pumpover effect (Kerner, Osipov, 1979, 1980, 1981). The instability criterion determining the minimum period of stable strata $\mathcal{L}_p = \mathcal{L}_{min}(A)$ is practically reduced to (53) and (54) (Kerner, Osipov, 1979, 1980, 1981). Generally, in N- and И -systems $|q'|$, $|\Theta'_\Theta|$, $Q'_\eta < 1$ therefore $a_0 \sim 1/\mathcal{L}_p$, and (Kerner, Osipov, 1979, 1983 b)

$$\lambda \sim -\varepsilon(\mathcal{L}_s/L) - \exp(-\mathcal{L}_s/\ell). \tag{64}$$

Equations (53), (54) and (64) at $\mathcal{L}_p < L$ provide an estimate for the width of periodic strata (Fig. 1 b) at which stability is lost (Kerner, Osipov, 1985 a)

$$\mathcal{L}_s(A_\ell^{(N)}) \equiv \mathcal{L}_\ell^{(N)} \sim \ell \cdot \ln(\varepsilon^{-1}(L/\mathcal{L}_p)), \tag{65}$$

where $A = A_b^{(N)}$ is the bifurcation value of A for N strata of period $\mathcal{L}_p^{(N)} = \mathcal{L}_x/N$. From (65) and the relation between $\mathcal{L}_s$ and $\mathcal{L}_p$ resulting from (63), we can find the value $\mathcal{L}_p = \mathcal{L}_{min}$ corresponding to $A = A_b^{(N)}$. It follows from (63) and (65) that, as a rule, strata of period $\mathcal{L}_o = 2\pi k_o^{-1} \sim (l/L)^{1/2}$ in models with N- or И -shaped LR are unstable. Therefore, it is most probable that in the layering of the homogeneous state, i. e. at the point of growth of the critical fluctuation with $k = k_o$, a large-amplitude structure of period $\mathcal{L}_o$ in the form of $N = \mathcal{L}_x/\mathcal{L}_o$ periodically spaced broad strata will occur jumpwise (jump 0 → 1, Fig. 12).

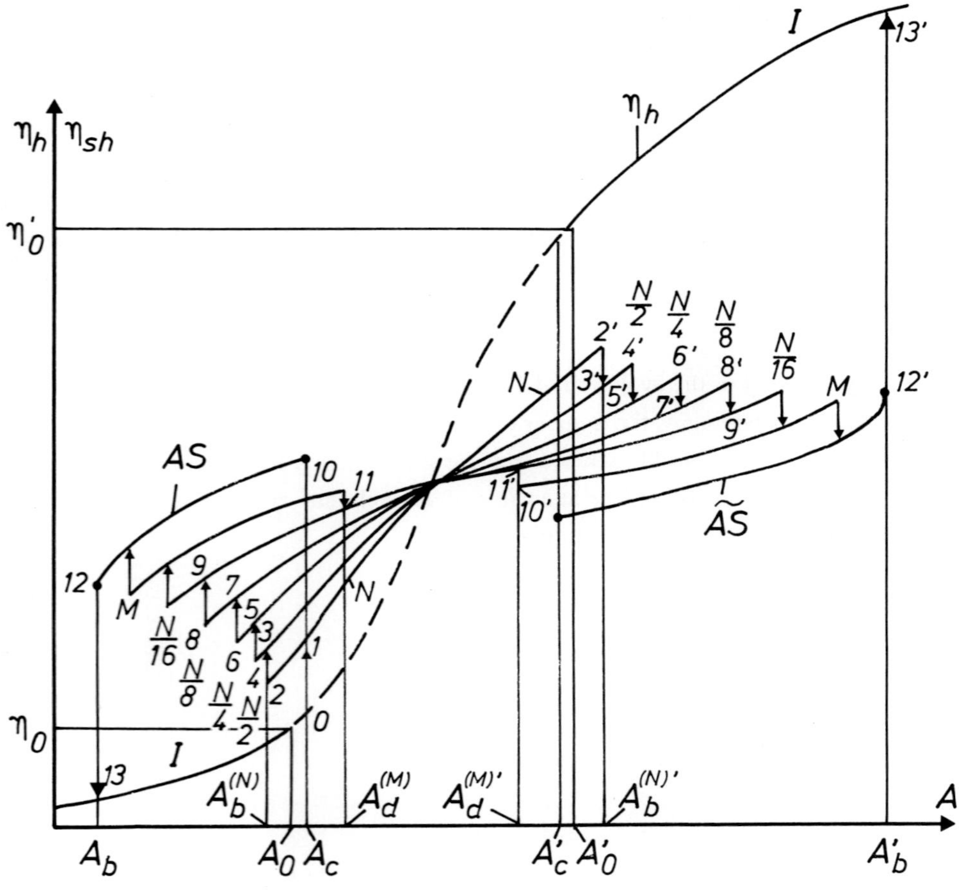

Fig. 12 Illustration of the evolution of periodic strata. Curves I correspond to the homogeneous state, AS and AS to the autosolitons shown in Fig. 2 b and 2 c, respectively, and numbered curves (N, N/2, ... N/16, M) to the periodic strata shown in Fig. 1 b.

The width of the strata decreases with A and at a certain $A = A_b^{(N)}$ (Fig. 12) they become unstable to the pumpover effect. As a result of a series of such period-doubling bifurcations (jumps $2 \rightarrow 3, ..., 8 \rightarrow 9$, Fig. 12), the number of strata in the system decreases jumpwise, and a single stratum, AS, may spontaneously arise (next section).

With A growing from $A = A_c$, the width of the strata $\mathcal{L}_s$ increases and they become more and more like narrow strata with a decreased activator value (Fig. 1 c), i. e. strata of width $\tilde{\mathcal{L}}_s \equiv \mathcal{L}_p - \mathcal{L}_s \ll \mathcal{L}_s$. At $A = A_b^{(N)}$, where their width $\tilde{\mathcal{L}}_s = \mathcal{L}_p - \mathcal{L}_s(A_b^{(N)}) \approx \mathcal{L}_b^{(N)}$ (65), these latter lose stability to the pumpover effect (jump $2' \rightarrow 3'$, Fig. 12). The succession of such period-doubling bifurcations (jump $4' \rightarrow 5'$, $8' \rightarrow 9'$, Fig. 12) reduces jumpwise the number of strata in the system.

The above results, which follow from the qualitative theory (Kerner, Osipov, 1979, 1980, 1981, 1985 a) were confirmed in detail by study of an axiomatic model (Kerner et al., 1984 a, b, c).

Note the peculiarity of the evolution of M strata of period $\mathcal{L}_p > L$ ($M << N = \mathcal{L}_x/\mathcal{L}_o$. With varying A, the number of such strata in the system may increase jumpwise as a result of local break between the strata (jumps $10 \rightarrow 11$ and $10' \rightarrow 11'$, Fig. 12). This situation takes place at a value of A such that η_t reaches η_o or η_m reaches η'_o (Fig. 11) (Kerner, Osipov, 1979).

Autosolitons in one-dimensional systems

In a stable system at $A < A_o$ ($\Theta_h < \Theta_o$) the extremum of the potential U_η (42) is on branch I (see above). $q'_\Theta > 0$ corresponds to a stable system, therefore, by (45) this extremum ($\eta = \eta_h$, $\Theta = \Theta_h$) is maximum. It corresponds to the saddle point of equation (42). The trajectory of a particle closing at this point, corresponds to the solution $\Theta(x)$, $\eta(x)$, which at $x \longrightarrow \infty$ tends to Θ_h and η_h for the stable homogeneous state of the system. The AS is symmetrical relative to the point $x = 0$. Thus, using the arguments from above for the range $0 < x < \mathcal{L}_s/2$ (Fig. 11), we arrive at the distribution $\Theta(x)$ and $\eta(x)$ in the AS shown in Fig. 2 b.

The AS (Fig. 2 a, 2 b) exists in the range $[A_b, A_c]$ (Fig. 12, curve AC). It vanishes jumpwise at $A = A_b$, when $d\eta_{sh}/dA = \infty$ and the system goes over into a homogeneous state (jump 12 → 13, Fig. 12). The critical width of AC at the point $A = A_b$ can be estimated from (53) (Kerner, Osipov, 1979, 1980, 1981), using the values of λ from (64) and $P \sim \varepsilon$:

$$\mathcal{L}_s(A_b) \equiv \mathcal{L}_b \sim \ell \ln \varepsilon^{-1} . \qquad (66)$$

The form and the properties of AS described by Kerner and Osipov (1979, 1980, 1981) are confirmed by studying axiomatic N- and $\mathcal{N}$ -models, which permits an analytical solution of the problem (Chernavskii, Ruijgrok, 1982; Kerner et al., 1984 b; Koga, Kuramoto, 1980; Romanovskii et al., 1984). At $A'_c < A < A'_b$ (Fig. 12, curve $\widetilde{AS}$), it is possible to excite an AS with a decreased activator value (Fig. 2 c). Such an AS having the width $\widetilde{\mathcal{L}}_s = \widetilde{\mathcal{L}}_b \sim \mathcal{L}_b$ (66) vanishes jumpwise at $A = A'_b$ (jump 12' ⟶ 13', Fig. 12).

The simplest structures in two and three dimensions

In two- and three-dimensional systems a one-dimensional broad AS (Fig. 2 b) exists in a smaller range $[A_{b1}, A_{c1}]$ of variation of A than in one-dimensional systems. At the limits of this range, the one-dimensional AS loses stability with respect to the growth of inhomogeneous fluctuations with a wave number $k_\perp \sim \varepsilon^{1/4} (\ell L)^{-1/2}$ of the activator (Kerner, Osipov, 1980); these are localised in the AS walls. The growth of such fluctuations in a two-dimensional system leads to a goffering of the AS walls, and in a three-dimensional system to the appearance on the walls of a cellular structure or to a break up of the AS into smaller regions (Kerner, Osipov, 1980). The effect of layering of the walls is characteristic of broad periodic strata (Fig. 1 b) and of AS with a decreased activator value (Fig. 2 c). The relevant critical values for the width of AS and periodic strata were described by Kerner, Osipov (1985 a).

In a three-dimensional system at $A < A_0$, it is possible to excite a radially symmetrical hot AS in the form of a drop or hollow sphere, and in a two-dimensional system in the form of a spot or ring (Kerner, Osipov, 1980). In a hot drop of radius $\rho_o \sim L$, the extremal values $\Theta_{max} = \Theta_{s3}$, $\Theta_{min} = \Theta_{s1}$ and the value of η_{sh} in the drop wall, $\eta(\rho_0) = \eta_s$, are determined (accurate to ε) from equations (58) (Kerner, Osipov, 1985 b). In the systems considered there may occur radially symmetrical structures in the form of concentric hollow spheres (rings) and also of periodically spaced drops (spots). The distance between the drops R_p is limited by the values $R_{min}(A)$ and $R_{max}(A)$ which are determined, as in the one-dimensional case, from the conditions for instability of the type of pumpover effect or local break between the drops (spots) (Kerner, Osipov, 1980, 1981, 1982 a, 1985 a).

The general form and properties of structures in two and three dimensions

Structures in two- and three-dimensional systems may take an extremely complicated form. The inhibitor $\eta(\vec{r})$ varies smoothly in space with the characteristic length of the order of L, while the distribution of the activator $\Theta(\vec{r})$ shows quite a different pattern: in some regions Θ varies sharply from $\Theta \simeq \Theta_{s1}$ to $\Theta \simeq \Theta_{s3}$ (58) along the length $l << L$ (Kerner, Osipov, 1980). In regions of smooth variation of $\eta(\vec{r})$ and $\Theta(\vec{r})$, these values satisfy (accurate to $\varepsilon = l/L << 1$) equation (42) and correspond to branches I ($\Theta(\vec{r}) < \Theta_0$) and III ($\Theta(\vec{r}) > \Theta'_0$) of the LR plot (Fig. 9 a, 9 b). The unstable region of the activator parameters ($\Theta_0 < \Theta(\vec{r}) < \Theta'_0$, Fig. 11 a) is located over a narrow range of size, $\sim l$, in the surface layers where there is abrupt variation in $\Theta(\vec{r})$ between different regions ($\Theta < \Theta_0$ and $\Theta > \Theta'_0$) of its smooth variation. A stable structure cannot contain a fragment of the unstable region ($\Theta_0 < \Theta < \Theta'_0$) within one of the stable regions, whose size greatly exceeds l in at last one direction (Kerner, Osipov, 1980). The requirements of the form of a stable structure (Kerner, Osipov, 1980) explain the results of numerical studies of two-dimensional structures in an axiomatic model (Balkarei et al., 1980, 1983). A structure of a given form exists over a certain range of variation in the parameter A (Kerner, Osipov, 1980). At the limits of this range the structure disappears as a result of different types of instability (pumpover effect, goffering of the surface layers of the abrupt variation of $\Theta(\vec{r})$ or local break between the surface layers).

Connection between the type of a structure and model parameters

In the above analysis of structures we used models in which the LR plot has a V-, Λ -, N- or И -shaped form (Fig. 3 a, 3 b, 9 a, 9 b). Such forms are not accidental, but are inherent in the very nature of the Turing instability and can be deduced from general considerations (Kerner, Osipov, 1980, 1981). It is convenient to classify models by the form of their LR plots as V-, Λ -, N- and И -systems. The N- and И -systems in which the LR and SE plots intersect at three points, are generally called trigger systems. The type of structure is also determined by the values $\varepsilon = l/L$ and $\alpha = \tau_\Theta/\tau_\eta$. On this basis, all the models can be divided into three qualitatively different groups (Kerner, Osipov, 1985 a, c):

K-systems,	for which	$\varepsilon << 1,$	$\alpha > 1$
Ω -systems,	for which	$\varepsilon > 1,$	$\alpha << 1$
$K\Omega$-systems,	for which	$\varepsilon << 1,$	$\alpha << 1.$

In K-systems[7)] at $A = A_c$ there occurs layering of the homogeneous state, and static large-amplitude structures (in one-dimensional case, strata) arise jumpwise (jump $0 \rightarrow 1$, Fig. 6, 12) (Kerner, Osipov, 1978, 1979, 1980, 1981). Small amplitude quasi-harmonic structures never arise in such systems. Large-amplitude structures also exist at $A < A_c$; the critical value $A = A_b < A_c$, below which these structures are not realised, corresponds to that point where $d\eta_{sh}/dA = \infty$ (Fig. 6, 12) (Kerner, Osipov, 1978, 1981). In $K\Lambda$ - and KV-systems the value of A_b decreases with ε . At $A < A_c$, a short-term local perturbation of finite amplitude can excite, in homogeneous stable systems, a stationary solitary state, a static autosoliton (Fig. 2) (Kerner, Osipov, 1978, 1979, 1980, 1981, 1985 c).

It is possible to excite a great many different structures in a system at the same parameter values - not necessarily periodic, but also complicated (Fig. 1 a, 1 e) and stochastically inhomogeneous structures (Kerner, Osipov, 1978, 1979, 1980; Kerner et al., 1984 b, c). In K-systems, along with static structures, turbulence may also occur spontaneously (Kerner, Osipov, 1982 b, 1983 a).

In the one-dimensional $K\Lambda$ - and KV-systems, including the GM model (3) and the Brusselator (4), narrow (of size $\sim l$) large-amplitude strata (Fig. 1 a) are formed (see, however, the footnote 5). At the point of layering of the homogeneous state there appear jumpwise (jump 0 1, Fig. 6) strata of period $\mathcal{L}_p \simeq L$. The period may appreciably exceed the period of critical fluctuations $\mathcal{L}_o = 2\pi k_o^{-1} = 2\pi(l\,L)^{1/2}$ that give rise to an instability (see above). With growth of the bifurcation parameter A, the number of strata increases jumpwise (Kerner, Osipov, 1978), as a rule, because of dynamic rather than fluctuational rearrangement of the system, i. e. due to burst-like

---

7) Systems with nondiagonal diffusion in which structure formation is due to the nondiagonal coefficients $L_{ij}(i \neq j)$ in (1) being dependent on X_i also belong to K-systems.

appearance of a new stratum between two strata (Kerner, Osipov, 1982 b). Thus, in contrast to existing opinion (Nicolis, Prigogine, 1977), self-organisation may take place in a strictly deterministic system (Kerner, Osipov, 1982 b) unconnected with the growth of the critical fluctuations. When A decreases, the number of strata decreases jumpwise due to fluctuational rearrangement, i. e. growth of critical fluctuations of double period (Kerner, Osipov, 1978).

In two- and three-dimensional strata, narrow strata (Fig. 1 a) and AS (Fig. 2 a) are unstable. Therefore, structures in two- and three-dimensional V- and Λ-systems can only be of the form of point regions of large values of Θ with a size ~ 1 (Fig. 7) resembling in cross-section a narrow stratum and spaced at a distance $\sim L$ (Kerner, Osipov, 1980, 1982 c).

In one dimensional KN- and K$\mathcal{N}$ -systems, broad and narrow strata are formed (Fig. 1) (Kerner, Osipov, 1979, 1980, 1981). Strata exist in the range $A_b < A < A'_b$. Where $A = A_b$ and $A = A'_b$ there exist points at which $d\eta_{sh}/dA = \infty$ (Fig. 12). The size of broad strata $\mathcal{L}_s$, the values Θ_{max}, Θ_{min} and $\eta = \eta_s$ in strata walls can be found from simple algebraic equations (58), (62), (63). In the two- and three-dimensional cases, broad strata are stable (Kerner, Osipov, 1980, 1985 a). In two- and three-dimensional KN- and K$\mathcal{N}$ -systems, it is possible to excite, in addition to broad strata, structures of extremely complicated form (see above) (Kerner, Osipov, 1980).

In the trigger KN- and K$\mathcal{N}$ -systems, two reversible waves of switching can be excited, which switch the system from one stable state into another: from state (Θ_{h1}, η_{h1}) to state (Θ_{h3}, η_{h3}) or vice versa. Such alternately switchable stable states allow excitation in the system of quite arbitrary successions of travelling pulses with different widths (Kerner, Osipov, 1985 b).

The homogeneous state of a heated Ω-system at $A = A_c$ loses stability to homogeneous fluctuations (with condition (7) satisfied) with a dominating frequency $\omega \simeq (\tau_\Theta \tau_\eta)^{-1/2}$ and, as a result, relaxation oscillations are set up jumpwise (Nicolis,

Prigogine, 1977; Andronov et al., 1981, Ebeling, 1976; Haken, 1978). At $A < A_c$ over a certain range of stability of the homogeneous state in Ω-systems, it is possible to excite a travelling AS (pulse) or an AS in the form of more complicated autowaves, including a reverberator (spiral wave) (Ivanitskii et al., 1981; Krinskii, Zhabotinskii, 1981; Polak, Mikhailov, 1983; Romansvskii et al., 1984; Vasil'yev et al., 1979).

In $K\Omega$-systems, there can be driven or arise spontaneously static, pulsating, travelling and other, more complicated, structures (Kerner, Osipov, 1982 b, 1983 a, b, 1985 a, b, c). The properties of static structures in $K\Omega$-systems with $\varepsilon << \alpha << 1$ and in K-systems coincide. In $K\Omega$-systems for $\alpha \leqslant \varepsilon << 1$ static structures turn out to be less stable than those in K-systems. They lose stability with respect, to pulsations, i. e. drops, spots and other fluctuations localised in the strata walls, whose amplitudes vary periodically with time. When the amplitude grows, there may appear pulsating structures, for instance, a pulsating AS (PAS) whose volume or form varies periodically with time (Kerner, Osipov, 1982 b, 1985 c). In addition to pulsating and static structures, travelling AS (TAS) and other types of autowaves can be excited in $K\Omega$-systems with $\varepsilon^2 < \alpha \leqslant \varepsilon << 1$. Thus, the $K\Omega N$- and $K\Omega И$-systems with $\varepsilon^2 < \alpha \leqslant \varepsilon << 1$ exhibit the greatest variety in pattern formation: with the same value of A, it is possible to excite in them a static AS (SAS), PAS, TAS and other more complicated structures (Kerner, Osipov, 1983 b, 1985 a, c). The theory of static, pulsating and travelling structures in $K\Omega$-systems was developed by Kerner and Osipov (1982 b, 1983 b, 1985 a, b). Pulsating AS have recently been found and studied numerically by Gafüchuk et al. (1986) and Osipov et al. (1987). A single pulsating drop (spot or stratum) can show up as a source of radiating waves from it without attenuation (Kerner, Osipov, 1982 b), i. e. this can be a leading centre such as that discovered experimentally in homogeneous systems (Romanovskii et al., 1984; Vasil'yev et al., 1979).

Travelling AS in $K\Omega$-systems may differ in their properties from those in Ω-systems. While in the latter two travelling AS annihilate on collision (Ivanitskii et al., 1981; Polak et al., 1983; Romanovskii et al., 1984; Vasil'yev et al., 1979), in $K\Omega$-systems they may instead change into a static or pulsating AS or a more complicated structure. This is due to the fact that at the forewall of the travelling AS in such systems $\eta \neq \eta_h$ and

$\Theta = \Theta_h$ (η_h and Θ_h are the homogeneous values of the inhibitor and the activator), i. e. there exists a region of size of L, where the values of η and Θ vary smoothly to η_h and Θ_h (Kerner, Osipov, 1981, 1983).

If an inhomogeneous fragment is excited in a certain region of a K- or $K\Omega$-system, it will either attenuate or will give rise to a stable structure in the entire system (self-adjustment effect) (Balkarei et al., 1983; Krinskii, 1984). The form of stable structures for given 'external' parameters A, B, ... is determined by the integral relations, by the type of the nonlinearity inherent in the system (LR and SE plots), and by the stability conditions. Thus, specification of the homogeneously distributed parameters A, B, ... means storage of an entire set of definite patterns, which can be reconstructed, with some probability, by one of their fragments. In this sense the systems under study may be considered as having associative memory (Kerner, Osipov, 1980).

Conclusion

This work presents the results of a nonlinear theory of large-amplitude dissipative structures being driven or arising spontaneously in various models of morphogenesis.

The general character of Turing's mechanism for the formation of static structures lends the theory great weight and attraction: the same basic models can explain the occurrence of order and structures in a variety of natural and technical phenomena. Turings's structures are considered by theorists as basic mechanisms for morphogenesis. It is generally accepted that a great variety of morphogenetic processes can be explained by the hierarchy of structures generated by this mechanism. However, facts to support this view are lacking. No experiments have been performed (or, at least, proposed) that could serve as a single-valued proof of this mechanism (Belintsev, 1983).

At the same time, experimental data are being accumulated which suggest an alternative approach. The most familiar example is morphogenesis in *Dictyostelium discoideum*. This slime mould can exist both as free living amoeba and as a colony forming multicellular organism. The initial stages of its formation, as shown by the splendid experiments of Gerisch and Durston are determined by interaction of cAMP waves propagating in the aggregation field from different sources (Durston, 1974; Gerisch, 1968). These waves belong to the class of autowaves (Ivanitskii et al., 1978, 1981; Krinskii, Zhabotinskii, 1981), strongly nonlinear waves that greatly differ in their properties from the types of waves studied so far. As is known, autowaves do not posess such properties as reflection and interference nor do they penetrate one another on collision but annihilate, in contrast to ordinary waves. As a result of the interaction between the cAMP waves and of their annihilation, some sources of these autowaves die away and the ones that persist become centres of aggregation fields and, later on, centres of growth of multicellular organisms.

Another fascinating example of pattern formation is furnished by colonies of *E. coli*. Budrene (1985) has observed that when two colonies of *E. coli* are grown on half-fluid agar, there appear rather wide ($\sim$5 mm) 'demarcation zones' between them. Within the framework of the existing concepts, it would be natural to explain their formation by one of the modifications of Turing's mechanism. Skulatchev, however, has hypothesised that in this case, as well as in *D. discoideum*, the mechanism can be linked to autowaves. There was no hypothesis as to how the propagating autowaves could lead to a static demarcation zone of considerable width. Agladze and Tzyganov in their experiments have shown that the formation of patterns is here based on the interconversion of two different population types of autowaves (Adler's waves (1966)) and on their annihilation (Agladze et al., 1987; Ivanitskii et al., 1987). It is in those places where the propagating autowaves annihilate that a static structure appears.

Note that autowaves, just as Turing structures, are quite a general phenomenon and are characteristic of a large class of basic models (Ω - and KΩ -models, see last section) (Ivanitskii et al., 1978; Krinskii, Zhabotinskii, 1981; Polak, Mikhailov, 1983; Vasil'yev et al., 1979).

Both Turing's mechanism and autowave mechanisms are now receiving extensive theoretical study. Experimental data on morphogenesis is also increasing. Planning and performance of experiments which could identify the mechanisms predicted theoretically, determine the relations between them, and new problems for the theorist are now of even greater importance.

Self-Organisation of Biological Morphogenesis: General Approaches and Topo-Geometrical Models

L. V. Beloussov, A. V. Lakirev

Introduction

The development of organisms may be considered as a self-organisation process, since it involves a spontaneous (lacking any pre-established external or internal blueprint) complication of macrostructure. The complication means here decrease in symmetry order, topological changes (loss of homeomorphism) and the appearance of new geometrical and structural properties. The latter processes, to which cell differentiation belongs, may be considered formally as 'colour' dissymetrisation.

The earliest firm empirical evidence of the self-organising properties of biological morphogenesis was embryonic regulation discovered by Hans Driesch almost a century ago (Driesch, 1892). This author argued that the self-organising properties should be the specific characteristics of only living systems. Recently it was proved that the same general properties belong to a number of nonbiological systems as well and a general physical-mathematical theory of self-organisation (often designated as synergetics) has

Thermodynamics and Pattern Formation in Biology

been created (Haken, 1978; Prigogine, 1980; Romanovskii et al., 1984). We believe that the description and interpretation of biological morphogenesis in the framework of synergic concepts is one of the most important and real tasks of modern developmental biology (see also Beloussov, 1984; Beloussov et al., 1985). In this paper we discuss first the general principles of such a description and later on suggest some particular models of morphogenetic processes. By stressing the mechano-geometrical properties of embryonic tissues our approach is somewhat similar to that of Odell (1984), Odell et al. (1981); Oster et al. (1983).

The dynamical structure of self-organising systems

Each system capable of self-organisation should contain variables of distinctly different characteristic times. Let us design such systems as multilevelled, the greater the characteristic time, the upper being the level of a given variable. The lowest level is that of dynamic variables, directly determining the visible system's behaviour. The upper levels belong to parameters (as a rule, there may be several hierarchically arranged parameters of different characteristic times) that are just the values which determine the set of potentially excited stable and nonstable states of a system, the general 'style' of its behaviour and its very capacity for self-organisation. Thus, the parametric (directed from the upper levels) regulation is of primary importance for self-organising systems. On the other hand, if for given parameter values there exist more than one stable state the ultimate state selection is determined (as a rule, with great degeneracy) by the initial conditions, that is, by the momentary values of dynamic variables (dynamic regulation).

The basic notions of classical embryology, such as competence, determination, differentiation and induction may be precisely and exhaustively expressed in the terms of self-organisation theory. Namely, the competent state may be defined as that possessing (depending upon the parameter values) more than one stable state;

determination means reaching an 'attraction basin' of a new (noninitial) stable state and differentiation the ultimate achievement of this state. Correspondingly, the appearance (or disappearance) of a competence should be considered as a parametrical event whereas determination and differentiation are dynamic events. Induction may be considered, in its broad sense, as either a parametric or a dynamic shift. There is some evidence in favour of treating at least primary (Spemann's) induction as a parametric event (Beloussov et al., 1985).

A wide range of self-organising systems is capable of generating autooscillations and space-deployed autowaves. Autooscillations are generated in two-component systems with their variables to be connected by "+", "-" feedback relations. Both variables should be of sharply different characteristic times. In addition, to obtain oscillating behaviour, defined parameters values are required. Thus, autooscillating systems should be no less than three-levelled. The space-deployment of self-organisation waves is usually created by introducing into the corresponding equation a diffusion member, describing short-range transmission of some activating substance or physical influence. More regular spatial patterns can be obtained by introduction, in addition, of certain long-range diffusing chemical or physical factors, possessing inhibitory properties (Meinhardt, 1982; Belintzev et al., 1985 a, b).

Levels in biological morphogenesis

At first glance, it seems impossible to subdivide the entire multiplicity of ontogenetic variables into a limited number of descrete levels. Meanwhile recent results in cell biology permits one to outline a single category of interrelated events with the small (in a biological scale) characteristic times (about 10^0 - 10^2 s) which may be defined as dynamic variables of morphogenetic processes. These are what one may define as *cytoskeletal-membrane transformations* (CMT): aggregate transformations of actin and actin-bound proteins, lateral mobility of membrane components, endo-exocytosis, recirculation of vesicles and others. It is exactly these processes that create real morphogenetic forces.

It is convenient to begin the analysis of the upper level variables (parameters) from those of greatest characteristic times. One of, and perhaps the most important of these, slowest variables is the genome: in most Eukaryota the reverse frequency of spontaneous mutability is 4 - 6 times greater than the life cycle time (Dobjansky, 1970). The parametric function of the genome has been suggested and substantiated by several authors (and especially by R. Goldschmidt, C. Waddington and D. Chernavskii; for review see Beloussov et al., 1985). However, this function is realised in an indirect way, via a number of chemical and physical mediators introducing into the system their own characteristic times of synthesis, degradation and/or relaxation.

As to chemical agents, the most important for early embryogenesis are embryonic inductors, triggers of genetic expression, cytoskeleton regulatory proteins etc. The characteristic times of their accumulation up to a functionally active level are on the order of hours to dozens of hours (10^3 - 10^5 s). We designate this group of variables as biochemical parameters.

A bilevelled system 'biochemical parameters - CMT' may be considered as one of the simplest capable of morphogenetic self-organisation and of the creation of new intermediate levels. This can be illustrated by a model of self-organisation of polarised cell domains (Belintzev et al., 1985 a, b). In this model the dynamic variable belongs to CMT and is the wide-spread process of cell polarisation with a characteristic time of a few minutes. The model equations also include a quasi-diffusional member expressing the transmission of polarisation capacity from one adjacent cell to another. This transmission process (the so-called *contact cell polarisation* - CCP) has been observed both in cell cultures (Middletone, 1977) and in intact embryonic tissues (Beloussov, 1980; Beloussov et al., 1975).

The parameters which regulate the model's behaviour and may be qualified as biochemical are the diffusion (CCP-) coefficient D, the ε value characterising the intercellular force directed towards further cell polarisation and a certain value measuring the stability of a cell's polarised state. If some combination μ of these parameters exceeds a threshold value an initially homogeneous epithelial layer consisting of nonpolarised cells will start to become segregated to a domain of the markedly polarised and to a domain of tangentially stretched cells. This leads to the formation of a regular *tensile field* consisting of these two domains. The tensile fields of such a basic topology may be traced in the normal development of amphibian embryos. They create a new important level which should be placed, according to the characteristic times of successive field existence, between the levels of biochemical variables and of CMT.

Tensile fields in amphibian development and their morphogenetic role

In the embryonic development of amphibians from late blastula up to the tail-bud stage several successively appearing tensile fields can be distinguished, each of them embracing the embryonic territory of about several dozens micrometers (10^{-4} m) and retaining its general topological properties over a few hours (Beloussov et al., 1975). Each new field is created because of a profound and relatively rapid (taking no more than few dozen minutes) transformation of the preceeding field. Further morphogenesis is spatially regular and structurally stable only if it takes place in a fully developed tensile field (Beloussov, 1979).

To reveal the structure of tensile fields we have employed two different methods:

1. Tracing rapid local relaxation deformation (contractions, bending etc.) of dissected embryonic tissues (Beloussov et al., 1975) and
2. calculating the relative tension of cell walls on the basis of cell geometry, as viewed in histological sections (Lakirev, Beloussov, 1986).

Both methods give comparable results but the second is more exact. It is based upon the following assumptions:

a) all the tensile forces in embryonic tissues are localised in cell membranes and/or subcortical cell parts;

b) in each moment a net of tensed cell walls is mechanically equilibrated.

This means that in each cell apex (cell wall intersection point) $\Sigma \bar{f} = 0$ where $\bar{f}$ is a tensile force. Under these assumptions the angles between cell walls are direct functions of the relative values of the applied tensile forces. Passage, via computer calculations, from one cell apex to another through the whole tissue region under study yields the

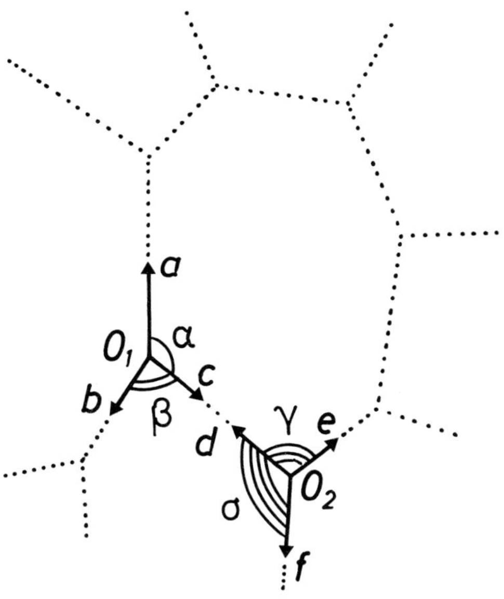

Fig. 1 A scheme of the geometrical methods of cell wall tension measurement. We start from an arbitrary cell apex 0_I and ascribe to a first tension force (say, a) an arbitrary value. Measuring α and β angles we can express the values of b and c tensions in relation to a. Then we pass to the next apex, say O_2, measure γ and δ angles and express e and f tensions in relation to c. In such a way we pass through all cell apices in a given zone and even return to the initial apex from another side and measure some tensions for the second time. If the second measurements do not differ significantly from the first one may conclude that the tensions located outside the given plane do not markedly influence the tensions oriented in this plane.

entire tensile pattern (Fig. 1). Presentation and further analysis of tensile patterns have been greatly facilitated by the markedly bimodal distribution of tensile forces (Fig. 2). This points to the nonlinear character of cell stretching (see also Belintzev et al. 1985 a, b). In our further constructions we have only used 'strongly tensed' cells (those on the right hand peak of the histogram, Fig. 2) to characterise the tensile patterns.

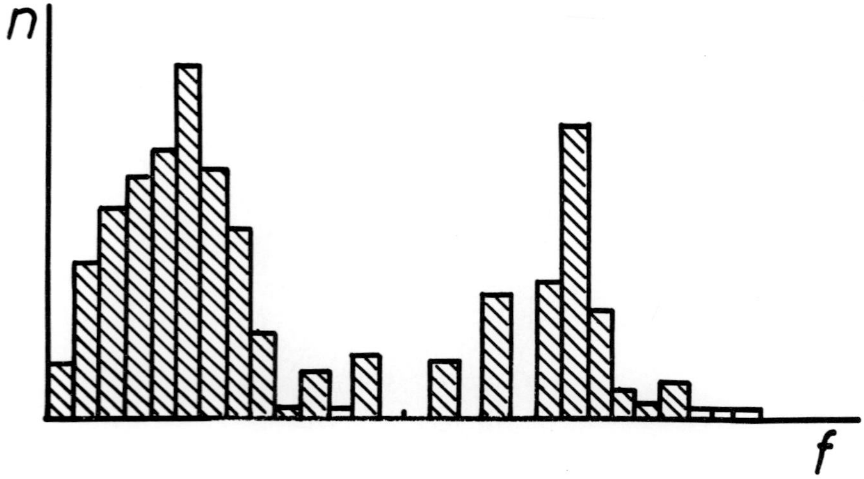

Fig. 2 A markedly bimodal histogram of tensile forces in cell walls of early neurula (Fig. 4 a), as detected by geometric methods (see text). Abscissa: tensile forces; ordinate: number of cell walls. The cell walls belonging to the right-hand peak are considered as strongly tensed (shown by solid contours on Figs. 3, 4).

Fig. 3 a gives the structure of the tensile fields of a suprablastoporal region of *Rana temporaria* early gastrula, and Fig. 4 a that of the same species early neurula obtained by this method. The main elements of both fields are the strips of stretched cells ('cross-lines') uniting in the first case the blastopore with the closest blastocoel region and in the second case the dorsomedial plane with the lateral embryo walls. These tensile patterns qualitatively coincide with those previously revealed by the dissection technique (Beloussov et al., 1971).

Experimental evidence has been presented elsewhere (Beloussov, 1979, 1980) that morphogenetic movement of gastrulation in the suprablastoporal region (longitudinal extension - lateromedial compression of cell material) creates the initial conditions for the formation of neurulation tensile fields and those of neurulation for the appearance of tensile fields in the anterior embryo region. The problem remains, however, what are the relations between an initially established tensile field and the immediately following cell movements in the same region. Can the latter perhaps be directly determined by the structure of the tensile field? If the answer is positive we could achieve a closed description of the long periods of embryonic morphogenesis as a function of tensile fields. We have investigated this problem by means of computer simulation.

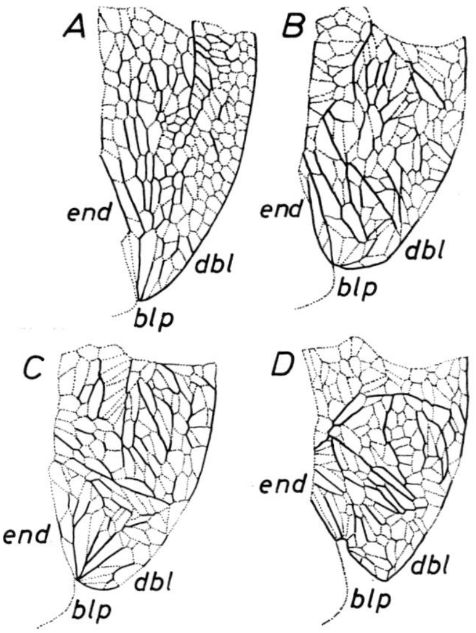

Fig. 3 Tensile fields in the suprablastoporal region of early gastrula (saggital sections). Dotted lines are for weakly tensed cell walls and filled lines for strongly tensed walls. In A the real (slightly schematised) cell arrangement and the corresponding tensile field are shown. B, C, D are the computed configurations of successive steps (2nd, 5th, and 7th step correspondingly) of the model. blc - blastocoel, blp - blastopore, dbl - dorsal blastoporal lip, end - the area occupied by endoderm. Note the imitation of blastopore invagination and blastoporal wards shift of the dorsal noninvoluted material (approach of the asterisk to the blastopore).

A model of stretching-dependent active cell contractions during gastrulation and neurulation in amphibian embryos

Quite different contractile systems, from the cytoplasmic strands of myxomycets Physarum to vertebrate heart are known to respond to artificial stretching by active contraction (Fleischer, Wohlfahrt-Bottermann, 1975; Kitabatake, Suga, 1978). This seems to be true for many embryonic cells as well: from observation actively polarised (stretched) cells are actively contracting (within a few minutes or, at least, dozens of minutes). We attempted to employ this observation to the model assuming that the stretching-contraction dependence for embryonic cells can be described by an equation similar to that proposed for heart muscle, namely

$$\ln \sigma = \ln a + b\varepsilon \tag{1}$$

where σ is stretching and ε contraction (Kitabatake, Suga, 1978). We suggested also that (1) each cell is contracted along the direction of its maximal stretching determined by the method described above; (2) a contracting cell is pulling its neighbours thus stretching them; at the next step of the process the latter cells are also actively contracting themselves because they are now stretching (see Lakirev, Beloussov, 1986, for details). Taking as the initial state the real tensile patterns established at the beginning of gastrulation and neurulation (Fig. 3 a, 4 a) we have computed, according to the above assumptions, several steps in further morphogenesis (Fig. 3 b - d, 4 b - d). The construction obtained satisfactorily imitated not only the external shaping but real cell rearrangements, too. In particular, blastopore invagination, in the gastrulation process, the animal-wards shift of involuted cell material (in relation to the noninvoluted one) and the appearance of circular cell structures inside the blastoporal lip have been imitated. In the neurulation process not only has the neural tube closure been reproduced but also the rounding of the neural tube external surface. These first attempts show that tensile fields created at the beginning of certain developmental periods may contain in themselves a great part of the 'spatial information' required for the immediately following period of morphogenesis. This information is realised by stretching-dependent cooperative cell contractions which gradually reconstruct the initial field and prepare the topological transformations which lead to the creation of a new tensile field.

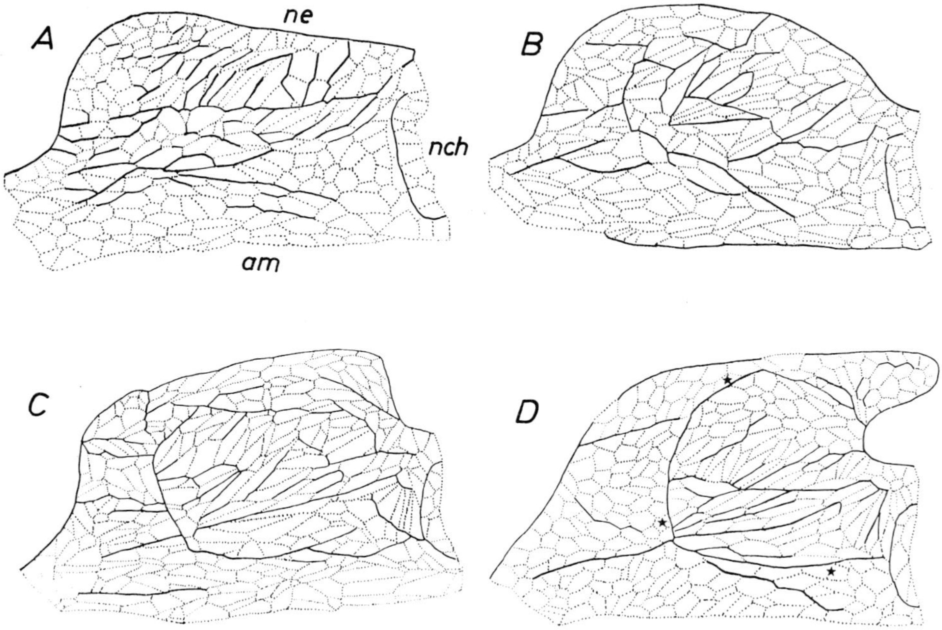

Fig. 4 Tensile fields during neurulation in amphibian embryos. Left-hand parts of the transverse sections through the dorsal embryo wall are represented. Dotted and filled contour lines have the same meaning as in Fig. 2. In a the real (slightly schematised) cell arrangement and the computed tensile field at the early neurula stage are represented. b, c, d are the computed configurations (2nd, 4th and 9th step correspondingly) of the neurulation process. Note the imitation of neural tube closure (in d the 'neural groove's' edges have crossed the saggital line of a computed configuration) and the rounding of neural tube contours (shown in d by asterisks). am - axial mesoderm, nch - notochord, ne - neuroectoderm.

A model of local curvature increase in the epithelial rudiments of hydroid polyps

The model presented above was related to vast multilayered rudiments within the fairly complex tensile patterns which were taken as the initial state. There exists meanwhile a large class of more simply constructed rudiments consisting of one or two more or less homogeneous cell layers but fully capable of autonomous morphological complication. A good example of this class of rudiments is the vegetal generation of hydroid polyps. The aim of the following model is to represent the morphogenesis of hydroid buds as a mechano-geometrical self-complication process based upon the deformation interactions of neighbouring cells.

Growth and morphogenesis of the vegetal outgrowths of hydroids consists of a series of so-called *growth pulsations* (GP) with periods of several minutes (Beloussov, 1973; Beloussov, Dorfman, 1974; Beloussov et al., 1984; Zaraiskii et al., 1984). During the 'upward phase' of each GP the cells ar turning themselves from the oblique to a perpendicular orientation (in relation to the outgrowth axis), and during the opposite, 'downward phase' back again. The mechanical force required for elongation and shaping is generated only during the upward GP phase. In each GP the cells of the same outgrowth are turning nonsimultaneously and, as a rule, in a proximo-distal succession (Zaraiskii et al., 1984). It was also observed that the greater the relative duration of the upward phase the greater should be the width-length ratio of a fully developed rudiment (Beloussov et al., 1984).

Proceeding from these data we suggested a model of hydroid morphogenesis based upon the following assumptions:

1. the shaping (local curvature changes) of a rudiment is caused by local mechanical cell interactions repeating during each GP;
2. the first step of these interactions consists in the passive compression of a j cell by a force f_{j-1} exerted by its upwardly turned proximal neighbour, $(j - 1)$ cell, and by the equal force f_{j+1} of elastic resistance exerted by the more distal $(j + 1)$ cell; the resulting force is F_j (Fig. 5);

3. the second step is the active shift D_j of a j cell which is proportional to F_j and oriented in the same direction:

$$\bar{D}_j = k\,\bar{F}_j\,. \qquad (2a)$$

The coefficient of proportionality k may be considered as a measure of an internal mechano-chemical activity of a j cell.

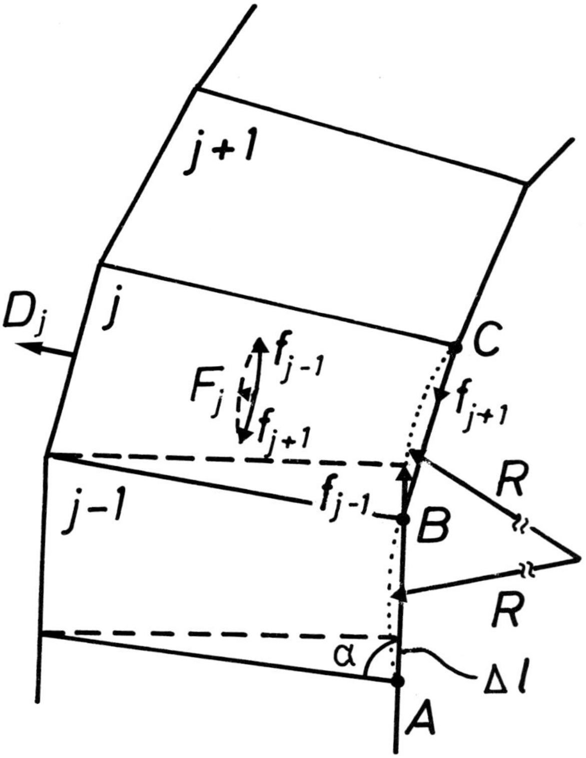

Fig. 5 A scheme of the forces leading to curvature increase in the ectoderm of a bud of hydroid polyp. Inside j cell a parallelogram of f_{j-1}, f_{j+1} forces is shown.

From the geometric construction one obtains

$$F_j = f_{j-1} \cdot L/R \cdot \sqrt{1 - (L/2R)^2}, \qquad (2\,b)$$

where L is the length of a basal wall of a cell (the lengths of all the cells considered are taken as equal) and R is a radius of ABC arch and, consequently, a radius of a layer curvature in the given area. As in the real case $R >> L$, we can simplify (2 b) and take

$$F_j \sim 1/R . \qquad (2\,c)$$

Finally, we assume that f_{j-1} is proportional to the upward shift $\Delta \ell$ of the basal surface of a $(j - 1)$ cell and

$$\Delta \ell \sim \cos \alpha . \qquad (2\,d)$$

Combining (2 a), (2 c) and (2 d) we obtain the main algorithm of the 'curvature increase' model:

$$D_j = k \cos \alpha \,/\, R . \qquad (3)$$

It is worth mentioning, that the biological meaning of this equation is rather similar to that of (1). The tangential compression of a j cell is indeed appropriate to its radial stretching and its subsequent shift away from the curvature centre is homologous to the cell's active contraction response, as suggested in the previous model. In both cases the same biphasic succession 'passive deformation by an external mechanical influence - active shift in the direction of the preceeded deformation' takes place.

Equation (3) has been used for modelling in two different regimes:

1. $k(t) = const$
2. k is a periodic function of t corresponding either to a 'narrow pulse' (Fig. 5 g) or to a 'broad pulse' (Fig. 5 h) regime.

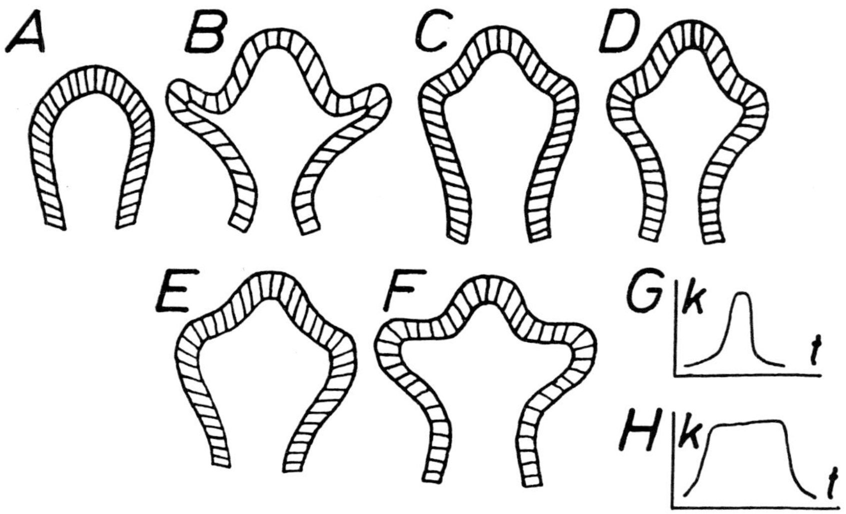

Fig. 6 Modelling of Hydrozoa buds shaped according to equation (3). a - a real shape of Obelia bud which is taken as initial for all the successive constructions. b - a shape modelled in $k(t) = const$ regime, 20th step. c - a shape modelled in the periodical 'narrow pulse' regime (see g), 15th step. d - same regime, 20th step. e - a shape modelled in the periodic 'broad pulse' regime (see h), 15th step. f - same regime, 20th step. In all the constructions $k = 16$. g - a scheme for a 'narrow pulse' h - a scheme for a 'broad pulse'.

The results obtained are briefly as follows (for details see Lakirev et al., 1987). Under all the regimes employed the initially smooth nondifferentiated shape of a hydroid bud (Fig. 6 a) is transformed into the three-foiled differentiated shape closely resembling the normal shape: the central evagination corresponds to a hypostome and the lateral ones to the transverse sections of a tentacle whorl (Fig. 6 b, d, f). In the 'narrow pulse' regime the correspondingly narrowed rudiments have been obtained (Fig. 5 c, d) and in the 'broad pulse' regime the transversely extended rudiments (Fig. 5 e, f). This correspondence of the models with the real GP-pattern, the rudiments shapes correlation (Beloussov et al., 1984), supports the models' assumptions.

Conclusions

A study of embryonic morphogenesis as a self-organising process is just beginning. Now it seems of particular importance to establish the close and heuristically valuable connections between biological experiment and self-organisation theory. In this paper we attempt to outline some promising areas for study. The first, and the most general one may be the analysis of experimental data in the framework of a concept of a multilevelled hierarchy of processes coupled parametrically and dynamically. The second, more particular one is the elaboration of mechano-geometrical models which are, we believe, closer to real morphogenesis than purely chemical (diffusional) models. In the latter aspect, one of the most important tasks should be revelation of the exact mechano-chemical nature of the morphogenetic algorithm, which remains at present more or less formal.

Topological and Thermodynamic Structures of Morphogenesis

E. V. Presnov, S. N. Malyghin, V. V. Isaeva

Introduction

The development of topology, thermodynamics and the theory of biological morphogenesis proceeded separately. Therefore it is interesting to establish inter-relations between them and to derive from these inter-relationships consequences which can shed new light on each of these distinctly separate sciences. More specifically, our purpose is to give a formal definition of the established but uncertain notion of a morphogenetic field. The importance of this notion will be demonstrated in the first part. Then, in the second part, we present a mathematical basis for its introduction. Finally, in the third part we give a definition of the morphogenetic field and reveal its thermodynamic content. In the last part another example is discussed.

We are inclined to consider our theoretical picture as a fragment of the more comprehensive and significant subject: topological structures of positional information (Presnov, Isaeva, 1985).

Thermodynamics and Pattern Formation in Biology

Origin of developmental patterns in early embryogenesis

In the history of embryology there have been many attempts to solve the problem of the origin of egg promorphology which determines the following determination of blastomeres. One attempt was made by Child whose book (Child, 1941) deals with the spatial organisation of development. Child's physiological gradients or axial gradients of metabolism intensity and the corresponding damage gradients found in the eggs and embryos, as well as in the organism of many adult animals, determine, according to Child spatial organisation of differentiation, various portions of the gradient axis. Child's ideas greatly influenced the further development of spatial organisation. hypotheses in embryology and general biology and still remain of interest up to now: Wolpert's (1969) more modern arguments of positional information theory directly develop Child's ideas.

Child's concept relates the presence and orientation of gradients to the heterogeneity of the environment. Egg promorphology appears as though imprinted from environmental nonhomogeneities. Thus animal-vegetative differences of the oocyte and egg are also explained by connection of one pole of the oocyte to the gonad wall; formation of the dorsoventral axis in some *Chordata* is ascribed to local penetration of the sperm. However, there are some objections to such explanation of egg polarisation: often heterogeneity of the environment cannot be discovered or morphogenesis is not oriented in accordance with this heterogeneity. For example, the sagittal plane in the amphibian egg is found under parthenogenetic activation and in the case of sperm penetration exactly in the animal pole of the egg (Gasaryan, Belousov, 1983).

Our proposition is based on the statement that each smooth vector field on a two-dimensional sphere has at least one singular point (i. e. a point at which the vector is equal to zero). Less pedantically this can be formulated as follows: the hedgehog cannot be combed - a tuft is always produced. Additionally, the singularity of the vector field has an integral index equal to the algebraic number of rotations of the field vectors when rotated around a sufficiently small positively oriented circle. The total number of indices over all singular points on each compact manifold does not depend on the choice of the field and is determined only by the topological properties of the manifold itself. This number, named the Euler characteristic, is a topological invariant of the manifold. For the two-dimensional sphere, the Euler characteristic is equal to 2; but for a torus it is equal to 0, i. e. the torus can be smoothed and the summit is absent (Milnor, 1965).

On the other hand it has been proved experimentally that the main carrier of egg promorphology is its cortical layer; the thin peripheral layer of the ooplasm adjacent to the plasma membrane. Consider the egg cortex as a two-dimensional sphere on which a vector field of any nature is located. The concept that the biological field is a system of some vectors usually considered as gradients of some morphogenes is widely held. Retaining the idea of diffusion of substances, so-called morphogenes, we assume that structural microheterogeneity of the egg cortical layer serves as a carrier or indicator (like iron filings in the magnetic field) of the vector field on the egg surface. In reality, the cortical cytoskeleton, whose main structural elements are actin microfilaments (see reviews by Vacquier, 1981; Presnov, Isaeva, 1985) is located in the egg cortex. From this proceeds the sought for structural microheterogeneity of the egg cortical layer consisting of individual microfilaments or their bundles, and also, possibly, of the cortical microtubuli found in the eggs of some animals. The linear elements a of naturally orientated cortical cytoskeleton can be regarded as visualisers of the field vectors. In its turn, the theorem of the sum of indices provides at least one singular point in any vector field on the sphere, i. e. egg polarity. This conditions primary polarity as animal-vegetative. The singular point can serve either as a vegetative or as animal pole, thereby determining animal-vegetative differences.

During spermium penetration a modification of the vector field takes place, and one more singular point is formed, i. e. the fertilisation cone at the site of spermium penetration. This disturbs the balance of singular point indices and automatically induces formation of another singular point; in the region of the grey or yellow crescent in amphibia and ascidia for example (Fig. 1). This predetermines the sagittal plane of the embryo.

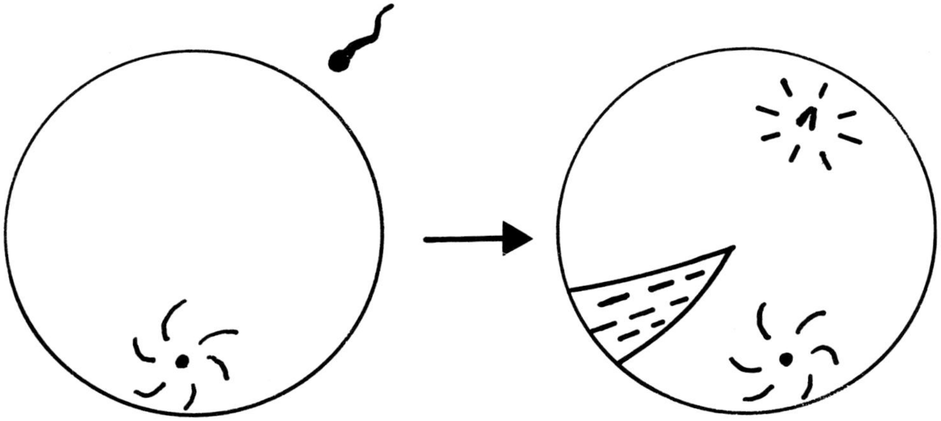

Fig. 1 See text.

When the sperm in amphibia penetrates at the animal pole, a change of the index of this singularity is possible as a result of modification of the cortical field after the sperm intrusion. Or, if the pole is not a singular point, a singularity in it is created by such a fertilisation variant. Both these alternatives again initiate a new singularity with index balance preservation (Σ ind = 2) and dorsoventral polarisation.

In case of parthogenetic activation induced by nonlocalised action of some agent development begins only when the total index of the preexisting points (or point) deviates from 2, and this results in the appearance of an additional singular point (or points) which determines the sagittal plane.

Thus the theorem involving vector fields on the sphere explains heterogeneity, polarity of the egg cortex, i. e. it points to obligatory self-organisation of the oocyte independent of environmental nonhomogeneities. External heterogeneity can only influence the localisation of the singular point (or points), existence of such a point (points) is an inevitable consequence of microheterogeneity in the organisation of the egg surface layer.

It appears from these conclusions that no explanations of the environmental effects on the establishment of oocyte polarity are required. But a more subtle problem is how to understand and reveal the effects of the environment in this case. It is known that the dimension of physical space equals 3. If it were different, by a unit higher (4) or lower (2), the vector field would be analysed not on an even-dimensioned sphere, but on an odd-dimensioned sphere. In the latter case singularities are not obligatory for the vector field. Thus the geometrical arrangement of our world is reponsible for manifestations of the multi-valued phenomenon of positional information (Presnov, Isaeva, 1985)

Topological modelling of biological shapes and pattern formation

Topological model of the biological shape

Firstly we define *the space Biol of the biological shapes*. Its elements or points are the compact domains Φ lying in the three-dimensional Euclidean space $\mathbb{R}^3$ and having a smooth (or piecewise smooth) boundary $\mathcal{F} = \partial\Phi$. We shall identify the points of this set as biological shapes. Indeed, any living organism occupies some volume at each time, i. e. we have some point of the set *Biol.* We do not claim, however, that an organism can be placed in any domain. A subset of true biological shapes could be distinguished in *Biol*, but we shall refrain from doing it for convenience.

Now we introduce a metrics on the set *Biol*, determining the sphere $\mathcal{U}_\varepsilon^\Phi$ with centre Φ and radius ε as

$\mathcal{U}_{\varepsilon}^{\Phi} = \{\Phi' | \partial\Phi' \subset \varepsilon\text{- neighbourhood of } \partial\Phi\}$.

It is clear that the process of organism development is a continuous map $\gamma: t \rightarrow \Phi_t$ from the half-line $\mathbb{R}^+ = \{t | t \geq 0\}$ into the space *Biol*. The point t = 0 is mapped under γ to the sphere Φ_0 (zygote). The family of these maps for all organisms allows endowment of the space *Biol* with a richer structure.

In particular, we say that shapes Φ_1 and Φ_2 are connected by the morphism $\Phi_1 \rightarrow \Phi_2$, if there exists an organism such that it occupies the domain Φ_1 at a time t_1 and the domain Φ_2 at a time $t_2 > t_1$. It is clear a priori that the presence of morphisms $\Phi_1 \rightarrow \Phi_2$ and $\Phi_2 \rightarrow \Phi_3$ does not imply a presence of morphism $\Phi_1 \rightarrow \Phi_3$, since while one organism transfers Φ_1 to Φ_2 and another organism transfers Φ_2 to Φ_3, there may not occur any third organism that would transfer Φ_1 to Φ_3. It should be noted, however, that we do not know any real example of this kind. If the existence of such a third organism is admitted as a hypothesis, the subspace of the true biological shapes will be converted into a (topological) category. If this hypothesis is not admitted, there will still be a structure of the oriented graph Γ on the set *Biol.*

According to the well known thesis of Thompson (1942), the shapes of any two closely related organisms can be produced one from another by a continuous deformation. This is a pure phenomenological hypothesis. The definition of equivalence in the space *Biol* mentioned below is a result of this thesis.

We determine the equivalence and the corresponding factorisation in two steps.

<u>First step.</u>

We say that two points Φ_1 and Φ_2 of the space *Biol* are equivalent, if there exists a diffeomorphism $\mathbb{R}^3 \longrightarrow \mathbb{R}^3$ (i. e. one-to-one map, smooth in both directions), transferring the domain Φ_1 onto the domain Φ_2. We also specify that the sign of Gaussian curvature (the signs of main curvatures for the elliptic points) is preserved in the boundary points, as well as the fracture type on the boundary.

Second step.
We say that the shapes Φ_1 and Φ_2 are equivalent, if there is a morphism $\Phi_1 \to \Phi_2$, which is reversible in the sense that there exists a morphism $\Phi_1 \to \Phi_2$.

Let *Sh* denote the factorspace of the space *Biol* corresponding to this equivalence. The above factorisation formalises the concept of shape in the closely related organisms. For this reason the space *Sh* is called *the space of standard biological shapes.*

Under the factorisation ψ : *Biol* → *Sh*, the curve $\gamma : \mathbb{R}^+ \longrightarrow$ *Biol* (individual organism development) is passed to the curve $\omega = \psi \circ \gamma : \mathbb{R}^+ \to Sh$. Unlike γ the curve has no self-intersections. We call ω *the process of individual development of closely related organisms*.

The problem of shape determination is closely related to the problem of shape recognition in the concept of tolerance (Zeeman, Buneman, 1968)

Topological classification of biological shapes

We classify biological shapes using the known classification of two-dimensional closed manifolds. The two-dimensional closed manifold is a compact Hausdorff topological space, each point of which has a neighbourhood homeomorphic to the open disc on the plane. We shall assume that this manifold is orientable. Because of their orientability these manifolds are indeed embedded in three-dimensional Euclidean space $\mathbb{R}^3$. An example of such a manifold is the sphere to which p handles are attached (Fig. 2).

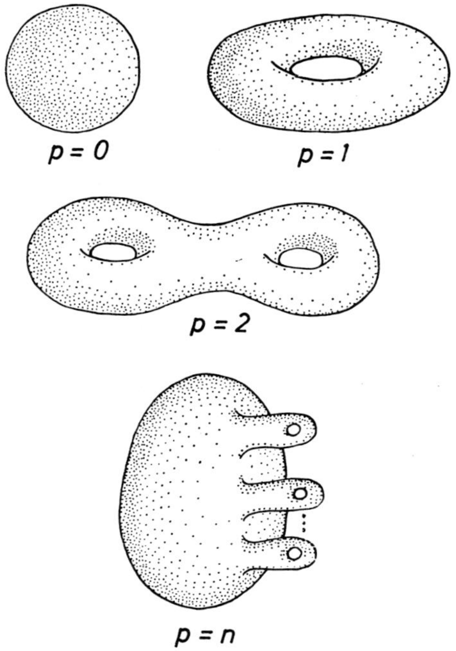

Fig. 2 See text.

The number p is invariant, i. e. with $p \neq q$ the attached sphere with p handles is nonhomeomorphic to the sphere with q handles. It is proved in topology (see, for instance, (Wallace, 1968)), that any two-dimensional closed orientable manifold M^2 is homeomorphic to the sphere with p handles. (It should be noted that they are homeomorphic as abstract spaces rather than those embedded into $\mathbb{R}^3$. In the latter case, the 'knottedness' of handels may serve as basis for a more sophisticated classification of these manifolds.) Number p is called as *the genus* of the manifold M^2.

Another invariant of the space X is presented by *the fundamental group* $\pi_1(X)$. Its elements are the homotopy classes in X, i. e. of maps $S^1 \to X$ of circle S^1 into the space X such that the base point in S^1 maps to the base point in X. The genus p and the fundamental group $\pi_1(M^2)$ define each other in a single-valued way for any two-dimensional manifold M^2.

It is clear that the boundary $F = \partial\Phi$ of the living organism shape (the shape is assumed here as an element of the space *Biol*), provided it is not a manifold (for example, that produced by hedgehog spines) is at least homotopically equivalent to some manifold M^2. We shall identify the genus of this manifold p as *the genus of the living organism* (Maresin, Presnov, 1985).

Topological operations in embryogenesis

The development of each organism begins with a zygote whose boundary is spherical (the genus of a sphere is equal to zero). In most multicellular organisms the genus of an adult organism differs from zero. Since the genus p is an integer it changes discretely. In the course of changes the two-dimensional manifold, modelling the surface of the organism, undergoes topological operations.

Two types of topological operations are distinguished: type 0 (φ_0) and type 1 (φ_1) (Fig. 3). The operation cannot follow this pattern during embryo development, since the embryo can utilize only its own cellular resources in the course of operation. Changes of the magnitude of the genus p in embryogenesis occur through fusion and ruptures of cell layers, the latter being equivalent to adding new handles to the initial sphere, i. e. the blastula. The anatomical structures arising in the course of such developmental operations are the formations connecting the gut cavity with the outside:

a) proctodaeum and stomodaeum
b) branchial clefts (usually five pairs)
c) ear cavities (modification of one branchial cleft pair)
d) nasal cavities.

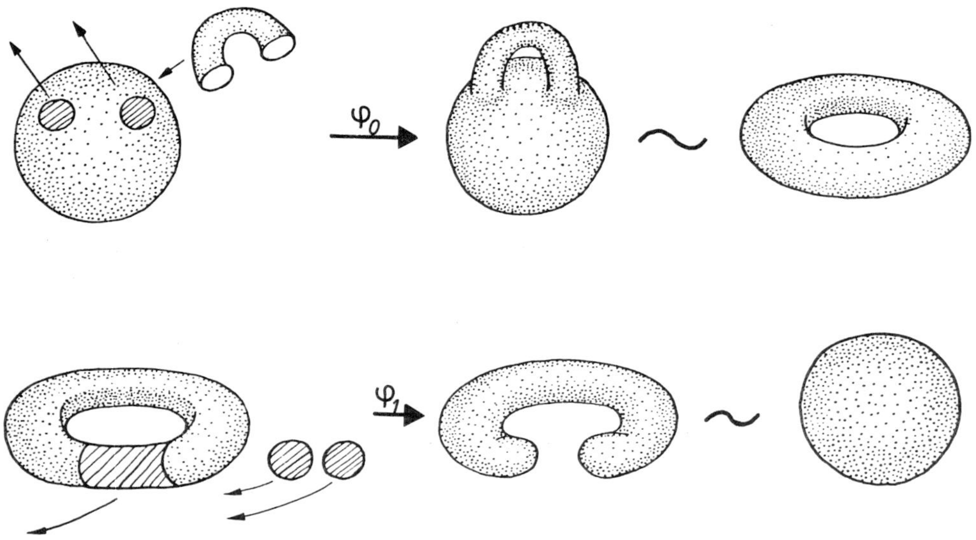

Fig. 3 See text.

The adaptive changes and supplementary developmental operations in the course of embryo development precludes assignment of a definite value p to each taxonomic animal group, although the relationship between the increase in value p and the complexity of animal organisation is apparent. Thus, the Chordata should be classified as a type with $p = 13$ (Fig. 2). The value of p may decrease within certain classes, for example, by subsequent embryonic closure of branchial clefts or fusion of paired nasal cavities into one. On the contrary, the value of p may increase because of subsequent addition and splitting of branchial clefts, resulting in some groups of the lower Chordata to an increase in p of a few hundreds (Fig. 4). Hence, there is sense in considering the above developmental operations as *fundamental*.

One experiment

Changes of the shape Φ of a living organism induce changes of the group $\pi = \pi_1(\partial\Phi)$ (see second section). Unfortunately, the construction of the fundamental group is not functional in relation to the morphisms determined in the first section. Indeed, the change of surface topology $F = \partial\Phi$ is reduced to a series of its developmental operations (third section). The homotopic loops may become nonhomotopic and vice versa. Nonetheless, it would be interesting to trace the changes of group π in ontogenesis, or at least of one of its elements. Thus, it is preferable to select a canonical loop on F. This could be the one-dimensional boundary of the surface contact between two blastomeres after the first cleavage division. The loop should be oriented to produce an element from π. This procedure is, however, not always simple. If the selected loop resulted from meridional division (via the animal and vegetative egg poles), the canonical orientation can be set only in the case where the direction of the dorsoventral axis or the bilateral symmetry plane of the future embryo is known (Fig. 5 a). If the plane of the first division is equatorial, the loop can always be oriented canonically (Fig. 5 b). However, one of the three cleavage divisions is always equatorial, and thus, the first equatorial loop can be taken.

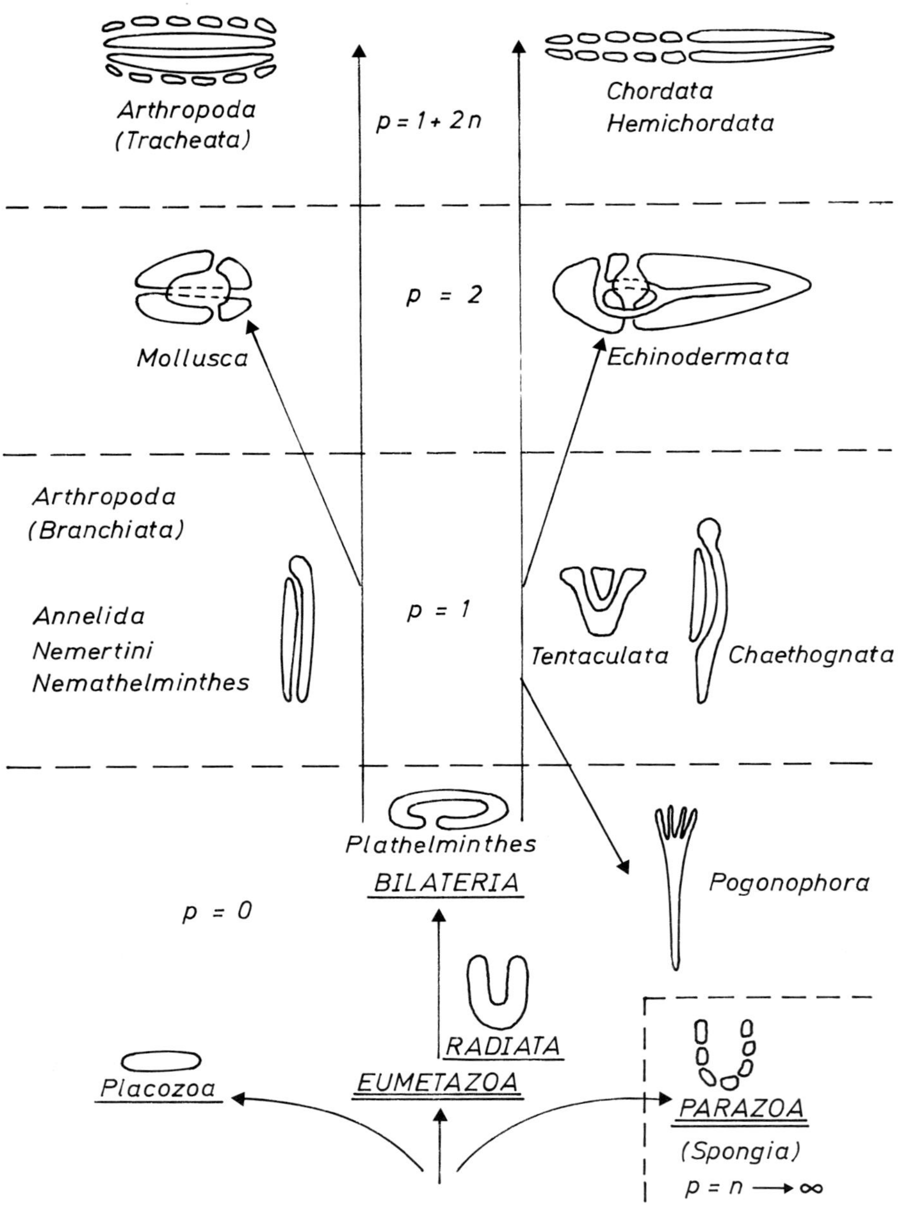

Fig. 4 See text.

The suggested experiment is the description of the fate of the canonical loop d in ontogenesis. It would be extremely interesting if with $p \neq 0$ the image of the loop d were nonhomotopic to zero. Then the question of the algebraic number of its turns round each hole of the domain Φ could be raised (in other words, the question concerning decomposition of cycle $h(d)$ according to the standard basis of the homology group $H_1(F)$, where $h: \pi_1(F) \rightarrow H_1(F)$ is a Hurewicz-homomorphism (Fuks et al., 1969), that could be identified, according to the Poincaré theorem, with factorisation by the commutant of the group $\pi_1(F)$.

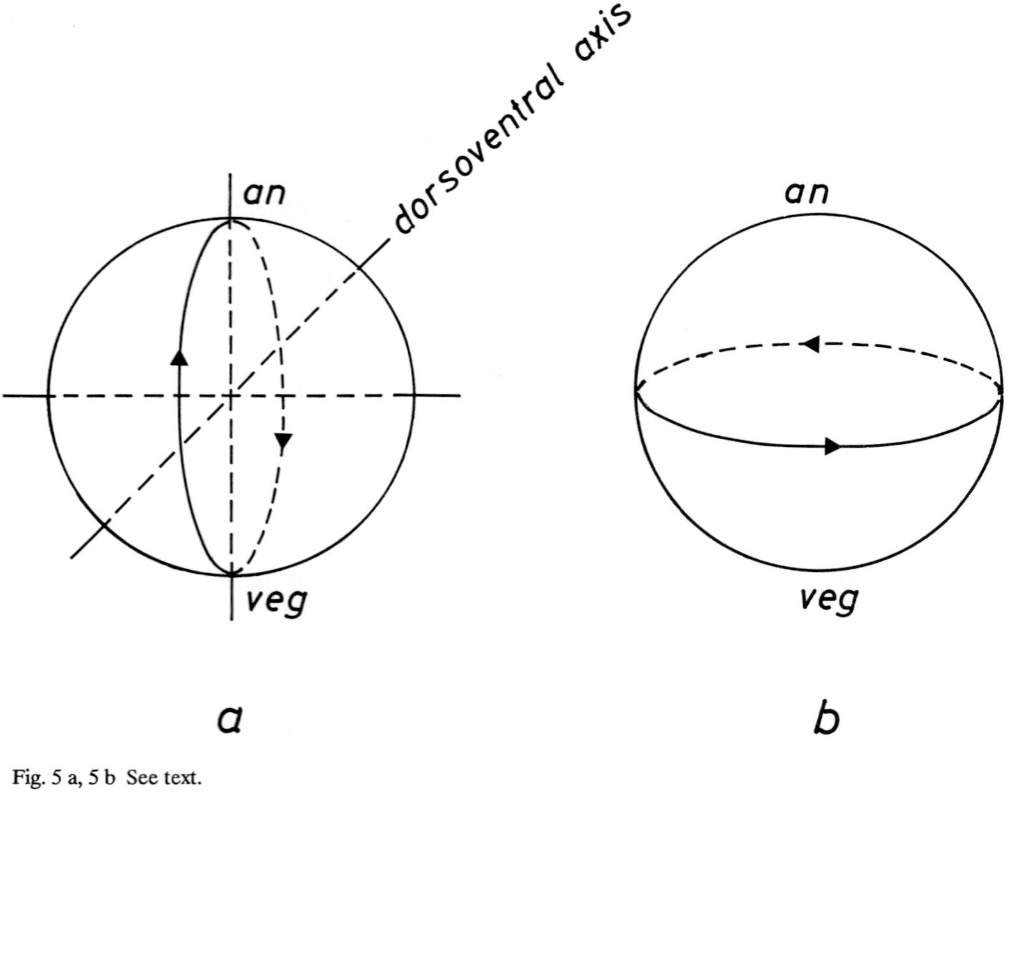

Fig. 5 a, 5 b See text.

Cell structure

Here we shall model a living organism using a cell complex (compare with Hoffman's axiomatics (Hoffman, 1973)). *The cell complex* (or *CW-complex*) (Fuks et al., 1969) is the topological space K, presented as a disjoint union of its subsets e^q_i (*cells*). The continuous map $f: D^q \rightarrow K$ (*characteristic map*) from the closed q-dimensional sphere D^q to K is present for each cell. Its restriction on interior *Int* D^q of sphere D^q is a homeomorphism onto e^q_i. To avoid pathologies it is assumed that the boundary of each cell lies in the finite union of cells of smaller dimensions, and that the topology of K is weakest among others, with regard to which all characteristic maps are continuous.

The applicability of this definition to cell structure modelling of the living organism tissue is apparent. We shall present the organism as a three-dimensional cell complex K lying in $\mathbb{R}^3$. It would be natural to consider in K the connected subcomplexes T_j, consisting of cells. The subcomplexes T_j will be identified as *tissues* here.

When the physiological activity of the organism is manifested its cell model K undergoes cell homeomorphic transformation $K_1 \rightarrow K_2$. The tissue structure is retained.

In the third section, only the fundamental operations, changing the genus p of the external organism surface, were described. The development is also characterised by other developmental operations resulting in separation of the initially connected cell layer into a few tissues, that may cause inversion of the initially outer part of the embryo toward the interior. Such tissues can undergo additional operations which do not change the topology of the external organism 'envelope'.

The classification of real developmental operations, occurring because of fusion and rupture of cell layers, is presented in Fig. 6. In real morphogenesis it is convenient to differentiate operations changing the genus but not breaking tissue connectedness (Fig. 6.1, 6.2) and those that leave the genus unchanged but give rise to changes in connectedness (Fig. 6.3, 6.4).

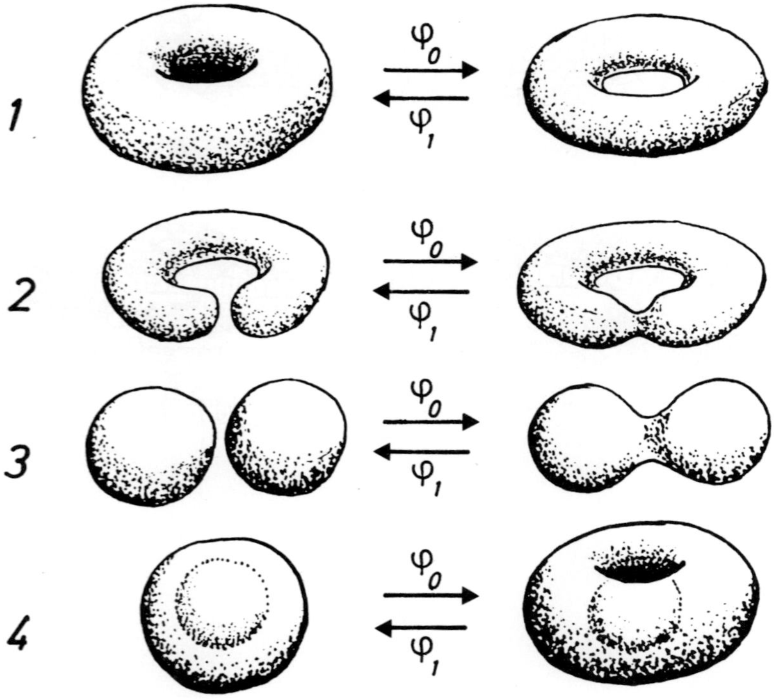

Fig. 6 See text.

Locally types of developmental operations occur in the following way (Fig. 7):

a) contact of two layer areas
b) fusion of areas
c) disintegration of the fused area into individual cells (rupture)
d) overgrowing of the fused area.

The analysis of developmental operations has shown that not all types of fusion really occur in the course of development, as it is essentially, which sides (internal or external) of the surface of embryonic layers come in contact with each other (Fig. 8).

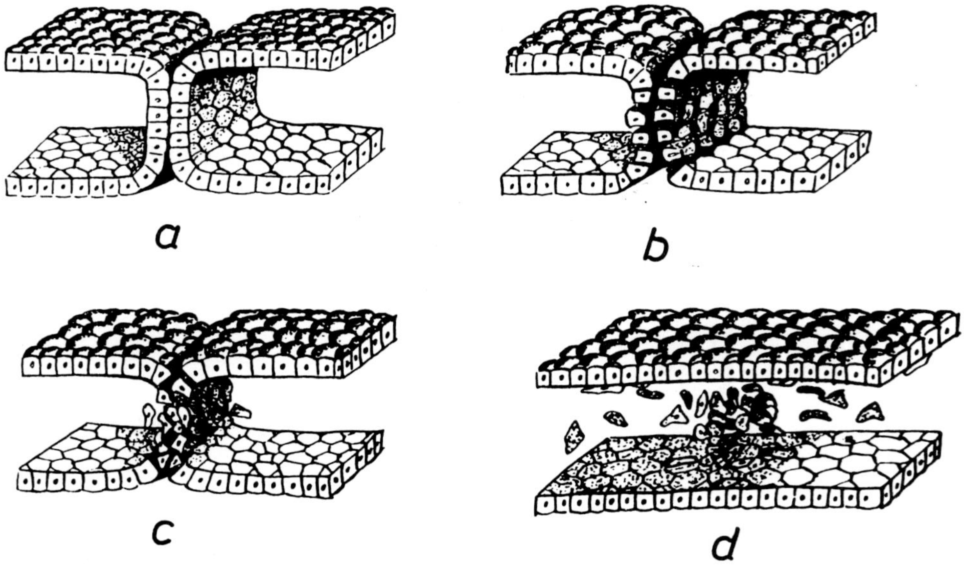

Fig. 7 See text.

Morphogenetic transformations in individual organism development can be presented as the following serial procedures:

1. cell division: circle $S^1 \subset \partial \bar{e}^3$ is distinguished on the mother cell, shrinks along $\bar{e}^3$ into a point, and the tubular neighbourhood of S^1 in $\bar{e}^3$ is attached (fused) symmetrically, forming a contact membrane of daughter cells
2. cell death: $e_{ij} \longrightarrow T_j^2$ (a map on the two-skeleton)

3. handle attaching to one or two tissues (in the second case it is a connected sum of two tissues) (oral or anal plate rupture, branchial cleft formation)
4. handle cutting; in this the tissue can be cut into two tissues followed by constriction (closing of the branchial clefts in terrestrial vertebrates)
5. cavity formation (schizocoel cavity formation)
6. cavity agglutination: opposite of 5.
7. tissue formation from individual cells (the formation of endothelium and mesothelium from mesenchyme-like cells)
8. tissue loosening (disintegration of the neural crest into individual cells at neurulation).

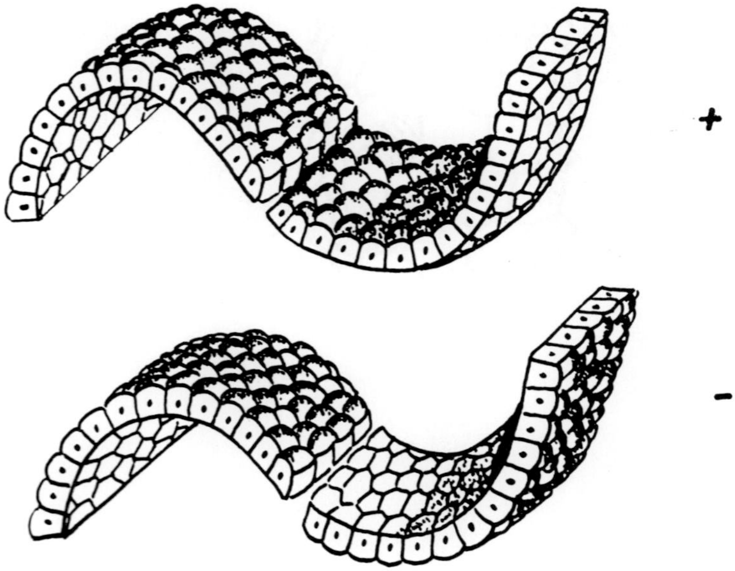

Fig. 8 See text.

Morphogenetic fields

Discrete morphogenetic fields

The cellular structure of tissue determines the numerical characteristics of its constituent cells. The association of some number with each cell is noncanonical. For example, the number of neighbours of a given cell (Fig. 9) or its volume or something else can be taken as such a characteristic.

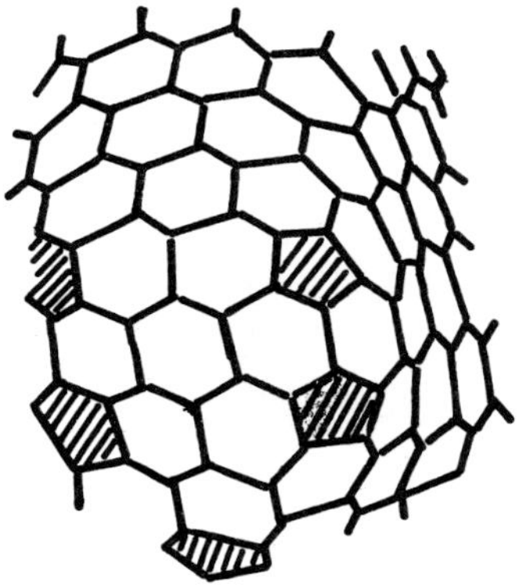

Fig. 9 See text.

Let some (most frequently integral) number be associated with each cell in some way. Then we shall say that a *discrete morphogenetic field* is present on the tissue. Let there also be rules altering the cellular structure itself (cells are born and die) and, consequently, its numerical characteristics. Such rules determine the development of our structure and its numerical characteristics. This development is a temporal variation of the discrete morphogenetic field.

An example of amazing beauty, illustrating such a dynamic situation, is furnished by J. Conway's game 'Life' (see Gardner, 1971). An analysis of this game forms an essential component of the theory of cell automata.

Scalar morphogenetic fields

Now consider, as before, the living organism as a three-dimensional body with a two-dimensional boundary M^2. Let some numerical function $f: M^2 \to \mathbb{R}^1$ be given on M^2. As f it is possible to take, for example, the total (Gaussian, Riemannian) curvature of the surface M^2 or its mean curvature. We shall consider not all the functions but only the smooth ones (infinitely differentiable) and further only the so-called Morse functions. These functions are quite sufficient for our purposes since they are dense in the space of all functions. The Morse function, by definition, is a function whose critical points are nondegenerate (i. e. at those points $x \in M^2$ where $df = 0$ the form d^2f is nondegenerate). An illustration of the Morse functions is given by the height function (Fig. 10).

Morse functions are interesting just because of their critical points. According to the Morse lemma, at each critical point there exist local coordinates in which the function f can be written as

$$f(x) = -x_1^2 - x_2^2 - \dots - x_\lambda^2 + x_{\lambda+1}^2 + \dots + x_n^2 .$$

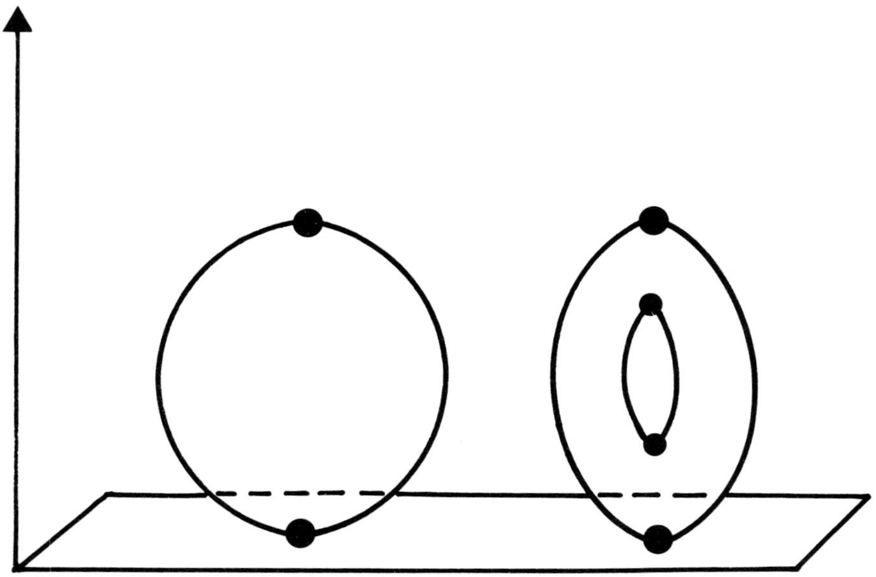

Fig. 10 See text.

The number of negative squares λ is referred to as *the index of a critical point*. It can be proved that the index is independent of the method of selection of the local coordinates. By definition, $\lambda \geqslant 0$.

Denote by $\mu_\lambda(f)$ the number of critical points of the index λ of the Morse function f. It appears that the alternating sum of these numbers

$$\sum_{\lambda \geqslant 0} (-1)^\lambda \mu_\lambda(f)$$

coincides with the Euler characteristic of a manifold M^2. Therefore the number written is independent of the Morse function f. Moreover, it is topologically invariant. It will be recalled that for the two-dimensional closed surface M^2 of genus p (Fig. 2) the Euler characteristic is $\chi(M^2) = 2 - 2p$. The treatment of Morse theory can be found in many textbooks (see, for example, Doubrovine et al., 1985).

Thus, the Morse function determines the topology of the whole manifold. To put it differently, what is local can be varied in a wide range but nevertheless determines the global properties of the manifold. Or, again, it signifies that each domain of the surface of the living organism knows its position in the whole and, moreover, the relative position of all the parts of the organism. Such a property was called by Wolpert (1969) positional information. Thus, the Morse function embodies some properties of positional information. The latter allows us to call the arbitrary Morse function on M a *scalar morphogenetic field.*

In some cases the scalar field determines not only the topology of a manifold but also its geometry. For instance, the curvature of a flat curve uniquely (up to motions) determines this curve. The spatial curve is determined by the curvature and torsion. The closed surface of a positive constant mean curvature is a sphere. The surface of a negative mean curvature is a Beltrami pseudosphere. If the surface of zero mean curvature is spanned on two sufficiently close equal and coaxial circles then it has one of the following three types:

- stable catenoid
- unstable catenoid
- two flat discs (Fomenko, 1984).

It is clear that for any change in the surface the scalar field on it must change, too. The reverse is not true, the scalar field can change but the surface in this case can be invariant, even metrically. Therefore one should distinguish radical changes in the scalar field, at which the topology of the surface is rearranged, from other changes (unimportant from our point of view). We believe that the local maxima and minima of the scalar field are centres, respectively, of evagination of the surface.

Vector morphogenetic fields

A vector field grad f, given in the local coordinates (x_i) by the vector of the partial derivatives $(\partial f / \partial x_i)$, can be associated with each function f on the Riemannian manifold. In our case the manifold is a closed two-dimensional surface and the function is the Morse function. The trajectories of the field grad f, as is known, are perpendicular (in the given Riemannian metrics) to the level lines of the function f. These trajectories represent lines of steepest descent. The critical points of the function f are no other than singularities of the vector field grad f, i. e. those points at which grad $f = 0$. For each point of the manifold with a defined vector field one determines *the index* as the number of turns of the field vector when carrying it along a sufficiently small circle around this point counter-clockwise (see first part). At the singularity of the vector field grad f, its index j and its index as a critical point of the function f are related by the relation $j = (-1)^{\lambda}$.

Now we shall discuss the inverse problem: to find for the given vector field v the function f whose gradient is the field v. In other words, is the equation grad $f = v$ always resolvable for f? The answer is obvious: the inverse problem is not always resolvable. This follows, if only, from the fact that the gradient of any function on the compact has at least two singularities since this function has both a maximum and a minimum. At the same time one can present an example of a vector field on a sphere with one singularity. In connection with this, the following theorem is interesting:

Theorem

(Presnov, 1973; Edelen, 1973). For any continuously differentiable map $\mathcal{F} : \mathbb{R}^n \to \mathbb{R}^n$ there exists, uniquely, a decomposition

$$\mathcal{F}(x) = \mathrm{grad}\, \varphi(x) + u(x)$$

such that $\langle x, u(x) \rangle = 0$ and $\varphi(0) = 0$.

Proof.

To begin with, we prove the existence of such a decomposition, describing it by formulae. For this purpose take an auxiliary function $\sigma(x) = \langle x, J(x) \rangle$ and put

$$\varphi(x) = \int_0^1 \frac{1}{\tau} \sigma(\tau x)\, d\tau .$$

Assume that $u(x) = J(x) - \operatorname{grad} \varphi(x)$. Check by calculation that $\langle x, u(x) \rangle = 0$:

$$\langle x, u(x) \rangle = \langle x, (J(x) - \operatorname{grad} \varphi(x)) \rangle = \langle x, J(x) \rangle - \langle x, \operatorname{grad} \varphi(x) \rangle =$$

$$= \sigma(x) - \sum x_i \frac{\partial}{\partial x_i} \int_0^1 \frac{1}{\tau} \sigma(\tau x)\, d\tau = \sigma(x) - \sum x_i \int_0^1 \frac{\partial \sigma}{\partial x_i}(\tau x)\, d\tau =$$

$$= \sigma(x) - \int_0^1 \frac{d}{d\tau} \sigma(\tau x)\, d\tau = \sigma(x) - \sigma(x) + \sigma(0) = \sigma(0) = 0.$$

Thus, the existence of the decomposition has been proved. To check, that it is unique, let there be two such decompositions

$$J = \operatorname{grad} \varphi + u, \qquad \langle x, u \rangle = 0,$$
$$J = \operatorname{grad} \psi + v, \qquad \langle x, v \rangle = 0.$$

Subtract the second from the first and use the linearity of the gradient and the scalar product:

$$0 = \operatorname{grad} (\varphi - \psi) + (u - v), \qquad \langle x, (u - v) \rangle = 0$$

Hence,

$$\langle x, \operatorname{grad} (\varphi - \psi) \rangle = - \langle x, (u - v) \rangle = 0.$$

But the solution of a homogeneous linear equation in first-order partial derivatives is a constant (see, for example, Arnold, 1980), i. e. $\varphi - \psi =$ const. And since $\varphi(0) = \psi(0) = 0$ we obtain $\varphi = \psi$, whence $u = v$. The proof has been completed.

Generalisation.
This theorem has been formulated and proved for the Euclidean space $\mathbb{R}^n$. It can be generalised to an arbitrarian Riemannian manifold M but we confine ourselves to manifolds without boundary.

Theorem.
For any smooth vector field J on the closed Riemannian manifold M there exists a unique decomposition

$$J = \text{grad}\,\varphi + u$$

such that $\langle x, u \rangle = 0$, and in this case the potential φ is uniquely defined up to a constant.

Explanation.
Here J and u represent maps from M into a space of tangent bundle TM. The scalar product $\langle x, u \rangle$ is defined in view of the fact that the Riemannian metrics from M is lifted to TM and M is canonically imbedded into TM (by means of the zero section). In the local coordinates, in which the Riemannian tensor is an identity matrix, the condition $\langle x, u \rangle = 0$ is written as $\sum_{i=1}^{n} x_i u_i = 0$. The same condition can be rewritten briefly in the same coordinates as $L_u x^2/2 = 0$, where L_u is the Lie derivative along the vector field u.

Proof.

Cover the manifold M with charts U_k (i. e. with domains which are homeomorphic to the Euclidean space $\mathbb{R}^n$) so that the images (under fixed homeomorphisms $\mathbb{R}^n \to U_k$) of open unit spheres $V_k \subset U_k$ also cover the manifold M. By virtue of the paracompactness of M it may be assumed that this coverage is locally finite. Let us choose, next, the partition of unity, subordinate to the covering $\{V_k\}$ (see, for example, Doubrovine et al., 1984). By definition, it represents such a set of the functions $M \to \mathbb{R}^1$ that

(1) $0 < \psi_k(x) < 1$ for all $x \in M$;

(2) $\operatorname{supp} \psi_k \subset V_k$ for any k (the symbol $\operatorname{supp} \psi_k$ denotes the support of the function ψ_k, i. e. a set of points $x \in M$ in which $\psi_k(x) \neq 0$); and

(3) $\sum_k \psi_k(x) = 1$ for all $x \in M$.

By multiplying the field J by the function ψ_k, we localise it in the domain V_k in which the decomposition theorem, we have proved in the Euclidean case, is valid. However, we apply this theorem to the vector field $J_k = J\psi_k$ not in the domain V_k but in the (also Euclidean) domain U_k:

$$J_k = \operatorname{grad} \varphi_k + u_k \qquad \langle x, u_k \rangle = 0$$

Outside the support $\operatorname{supp} \psi_k$ the filed J_k is zero. In particular, it is zero on $U_k \setminus V_k$ and any sphere $B \subset U_k \setminus V_k$. But the zero field has a trivial decomposition: $0 = \operatorname{grad} c + 0$, where c is a constant (however, the potential in general is defined up to a constant). According to the already proved uniqueness of the decomposition in B the function u_k is identically zero in B and the potential φ_k is a constant. Since this is true for any sphere $B \subset U_k \setminus V_k$ then it must be true too for $U_k \setminus V_k$ as well. Thus, the function φ_k and the field u_k are extended continuously and smoothly by the same constant and zero to the whole domain U_k and, moreover, to the whole manifold M. So, the local decomposition

$$J_k = \operatorname{grad} \varphi_k + u_k,$$

is essentially a global decomposition.

Having performed such a decomposition for each k, we sum these decompositions together. By virtue of the linearity of the gradient and in view of the fact that for each point $x \in M$ the sums are finite, we have

$$J = J\left(\sum_k \psi_k\right) = \sum_k J\psi_k =$$

$$= \sum_k (\operatorname{grad} \varphi_k + u_k) = \operatorname{grad} \sum_k \varphi_k + \sum_k u_k .$$

Putting $\varphi = \sum \varphi_k$ and $u = \sum u_k$, we obtain the required decomposition

$$J = \operatorname{grad} \varphi + u,$$

since $\langle x, u\rangle = \langle x, \sum u_k\rangle = \sum \langle x, u_k\rangle = 0$. The theorem has therefore been proved.

Note 1.

The decomposition theorem has found brilliant application in the thermodynamics of irreversible processes far from equilibrium. It gives a decomposition of the thermodynamic fluxes $J(X)$ into dissipative and nondissipative parts. The dissipative part is a gradient, it automatically satisfies the nonlinear Onsager reciprocity relations. We note that in this case no linearity requirements are imposed on the initial field of thermodynamic fluxes. It is noteworthy that this fundamental decomposition theorem remained unknown for a long time despite the abundance of papers on the Onsager relation in the linear region and near it. This theorem is indicative of the potentiality of the Glansdorff-Prigogine evolution criterion (Presnov, 1973, 1978; Presnov, Malyghin, 1982; see also Bataille et al., 1982). Independently this theorem was rediscovered by Merkin (1975) with a view to determining the structure of forces in mechanics.

Note 2.

The term 'dissipativity', which has been dealt with above, has a physical meaning. We shall explain it in more detail. Let there be a thermodynamic system (X, J), where X are thermodynamic forces and $J(X)$ are thermodynamic fluxes. As is known, the Lie derivative of the function $1/2 <X, X> = X^2/2$ along the vector field J

$$L_J X^2/2 = X_1 J_1 + \ldots + X_n J_n$$

is none other than the *entropy production* of this system. Since for the fluxes $u(x)$ the equality $L_u = x^2/2 = 0$ holds it is natural to call them nondissipative. It is appropriate to mention here that in mathematics there is also a notion of 'dissipativity'. Namely, the dynamic system

$$\dot{x} = J(x)$$

is called *dissipative* if $\operatorname{div} J < 0$. We emphasise that although the thermodynamic system (X, J) can be formally associated with the dynamic system

$$\dot{X} = J(X)$$

notions of 'dissipativity' will differ so that they should not be confused. It has been proposed (Presnov, 1984) that the theorem proved above applies to the stochastic behaviour of such systems.

So, our decomposition is associated with the physical notion of 'dissipativity'. It is interesting that there is a decomposition related to the mathematical notion of dissipativity.

The Hodge theorem

(see, for example, Marsden, McCracken, 1976). For any smooth vector field J on the closed Riemannian manifold M there exists a unique decomposition

$$J = \operatorname{grad} g + Y,$$

such that $\operatorname{div} Y = 0$.

We apply our theorem to morphogenesis. According to it, any vector field has two additive components. One of them, the gradient (dissipative), is related to a scalar morphogenetic field (the Morse function) which is its potential. The other nondissipative component has no potential. An arbitrary vector field on the surface of a developing ovum will be called a *morphogenetic vector field.*

An unfertilised ovum carries (see first part) a cortical morphogenetic vector field determined by its previous spatial relations with the (cellular and noncellular) environment in the maternal organism. We believe that the dissipative component of this field is determined by the maternal organism, coordinates the growth of limbs and determines changes in the topology of the organism, i. e. the surface of an embryo follows changes in the scalar field (invaginates, bends, etc.). Thus, the dissipative part of the cortical morphogenetic field represents a genetic component of morphogenesis.

What does the nondissipative component of the initial morphogenetic field control? Activation by or without a sperm includes the program of changes in this component. At first the nondissipative part overshadows the dissipative part, in other words, conceals the dissipative (genetic) component in the early stages of development. In the course of time the nondissipative part goes to zero and disappears, freeing the dissipative part which was from its influence concealed in the initial field of the egg.

Consider the process of development in detail. The potential of the dissipative component of the morphogenetic vector field on the surface of the egg is some Morse function. It has some singularities. Among them, in particular, are critical points of index 0 (minima) and index 2 (maxima). At first these points are not apparent because they are overshadowed by the nondissipative part of the field. In the course of time, however, with the disappearance of the nondissipative components these points show through and play their role. It is precisely in them that there occurs invagination and evagination of the surface of the organism, i. e. a detailed morphogenesis is accomplished. A set of the critical points of indices 0 and 2 is divided into pairs of conjugate points, altogether there is an even number of such points, namely $2p$. Their mutual annihilation occurs, during the process of development, at the meeting of these points and the topology of the surface changes. Topologically this change consists in the attachment of handles (see second part). In reality the picture can be more complex because of the appearance of pairs of conjugate critical points which are also described topologically by the attachment of handles. Thus, the spherical surface of the embryo is rearranged to the surface of genus p (the adult organism with a definite surface (Fig. 11).

Analytically the development process described can be represented, for example, as

$$J_\lambda = \mathrm{grad}\,(\lambda \varphi_1 + (1-\lambda)\varphi_0) + (1-\lambda)u_0.$$

The moment of activation corresponds to $\lambda = 0$ and the end of pattern formation to $\lambda = 1$. The function φ_1 is obtained from the mutual annihilation of the conjugate critical points of the function φ_0 .

Certainly, the representation of the developmental process as a linear homotopy J_λ is written for illustrative purposes. It would be interesting to consider, also, other similar representations of the homotopy of the field J and the corresponding potential φ . Here, probably, methods of the catastrophe theory are applicable.

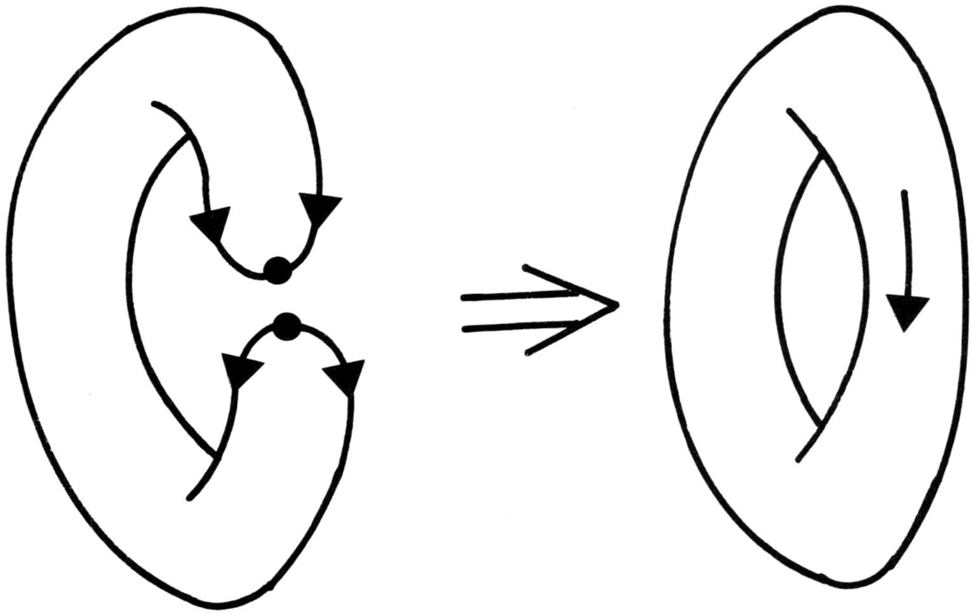

Fig. 11 See text.

Gastrulation as one of the aspects of the topological structure of positional information

Consider the blastula of any living organism. Obviously, its outer surface is a sphere or may be considered as homeomorphic to a sphere (it will be recalled that, from a topological point of view, we do not distinguish homeomorphic surfaces). Blastomeres, adjoining the environment, divide this surface into domains. If we disregard details of the interblastomere contacts then we obtain a graph located on our sphere. The edges of this graph represent an abstraction of interblastomere contacts. Thus, we consider the blastula as a sphere with a graph inscribed upon it. It is clear that the real structure of the blastula has been considerably simplyfied in this case. It is remarkable that such a simple interpretation enables one to express subtle statements about this structure.

We now pose the question as to how many types of regular patterns of such graphs can exist. Here we shall simplify the picture even more, considering as unimportant such details as the lengths of the edges of this graph and the angles between the edges. A graph on a sphere will be called *regular* (or *topologically homogeneous)* if, firstly, each domain on the sphere, obtained from the division of the sphere by this graph, is limited by the same number of edges and if, secondly, at each vertex of the graph the same number of edges meet. We give an example of such a graph from which it will be clear why it has been called regular. Consider a geometrically regular convex polyhedron and surround it with a sphere the centre of which is within the polyhedron. If a light source is placed at the centre of the sphere then the shadows of the polyhedron edges will just set on the sphere a regular graph (the process of casting a shadow can be thought of also as inflation of a rubber-made polyhedron till all of its vertices, edges and faces come in contact with the sphere). Even ancient Greeks knew that there are only five regular polyhedrons: a tetrahedron, an octahedron, a hexahedron (cube), an icosahedron, and a dodecahedron. But for the definition of a regular polyhedron much greater detail is required than that required for the definition of a regular graph. Perhaps there are other regular graphs? It is remarkable that the answer is: no. Namely, the Euler topological theorem states that there are also exactly five regular graphs on a sphere (see, for example, Appendix I to the book of Presnov and Isaeva (1985)).

Consider the regular polyhedrons in more details. Only a tetrahedron (4 faces), an octahedron (8 faces) and an icosahedron (20 faces) have triangular faces. Exactly three edges meet at each vertex of the tetrahedron and therefore it is structurally stable: small deformations of its edges do not change this number. The number of edges meeting at the vertex is called the *degree of the vertex.* In the octahedron the degrees of the vertices are equal to four and in the icosahedron, to five. Therefore the octahedron and icosahedron, are structurally unstable: small deformations can lead to break-down of their vertices into several vertices of smaller degree. The cube has 6 square faces while the degrees of its vertices are equal to three and therefore it is structurally stable. Finally, the dodecahedron has 12 pentagonal faces and the degrees of its vertices are equal to three, too, therefore it is also structurally stable.

There are, therefore, in total, five topologically homogeneous graphs on a sphere. The number of edges in such graphs is not great - no more than 20 and taking into account structural stability it is even smaller - no more than 12.

Now turn to the consideration of a cleaving embryo. Taking into account the foregoing purely arithmetical considerations it is clear that not later than after the fourth division of cleavage the pattern of the surface of the embryo becomes topologically inhomogeneous. We note here in passing that if we lived in a two-dimensional world then this statement would be wrong since a convex polygon can continue cleaving indefinitely, remaining regular in this case even in the metric sense. Thus, the existence of inhomogeneity in the patterns shown by blastomeres during cleavage results from the topological structure of our three-dimensional physical space. Our view of the topological structures of positional information is suggestive of the relation of this fact to gastrulation phenomenon. Incidentally, the obligatory presence of pentagons among the overwhelming majority of hexagons on the pattern of the intestine crypt was already attributed to the idea of the growth of the crypt precisely at the locations of such pentagons (Pyshnov, 1980, Fig. 9). In essence the pentagons and hexagons, derivable from the topological pattern of the graph, determine a discrete morphogenetic field which in the structurally stable case can be constant only for three patterns on the sphere.

Certainly, the gastrulation phenomenon is not directly related to the inhomogeneity of the patterns of blastomeres during cleavage because the invagination usually originates only in one place while the inhomogeneity can be observed in several places. We therefore think that the topological inhomogeneity of the blastula is the necessary cause of gastrulation but nevertheless it cannot be a sufficient basis for it. Hence, it is necessary to search for other causes which result in the gastrulation phenomenon.

In the egg there exists a vegetative singularity which determines the initial polarity of the ovum. This polarisation arises from topological factors (see part 1) according to which, on a sphere, there is necessarily a singularity of the morphogenetic vector field present. The identical topological character of the two phenomena encourages us to compare them. Namely, on the sphere there are singularities of the morphogenetic field and inhomogeneities of the blastomere pattern. Their coincidence will just, possibly, be the cause of gastrulation. This assumption is still to be proved through searches for a mechanism causing such a coincidence. The experimental fact, that the sea urchin happens to have monster-twins with two or three archenterons, our unpublished data, speaks in favour of its existence. We attribute this to the noncoincidence of singularities.

Conclusion

However, all these singularities of the cortical layer, egg polarisation and gastrulation do not pertain to biological details alone. Our sketch reveals also other meaningful aspects of a particularly mathematical world outlook. Not very long ago, at the end of the 19th century, a nonconstructive set-theoretical trend appeared in mathematics. It was developed in studies by, especially, K. Weierstrass, R. Dedekind and G. Cantor (Goldblatt, 1979). The difference between constructive and nonconstructive mathematics lies in the fact that the former provides statements about the existence of mathematic objects with rules for their visualisation and therefore rejects a number of principles of set-theoretical mathematics, i. e. abstraction of actual infinity for example. Meanwhile, this abstraction is responsible for the appearance of the so-called pure theorems of existence: in our case the theorem for the even-dimensioned sphere. The pure existence theorems underwent a period of blossom. They permeate the whole edifice of modern mathematics, postulating the existence of some objects, but at the same time giving no methods for the detection of these objects. There are grounds for believing that the objects whose existence is established without the abstraction of actual infinity but only in terms of a much more modest abstraction of potential feasibility are much closer to objective reality.

Thus we considered the interaction between the experimental and mathematical world (Bourbaki, 1948). And therefore if it is possible to explain the inherent polarisation peculiar to the egg, then this implies that the path leading to this understanding requires formal attention, if not thorough investigation. Who knows, perhaps the abstraction of actual infinity intuitively understood is the basis of theoretical constructions in biology, whilst the actual infinity hypothesis provides a more or less exact modelling of these phenomena.

Finally we would like to present a little fragment from the magnificent sketch 'L'Architecture de mathématiques' written by a well-known group of french scientists: 'Dans la conception axiomatique, la mathématique apparaît en somme comme un réservoir de formes abstraites - les structures mathématiques; et il se trouve - sans qu'on sache bien pourquoi - que certains aspects de la réalité expérimentale viennent se mouler en certaines de ces formes, comme par une sorte de préadaption. Il n'est pas niable, bien entendu, que la plupart de ces formes avaient à l'origine un contenu intuitif bien déterminé; mai c'est précisement en les vidant volontairement de ce contenu qu'on a su leur donner toute l'efficacité qu'elles portaient en puissance, et qu'on les a rendues susceptibles de recevoir des interprétations nouvelles, et de remplir pleinement leur rôle élaborateur.' (Bourbaki, 1948).

Fractal Shapes of Cell Membranes and Pattern Formation by Dichotomous Branching Processes

T. F. Nonnenmacher

Introduction

The design of pattern formation in biological systems challenged biologists, mathematicians and physicists to develop models in order to explain some fundamental features of nature's program for the architecture of complicated biological structures. The concepts of fractal dimensionality, renormalisation group theory, scaling theory and the principle of minimum entropy production have already been stressed and applied to chemical, physical and biological systems.

In this paper Mandelbrot's (1983) concept of fractal dimensionality is discussed and applied to different types of pattern formation in biological systems. In particular we will focus our interest on the fractal shape of cell membranes, and we will present an affine scaling law for dichotomous branching processes. The approach to biophysical problems presented here takes into account both, stochastic and deterministic methods to describe biological evolution processes. Power-law scaling plays a central part in our considerations. The mathematical background is the fractal concept of statistically self-

Thermodynamics and Pattern Formation in Biology

similar objects for our estimation of the Hausdorff-Besicovitch dimension D of the shape of cell membranes, and we modify some Weierstrass-type functions in such a way that they can be used to model dichotomous branching processes. An application to the human bronchial tree will be given and available experimental data will be interpreted within this context.

Already Thompson (1942) formulated the 'principle of similitude' stating that 'although biological systems may evolve according to different laws than do physical systems, they cannot violate basic physical laws in that process of development' (West et al., 1986). Meanwhile a number of scaling relations for biological systems have been derived and applied to interprete experimental data. Indeed, the concept of scaling has subsequently become a powerful strategy.

The analytical study of scaling in physical systems is essentially based on group-theoretical renormalisation techniques (Wilson, 1975) for thermodynamic systems. On the other hand the property of self-similarity or scaling is also one of the central concepts of fractal geometry (Mandelbrot, 1983). Indeed, Wilson's renormalisation group theory and Mandelbrot's fractal concept draw on the same inspiration, but the parallelism between both approaches is not quite complete. While the thermodynamic aspproaches to scaling properties based on Wilson's renormalisation technique try to 'explain' via models the underlying physical mechanism, Mandelbrot's fractal approach is a mathematical approach generalising the concept of Euclidean geometry with topological dimension $D = 1, 2, 3$ to fractal dimension D, which must not be an integer.

The concept of fractal dimensionality

A topological object (like a line-segment, a square in two-dimensional, or a cube in three-dimensional Euclidean space) possesses a scaling property: it can be divided into N identical (exactly self-similar) parts each of which is scaled down from the whole by a ratio $r = (1/N)^{1/D}$. Only in Euclidean space one has $D = 1, 2, 3$, respecitely. Mandelbrot's (1983) generalisation is straightforward: given a self-similar object, its fractal or similarity dimension is generally defined by

$$D = (\log N)/\log(1/r) \tag{1}$$

where D is in general not an integer (the von Koch curve (snowflake) has $D = 1.26$, for instance).

Not all objects in nature exhibit exact self-similarity. However, they often possess the property of 'statistical self-similarity' like coastlines or boarders between different countries, or even the surrounding curve of a cell membrane (Paumgartner et al., 1980); Baumann, Nonnenmacher, 1988). The closer one looks at a coastline, the more detail one sees; however, any attempt to divide the coastline into smaller parts results in the resolution of still more structure identical only in its statistical properties to the whole. The concept of statistical self-similarity can also be quantified in terms of a fractal dimension D. Using a yardstick (scale unit) of size r to measure the length of a coastline, the total length $L(r)$ is given by (Mandelbrot, 1983) $L(r) = rN(r)$ with $N(r) = \lambda/r^D$ leading to (see Fig. 1 a)

$$L(r) = \lambda/r^{D-1} \tag{2}$$

where λ is the length if $D = 1$.

An alternative approach (Fig. 1 b) to measure the fractal dimension D of a coastline is based on the box-dimension (Minkovski's method): One needs N(r) boxes (squares or circles etc.) of size r to cover the full object. Again one has $L(r) = \lambda / r^{D-1}$.

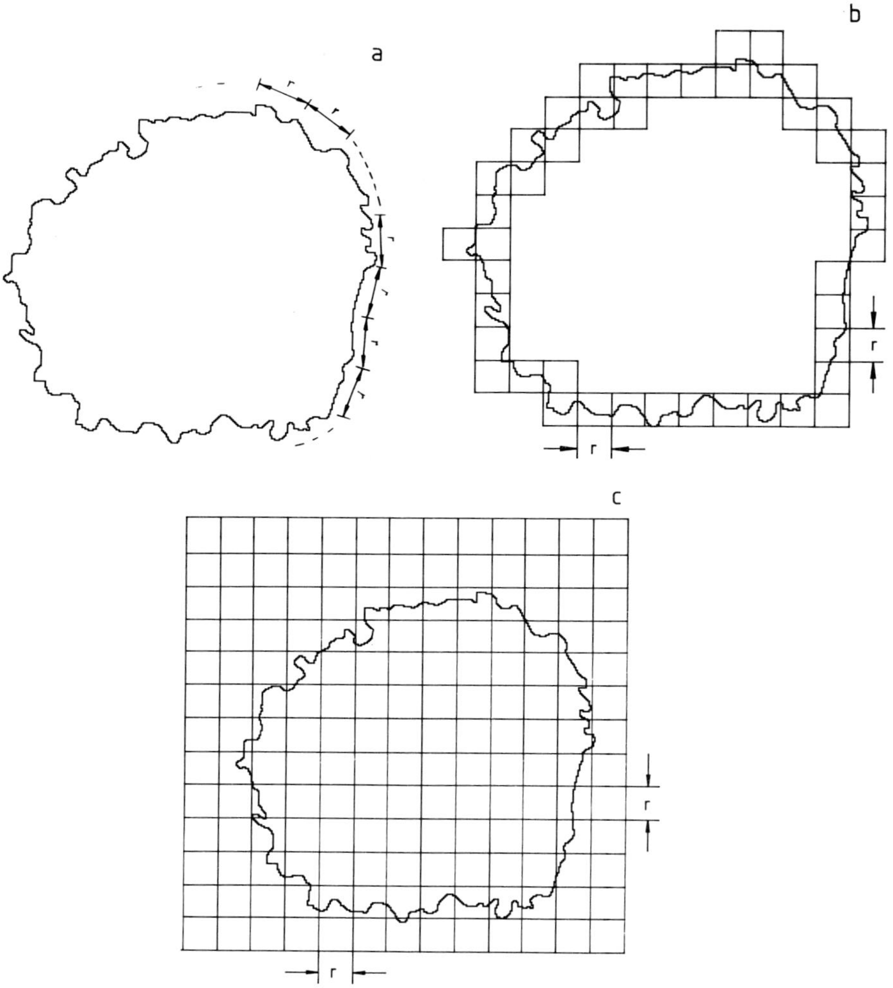

Fig. 1 Three alternative methods for the estimation of the fractal dimension D for the shape of cell membranes. a: the length L of the fractal curve is measured in units of a yard-stick-length r; b: the method of box-dimension, c: the concept of grid-dimension (see text).

A third method (practical alternative, Fig. 1 c) for the estimation of D is based on the concept of grid dimension: One needs N(r) grid squares. Count the number of squares which contain any piece of the curve, and use different origins for different scans. Again one finds $L(r) = \lambda / r^{D-1}$. All these methods lead - as it must be - to the same value for the fractal dimension D. The fractal curve shown in Figures 1 a to 1 c represents the 'coastline' of the human granulocyte membrane investigated in the next section.

Mandelbrot (1983) has already observed that one needs λr^{-D} intervals of length r, summing up to $L(r) = \lambda r^{1-D}$ for the coastline's length, where the dimension D may depend on the coastline that is chosen, and different portions of the same coastline, if considered separately, may produce different values of D. However, the value of D does not depend on the method used to estimate the length of the coastline. Published data on different coastlines lead to fractal dimensions D of about 1.2 to 1.3.

Sometimes it is of advantage to introduce the concept of 'mass-dimension' $m(L) \sim L^D$ (Mandelbrot, 1983). A pertinent example is the distribution of proteins on membrane surfaces. Investigations of such problems are usually based on the determination of the two-point correlation function g(x). If g(x) is known theoretically or determined experimentally (in many cases one has asymptotically a power-law dependence $g(x) \sim 1/x^{\kappa}$) then (by definition)

$$m(L) \sim \int^{L} g(x)\, d^E x \sim L^{E-\kappa} = L^D,$$

i. e. the fractal 'mass-dimension' D is given by

$$D = E - \kappa$$

where E is the Euclidean topological dimension E = 1, 2 or 3 and κ is the power-law exponent of the correlation function g(x). We note that the knowledge of g(x), of course, stores more physical information than is stored in the fractal dimension D, because integration over g(x) destroys some information. Thus, if an acceptable physical mechanism is already known it is not necessary in this case to go through the concept of fractal geometry (Krumhansl, 1986).

Application of the coastline concept to cell membranes

Recently Weibel (1979) observed that measurements of the human lung's alveolar area are conflicting: light microscopy yields ca. 80 m^2, while electron microscopy results in 140 m^2. Similarly, Paumgartner et al. (1980) found sharp disagreement between estimates of surface and volume density of endoplasmic reticulum and mitochondrial membranes in liver cells at different microscopic magnification. The same effect stating that the length (circumference) of a membrane - projected into a plane - is not a smooth line but a fractal curve with fractal dimension , was found very recently (Baumann, Nonnenmacher, 1987) by analysing the micrograph (in the image plane) of human granulocyte cell-membranes at different magnification.

Applying Mandelbrot's concept of the coastline's fractal dimension the disagreement between measurements at different magnifications disappear: like coastlines the circumference of a cell membrane is a fractal line, i. e. the smaller the scale of observation, the longer becomes the coastline which is observed (because more detailed structure can be measured). For the outer mitochondrial liver membrane Paumgartner et al. (1980) found $D = 1.09$, for the inner mitochondrial membrane $D = 1.54$ and for the endoplasmic reticulum the fractal dimension has the value $D = 1.72$. We found (Baumann, Nonnenmacher, 1987) $D = 1.10 \pm 0.05$ for the 'coastline' of the human granulocyte membrane in the image plane (see Fig. 2). For different portions of the same membrane, if considered separately, different (local) values of D may be found. A detailed discussion will be given in a forthcoming paper (Baumann, Nonnenmacher, 1987).

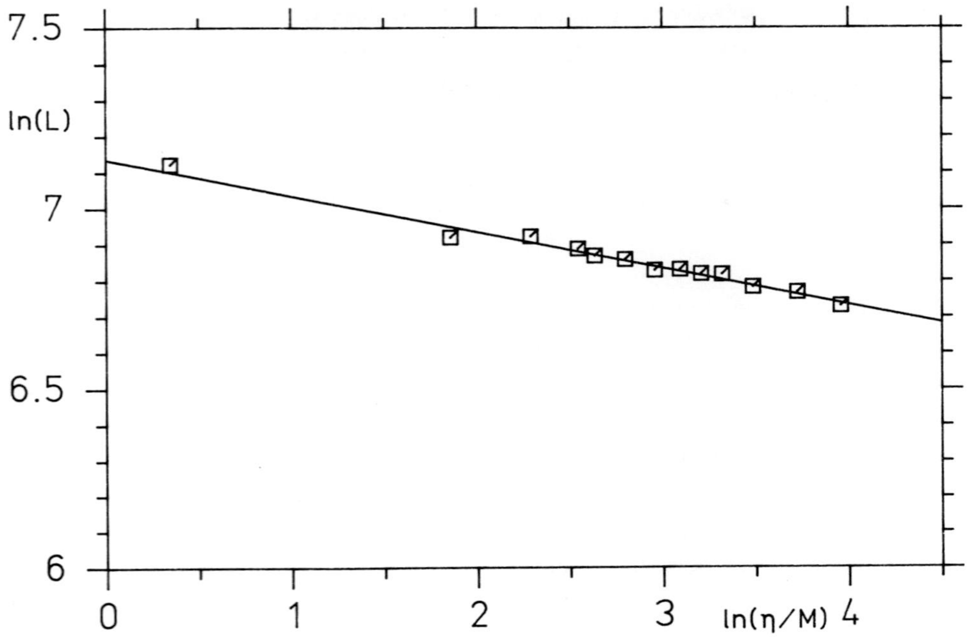

Fig. 2 log-log-plot for the fractal shape L(M) of a human granulocate membrane according to equation (4). The magnification M spans the range $8.000 \leq M \leq 12.500$. The slope of the straight line (mean square fit) gives the value of the fractal dimension D(= 1.10). The points represent experimental values for different magnifications M based on the yard-stick-method of Fig. 1 a.

As already pointed out above the concept of the coastline's fractal dimension D relates the length L of the coastline to D and to the yardstick unit r. However, assuming that the scale unit length r decreases proportionately as the magnification M increases ($r = \eta M^{-1}$) one obtains (Paumgartner et al., 1980) a formula relating L to M and D:

$$L(M) = \lambda \eta^{1-D} M^{D-1} \qquad (3)$$

where η is the 'resolvable scale unit' at $M = 1$, i. e. observing the magnified micrograph in the image plane carrying the test system. Using a log-log-plot for the data it is conventional to rewrite this result as

$$\log L(M) = (1-D)\log(\eta/M) + \log \lambda . \qquad (4)$$

This relation may be of help for the interpretation of experimental data obtained from different magnifications by light or electron microscopy. Figure 2 shows the graph of (4) for the human granulocyte membrane for magnifications in the range $8.000 \leqslant M \leqslant 12.500$. The upper limit $M = 12.500$ is dictated by our experimental device: For $M > 12.500$ the micrograph does not cover the full extension of the cell but only portions of it. However, one needs the complete circumference of the cell in order to determine its global fractal dimension D, since different portions may lead to different values of D (local fractal dimension).

Weierstrass series and related functions: affine scaling laws

Not all pattern formations exhibit the property of exact or statistical self-similarity which leads to a scaling behaviour characterised by only one number: the fractal dimension D (like the coastline concept, for instance). There are many examples requiring two scaling factors (or even more) for mathematical formulation of the scaling property. Such nonuniform scaling is known as self-affine rather than self-similar scaling. For self-affinity, however, the fractal dimension D is not as easily to define as for self-similarity. A prominent approach to model self-affine scaling properties mathematically is based on modified Weierstrass functions which have a discrete and geometric spectrum (Mandelbrot, 1983).

The series

$$W_0(\varsigma) = \sum_{n=0}^{\infty} \frac{e^{\gamma^n \varsigma}}{\gamma^{(2-D)n}} \tag{5}$$

represents for ς = it the complex Weierstrass function, which is (for $\gamma > 1$ and $1 < D < 2$) continuous but nowhere differentiable. Its formal extension, however, to $D < 1$ is both continuous and differentiable. t is usually the time variable. Mandelbrot pointed out that $W_0(it)$ is a fractal function in the sense that its graph is a curve whose Hausdorff-Besicovitch dimension exceeds unity. The frequencies γ^n form a discrete 'Weierstrass spectrum' spanning the range $1 \leqslant \gamma^n \leqslant \infty$ in geometric progression, i. e. the spectrum of the Weierstrass function is bounded below by $\gamma^0 = 1$. This lower bound $(n = 0)$ clashes with the scaling property. Hence, in order to eliminate this lower bound and to extend the spectrum γ^n down to $\gamma^n = 0$, i. e. to the full range $0 \leqslant \gamma^n \leqslant \infty$, Mandelbrot proposed the (Mandelbrot-Weierstrass) series (Berry, Lewis, 1980).

$$W_{MW}(t) = \sum_{n=-\infty}^{\infty} \frac{1 - e^{i\gamma^n t}}{\gamma^{nH}} e^{i\phi_n}, \tag{6}$$

where $H = 2 - D$, and ϕ_n represents an arbitrary phase. The added terms corresponding to $n < 0$ converge if $0 < H < 1$. The graph of $\mathrm{Re}W_{MW}(t)$ or $\mathrm{Im}W_{MW}(t)$ has a Hausdorff-Besicovitch dimension $D = 2 - H$, where H is the Hurst coefficient introduced by Mandelbrot (1983).

Two special cases already discussed in literature (Berry, Lewis, 1980) are of particular interest:

1. The choice $\phi_n = n\mu$ gives a deterministic $W_{MW}(t)$ with the affine scaling property

$$W_{MW}(\gamma t) = e^{-i\mu}\gamma^{H} W_{MW}(t). \tag{7}$$

2. Choosing ϕ_n as random phases gives a stochastic $W_{MW}(t)$ with statistically stationary increments $W_{MW}(t+\tau) - W_{MW}(t)$ and with variance

$$\langle |W_{MW}(t+\tau) - W_{MW}(t)|^2 \rangle = 2\sum_{-\infty}^{\infty} \frac{1-\cos\gamma^n\tau}{\gamma^{2nH}} = V(\tau), \tag{8}$$

which is a smooth function if $1 < D < 1.5$, and a fractal function of dimension D if $1.5 < D < 2.0$ (Berry, Lewis, 1980).

The variance $V(\tau)$ obeys the affine scaling law

$$V(\gamma\tau) = \gamma^{2H} V(\tau) \tag{9}$$

and may serve as a mathematical model for fractional Brownian motion and for 1/f-noise.

The power spectrum $P(\omega)$ of $W_{MW}(t)$ is the Fourier transform of the correlation function

$$C(\tau) = \langle W_{MW}(t+\tau) W^{*}_{MW}(t) \rangle$$

and is given by

$$P(\omega) = \sum_{n,m} \frac{e^{i(\phi_n - \phi_m)}}{\gamma^{(n+m)H}} \delta(\omega) + \sum_{-\infty}^{\infty} \frac{\delta(\omega - \gamma^n)}{\gamma^{2nH}} .$$

Application of the Poisson summation formula on the second sum (we neglect the first sum on the r. h. s. which contains only zero-frequency terms) 'transforms' the discrete Weierstrass spectrum γ^n to a continuous spectrum of frequencies $\omega > 0$:

$$P(\omega) = \frac{1}{\omega^{\beta} \ln \gamma} \left\{ 1 + 2 \sum \cos\left(\frac{2\pi m}{\ln \gamma} \ln \omega\right) \right\} \tag{10}$$

which is the affine scaling according to $P(\gamma\omega) = \gamma^{-(2H+1)} P(\omega)$. Here, we have put $\beta = 2H + 1$ $(0 < H < 1)$, and M is a cut-off depending on the range of frequencies $\omega_1 < \omega < \omega_2$ to be analyzed. Summation on m can be carried out leading to

$$P(\omega) = \frac{1}{\omega^{\beta} \ln \gamma} \left\{ 1 + 2 \cos\left[(M+1)\frac{\pi}{\ln \gamma} \ln \omega\right] \times \sin\left[M \frac{\pi}{\ln \gamma} \ln \omega\right] \operatorname{cosec}\left(\pi \frac{\ln \omega}{\ln \gamma}\right) \right\}.$$

The trend of (10) is

$$P(\omega) \sim 1/\omega^{\beta}, \quad \text{with } \beta = 2H + 1 \tag{11}$$

representing 1/f-noise. The second term in (10) represents harmonic oscillation around the trend. β is restricted to the range $1 < \beta < 3$, since $0 < H < 1$.
We note that the choice $H = 1/2 (\beta = 2)$ leads to the power-law spectrum of normal Brownian motion: $P(\omega) \sim 1/\omega^2$.

Thus, the usual Brownian formula $< (\Delta R(t))^2 > = ct$ (with $c = \text{const}$) relating the mean square displacement to the time-variable may be replaced by the generalisation (Mandelbrot, 1983) $<(\Delta R)^2) > ct^{2H}$. This extends the concept of normal Brownian motion (NBM) to fractional Brownian motion (FBM) falling into two sharply distinct subclasses:

1. the values $0 < H < 1/2$ correspond to antipersistent FBM, and
2. the values $1/2 < H < 1$ correspond to persistent FBM. The special case $H = 1/2$ represents just NBM.

Like the increments of the Mandelbrot-Weierstrass function the mean square displacement $< (\Delta R(t))^2> = ct^{2H}$ shows an affine scaling property: If the time scale t is changed by a factor γ ($t \rightarrow \gamma \cdot t$), then we obtain $< (\Delta R(\gamma t))^2 > = c\,\gamma^{2H} t^{2H} = \gamma^{2H} < (\Delta R(t))^2>$ 1 representing an affine scaling law for the increments ΔR of fractional Brownian motion. A model, discussed recently (Nonnenmacher, 1985 a, 1985 b), for chemokinetic movement of granulocyte cells shows good agreement with experimental observations for $H = 1/2$(NBM).

In addition to the stochastic properties of the Mandelbrot-Weierstrass function $W_{MW}(t)$ based on random phases ϕ_n there is a wide class of applications of modified deterministic ($\phi_n = 0$) functions $W(\xi)$. So, for instance, replacing (2-D) by H and adding terms $n < 0$ to the series (5) one obtains

$$W(\xi) = \sum_{-\infty}^{\infty} \frac{e^{\gamma^n \xi}}{\gamma^{nH}}, \quad \xi = x_1 + i x_2 . \tag{12}$$

The series converges for $x_1 < 0$, $\gamma > 1$ and $H < 0$. In addition, the 'frequencies' γ^n form a discrete Weierstrass spectrum spanning the full range $0 \leqslant \gamma^n \leqslant \infty$ with the consequence that (12) has the affine scaling property

$$W(\gamma \xi) = \gamma^H W(\xi). \tag{13}$$

Applying Poisson's summation formula to (12) one obtains (Nonnenmacher, Nonnenmacher, 1988)

$$W(v) = \frac{|v|^H}{\ln \gamma} \left\{ \Gamma(-H) + 2 \sum_1^{\infty} |\Gamma(v_m)| \cos\left(\frac{2\pi m}{\ln \gamma} \ln|v| - \arg \Gamma(v_m)\right) \right\} \tag{14}$$

where $v = -\xi = -x_1 - ix_2$, and $\nu_m = -H + i2\pi m/\ln\gamma$.

We will show that for $x_2 = 0$ the series (12) models the design of pattern formation for dichotomous branching processes and represents just the singular (scaling) part of the diameter law for the bronchioles derived recently by West et al. (1986) by going through the renormalisation group approach. We notice that (14) has, of course, the same affine scaling property (13) as the original series (12), because the Poisson summation formula leaves the scaling properties invariant.

Pattern formation by dichotomous branching processes

Dichotomous branching processes are of crucial importance, for instance, for the design of biological trees (botanical trees, neuronal trees, bronchial trees, arteries etc.). Mandelbrot (1983) reported some earlier observations and measurements on botanical trees and introduced the diameter exponent Δ in order to generalise Leonardo da Vinci's observation that the botanical tree's branch diameters before (d) and after (d_1, d_2) a bifurcation, satisfy the relation

$$d^{\Delta} = d_1^{\Delta} + d_2^{\Delta} \tag{15 a}$$

For certain botanical trees one has $\Delta = 2$ (Leonardo da Vinci's estimate), for arteries $\Delta \sim 2.7$ (Suwa, Takahashi, 1971) and for the bronchial tree $\Delta \approx 3$ (Weibel, 1963, 1980).

Concerning the lung's architecture the conventional interpretation of a variety of physiological tests of lung function is based on a symmetrical lung structure.

Asymmetries in lung anatomy have been discussed by Horsfield and Cumming (1968), and recently by Goldberger et al. (1985).

According to experimental findings (Weibel 1963, 1980) the airways of the human lung systematically scale down to about 23 generations of dichotomous branching, beginning with the trachea (generation number $z = 0$) to alveolar ducts and sacs ($z = 20$ to 23). Within each generation, the lengths and diameters of the airpipes have a characteristic range of size. At least, for the first ten generations ($z = 1, ..., 10$) Weibel's (1963) experimental values for the average diameter $\bar{d}(z)$ approximately follow an exponential law

$$\bar{d}(z) = d_o\, 2^{-z/\Delta} = d_o\, e^{-\alpha z} \tag{15}$$

with $\alpha = (1/\Delta)\ln 2$, and $\Delta = 3$ (i. e. $\alpha = 0.231$). d_o is the diameter of the trachea. Weibel (1980) calls (15) the fundamental law of design of the bronchial tree. Indeed, there are convincing arguments supporting Weibel's statement:

1. Wilson (1967) derived (15) as a consequence of the principle of minimum entropy production, and
2. Mandelbrot (1983) pointed out that the value $\Delta \approx 3$ is of crucial importance for the lung's architecture, because it guarantees a space-filling structure of the bronchial tree in contrast to certain botanical trees which have a 'surface-filling' ($\Delta = 2$) structure.

The experimental data (Weibel, 1963) are in agreement with the principle of minimum entropy production for mass-flow in a hydrodynamic pipe-system. Weibel (1980) observed that the bronchioles up to generation number $z \approx 16$ are pure conducting pipes, but the bronchioles beginning with generation number $z \approx 17$ up to $z = 23$ are occupied with alveoli in increasing number, and the transport process for gas molecules becomes dominated, not by mass-flow but by diffusion mechanisms which are dominant toward the periphery of the lung's branching system. Thus, there is indeed convincing evidence that (15) should represent - at least - a good approximation for air-flow in that portion of the bronchial network (up to $z \approx 16$) which is reponsible for pure hydrodynamic gas-flow mechanisms but not for diffusion processes.

Unfortunately, (15) already conflicts with the experimental data for generation numbers $z > 10$. Consequently, West et al. (1986) proposed a fresh approach for the design of the bronchial tree based on a renormalisation group theoretical concept which reflects a multiplicity of scales rather than a single scaling parameter α. Their model predicts a power-law dependency of the average diameter (and length) of type

$$\bar{d}(\bar{\alpha}z) = \frac{A(\bar{\alpha}z)}{(\bar{\alpha}z)^{\mu}} + B(\bar{\alpha}z) \tag{16}$$

where

$$A(\bar{\alpha}z) = \frac{C(\lambda p)}{\ln\lambda} \sum_{-\infty}^{\infty} e^{-i2\pi m\frac{\ln(\bar{\alpha}z)}{\ln\lambda}} \Gamma\left(\mu + i\frac{2\pi m}{\ln\lambda}\right) \tag{16 a}$$

and $$B(\bar{\alpha}z) = C(\lambda p) \sum_{\ell=0}^{\infty} \frac{(-\bar{\alpha}z)^{\ell}}{\ell!} \frac{1}{1-\lambda^{\ell+\mu}} . \tag{16 b}$$

$C(\lambda p)$ is some normalisation constant, Γ is a complex-valued Γ-function, and $\bar{\alpha}$ is a mean value over α-distributions.

This result is the Poisson sum representation of the infinite series

$$F(z) = C(\lambda p) \sum_{0}^{\infty} (\lambda p)^{n} e^{-\bar{\alpha}z\lambda^{n}} \tag{17}$$

which West et al. (1986) derived by using a physically motivated argument based on a scaling of the parameter α. They conjectured that their exact result (16) may be approximated by

$$\bar{d}(z) = (A_0 + A_1 \cos(2\pi \ln z / \ln \lambda)) / z^{\mu} \tag{18}$$

where A_0, A_1, μ and λ are four independent parameters. Indeed, experimental values from different groups could be fitted by (18) fairly well. However, there is a fundamental difference between results (16) and (18): Formula (18) satisfies an affine scaling law $\bar{d}(\lambda z) = \lambda^{-\mu} \cdot \bar{d}(z)$, while the exact result (16) does not show such a scaling property. The reason for this 'scaling defect' of (16) is already stored in the series (17), from which (16) has been derived by applying the Poisson summation formula to (17). Here, we are faced with a similar problem as Mandelbrot (1983) when he modified the original Weierstrass series in order to make it self-affine scaling.

To face this problem mathematically we start out with the series (12), which for $x_2 = 0$ represents a real-valued function and may be rewritten as

$$D(y) = \sum_{-\infty}^{\infty} \frac{e^{-y\gamma^n}}{\gamma^{nH}}, \quad \gamma > 1, \ y = -x_1 > 0, \ H < 0. \tag{19}$$

It incorporates all discrete Weierstrass frequencies γ^n down to zero, and has the scaling property

$$D(\gamma y) = \sum_{-\infty}^{\infty} \frac{e^{-\gamma^{n+1} y}}{\gamma^{nH}} = \gamma^{H} D(y) \tag{20}$$

i. e. if the y-values are stretched by a factor γ then the function $D(y)$ scales down by a factor γ^H. Convergency of the series (19) is guaranteed by requiring $y > 0$, $\gamma > 1$ and $H < 0$.

Identifying $y = \bar{\alpha}\, z$ (z is generation number) and $D(y) = d(z)/d_0$ we just recover (15) as the first term ($n = 0$) of the series (19). Neglecting all the terms with negative integers ($n < 0$) the remaining series restricts the eigenvalues γ^n to the range $1 \leqslant \gamma^n \leqslant \infty$. However, there is no a priori reason to restrict nature's evolution principle to such a limit. The added terms for $n = 0$ extend the Weierstrass spectrum to the range $0 \leqslant \gamma^n < 1$ allowing for large scale fluctuations.

Applying the Poisson summation formula to (19)

$$D(y) = \sum_{-\infty}^{\infty} f(n) = \sum_{-\infty}^{\infty} \int_{-\infty}^{\infty} f(x) \exp(i\, 2\pi m x)\, dx$$

with $f(x) = (1/\gamma^{xH}) \exp(-\gamma^x y)$, and substituting $\gamma^x = \exp(x \ln \gamma) = \xi$, one obtains

$$D(y) = \frac{1}{\ln \gamma} \sum_{-\infty}^{\infty} \int_0^{\infty} \xi^{\nu_m - 1} \exp(-\xi y)\, d\xi$$

with $\nu_m = -H + i\beta_m$ and $\beta_m = 2\pi m / \ln \gamma$. For Re $\nu_m = -H > 0$ and $y > 0$ the integration can easily be carried out (Gradsteyn, Ryzhik, 1980) with the result

$$D(y) = \frac{y^{H}}{\ln \gamma} \sum_{-\infty}^{\infty} y^{-i\beta_m}\, \Gamma(\nu_m)$$

where $\Gamma(\nu_m)$ is the complex-valued Γ-function. Identifying $y = \bar{\alpha} \cdot z$, $\gamma = \lambda$ and $-H = \mu > 0$, this result does just represent the scaling part (first term of r. h. s. of (16) of West's et al. (1986) diameter law (apart from the normalisation factor $C(\lambda\, p)$) .

D(y) may be rewritten as

$$D(y) = \frac{y^H}{\ln \gamma} \sum_{-\infty}^{\infty} \exp(-i\beta_m \ln y + \ln \Gamma(\nu_m))$$

with

$$\ln \Gamma(\nu_m) = \ln |\Gamma(\nu_m)| + i b_m, \quad b_m = \arg \Gamma(\nu_m).$$

Making use of the relations $\beta_{-m} = -\beta_m$ and $b_{-m} = -b_m$ we get the final result

$$D(y) = \frac{y^H}{\ln \gamma} \{\Gamma(-H) + 2\sum_{1}^{\infty} |\Gamma(\nu_m)| \cos(2\pi m \frac{\ln y}{\ln \gamma} - b_m)\}. \qquad (21)$$

An asymptotic estimation (large m-values) shows that (in leading order) $|\Gamma(\nu_m)| \sim \exp(-m)$ indicating fast convergency of the series in (21). Thus, taking into account only the first term (m = 1) of the sum and choosing $y = \bar{\alpha} z$ one obtains the following simple evolution law for the average diameter of dichotomous branching processes

$$D(z) = [(\bar{\alpha} z)^H / \ln \gamma]\{\Gamma(-H) + 2|\Gamma(\nu_1)| \cos(2\pi \frac{\ln \bar{\alpha} z}{\ln \gamma} - b_1)\}. \qquad (22)$$

Both, the exact result (21) and the approximation (22) satisfy the affine scaling relation $D(\gamma y) = \gamma^H D(y)$ in agreement with (20).

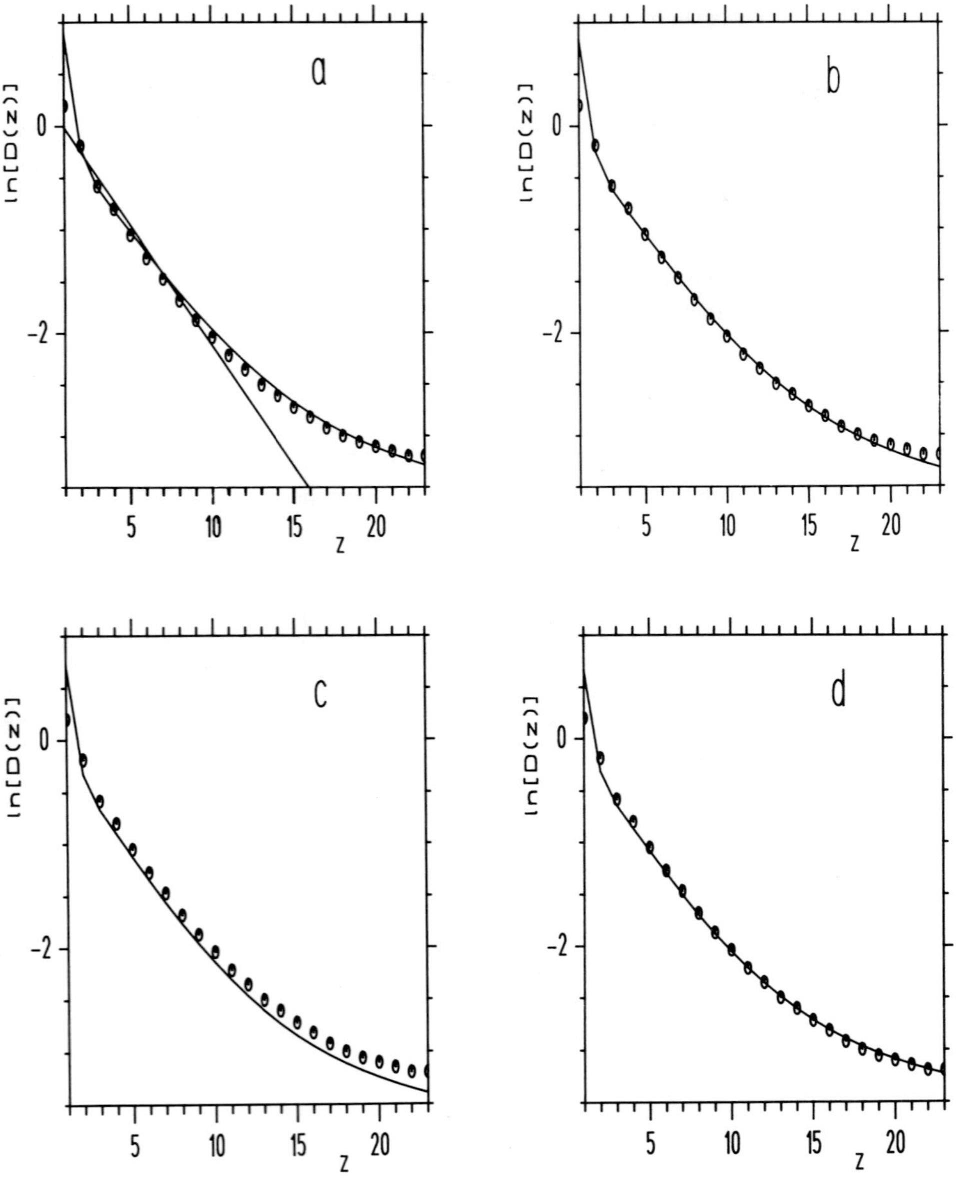

Fig. 3 lnD(z) versus generation number z. Circles represent experimental values from Weibel (1963). Equation (15) predicts a straight line (Fig. 3 a). The curve closely following the experimental data is the graph of equation (22) for the parameter choice: a: γ = 11, H = -1.26, $\bar{\alpha}$ = 0.231; b: γ = 11, H = -1.26, $\bar{\alpha}$ = 0.239; c: γ = 11, H = -1.26, $\bar{\alpha}$ = 0.2567; d: γ = 11, H = -1.20, $\bar{\alpha}$ = 0.2521.

Comparison with experimental data

In order to check the consistency of our result (22) with experiments we have only two fitting parameters (H and γ) available, if the constant α which appears in (22) is fixed as α = 0.231 (i. e. Δ = 3) in order to guarantee a space-filling lung tree (Mandelbrot, 1983). The choice H = - 1.26 and γ = 11 reproduces good agreement between Weibel's (1963) data for the human lung and the prediction of (22). We have also calculated contributions coming from higher-order terms (m = 2) in (21), but they are less than 1 %.

Figure 3 a shows the plot of the diameter scaling law (22). The straight line represents the law (15). To calculate the curve which closely follows the experimental data (circles) we used the parameter choice γ = 11, H = - 1.26 and $\alpha = (\ln 2)/\Delta$ = 0.231 (i. e. Δ = 3). We note that Δ cannot exceed the value Δ = 3 (space-filling lung tree). This ideal value (Δ = 3), however, can only be achieved if the branching process continues to infinitely thin bronchial pipes. But actually the branching process stops at a finite generation number of about z $\approx$ 23. Thus, the best one can achieve (Mandelbrot, 1983) is a value of Δ a little below 3 (Δ < 3). To test this hypothesis we varied $\bar{\alpha}$ in equation (22) for the parameter choice γ = 11, H = - 1.26 and $\bar{\alpha}$ = 0.239 (i. e. Δ = 2.9), Fig. 3 c for γ = 11, H = - 1.26, $\bar{\alpha}$ = 0.2567 (Δ = 2.70), and Fig. 3 d for γ = 11, H = - 1.20, $\bar{\alpha}$ = 0.2521 (Δ = 2.75).

Renormalisation group approach

As already pointed out by Mandelbrot (1983) there is some parallelism between the renormalisation group (RG) approach and the concept based on fractal dimensionality. Indeed, the fractal approach to the problem of dichotomous branching processes based on modified Weierstrass functions - as discussed in this paper - leads to the same (power-law) scaling part as West's et al. (1986) RG-approach, which, however, incorporates in additon a regular (non-scaling) part (B($\bar{\alpha}$ z) in (16). Hence, the exact RG-solution for the mean diameter takes the form

$$\overline{d(y)} = C_1(D(y) + B(y)), \quad y = \bar{\alpha} z \tag{23}$$

where

$$B(y) = \sum_0^\infty \frac{(-y)^n}{n!} \frac{1}{1-\gamma^{n-H}}, \quad H < 0 \tag{23 a}$$

and D(y) is given in (21). C_1 is an arbitrary normalisation constant.

Due to the appearance of B(y), the diameter-law (23) does not show the affine scaling property (20), which only holds for the 'singular part' D(y). For B(y) we find the relation

$$B(\gamma y) = \gamma^H (B(y) - e^{-y})$$

which, together with the scaling property (20) for D(y), leads to

$$\overline{d(\gamma y)} = C_1 \gamma^H (D(y) + B(y) - e^{-y}) = \gamma^H (\overline{d(y)} - C_1 e^{-y}). \tag{24}$$

It is of some interest now to ask for the contributions coming from the non-scaling (regular) part B(y). This question may be answered by giving a re-interpretation of the experimental data analysed in Fig. 3. However, before going through such a data analysis we note that West et al. (1986) did approximate the RG-result (23) by the scaling law (18) which contains four adjustable parameters A_0, A_1, μ and λ. We mention that the diameter-law (23) does also depend on four parameters $C_1, \bar{\alpha}, \gamma$ and

H. However, as we will demonstrate now, the specific model-parameters $\bar{\alpha}$, γ and H cannot be chosen arbitrarily. In order to derive such a law relating $\bar{\alpha}$, γ and H we go back to the definition of Mandelbrot's diameter exponent Δ, which is related to $\bar{\alpha}$ by $\bar{\alpha} = (1/\Delta)\ln 2$. Taking average diameters $\overline{d(\bar{\alpha}z)}$ Mandelbrot's relation (15 a) may be rewritten as

$$[\overline{d(0)}]^{\Delta} = 2[\overline{d(\bar{\alpha})}]^{\Delta} = \ldots = 2^N[\overline{d(\bar{\alpha}N)}]^{\Delta}$$

leading for $N \geqslant 1$ to

$$\ln\left(\frac{\overline{d(\bar{\alpha})}}{\overline{d(\bar{\alpha}N)}}\right) = \frac{N-1}{\Delta}\ln 2 = (N-1)\bar{\alpha}. \tag{25}$$

On the other hand, if we accept (for the current discussion) that the dominant contribution to the diameter-law (23) is stored in the scaling part D(y) or - as West et al. (1986) did - that (23) may be approximated by an affine scaling relation like (18), we find via (20)

$$D(\gamma\bar{\alpha}) = \gamma^H D(\bar{\alpha})$$

or

$$\ln\left(\frac{D(\bar{\alpha})}{D(\gamma\bar{\alpha})}\right) = -H\ln\gamma, \qquad -H = \mu > 0 \tag{26}$$

Approximating $\overline{d(y)} \approx C_1 D(y)$ and taking $\gamma = N$ we obtain upon comparing (26) and (25) the result

$$-H = \mu = \frac{\gamma - 1}{\ln\gamma}\bar{\alpha} \tag{27}$$

or

$$-H\Delta = \mu\Delta = \frac{\gamma - 1}{\ln \gamma} \ln 2 . \qquad (27\text{ a})$$

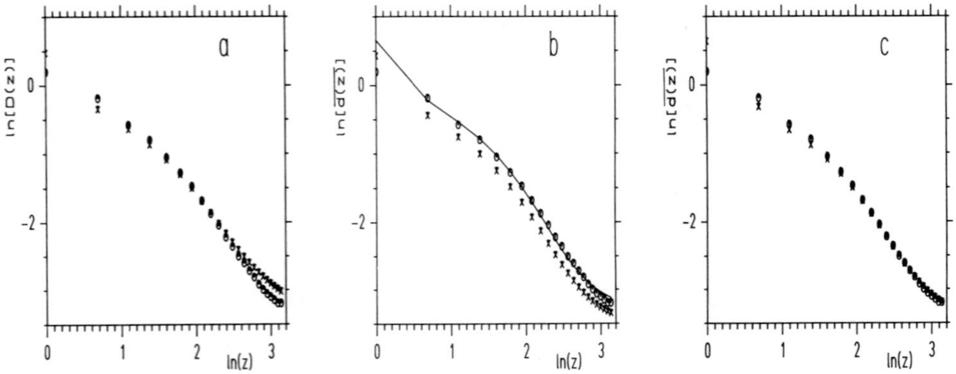

Fig. 4 In a log-log-plot the mean diameter law shows harmonic oscillations around the trend $H = -1.1$ with period $\gamma = 11$. Circles represent experimental values from Weibel (1963). The crosses in Fig. 4 a are values for $\ln D(z)$ calculated from (22) for $H = -1.1$, $\gamma = 11$, $\bar{\alpha} = 0.2638$ (i. e. $\Delta = 2.63$). Fig. 4 b shows, besides the experimental values (circles), crosses representing $\ln\overline{d(z)}$ calculated via (23) including 30 terms of the series for $B(y)$ with $C_1 = 1$ and same parameter choices as in Fig. 4 a. Full curve of Fig. 4 b represents (23) for $C_1 = 1.25$ and same data for H, γ and $\bar{\alpha}$ .The crosses in Fig. 4 c are values for $\ln\overline{d(z)}$ of (23) for $\gamma = 11$, $H = -1.1$, $\bar{\alpha} = 0.231$ ($\Delta = 3.0$), $C_1 = 1$ taking into account 30 terms of the series for B(y).

This relation restricts the choice of Δ , γ and H. Choosing $\gamma = 11$ (as above) we obtain $\mu\Delta = 2.89$. Since Δ is restricted to $\Delta \leqslant 3$ it follows $\mu \geqslant 0.96$. Fig. 4 shows a log-log-plot of the mean diameter for $\mu = -H = 1.1$, $\gamma = 11$, i. e. $\Delta = 2.63$. In order to investigate both the influence of the singular (scaling) part D(y) and the regular part B(y) on the mean diameter-law (23) we have plotted $\ln D(y)$ from (22) in Fig. 4 a and $\ln\overline{d(y)}$ from (23) in Fig. 4 b for the same data $\mu = -H = 1.1$, $\gamma = 11$ and $\Delta = 2.63$.

It is obvious that the most important contributions to the diameter-law (23) are coming from D(y), but it is also evident that contributions coming from the regular part B(y) should not be neglected.

In conclusion we notice that both, the RG-approach and Mandelbrot's concept of fractal dimensionality lead to a power-law dependence of the mean diameter for dichotomous branching processes. This part of the solution to the problem (i. e. D(y)) is self-affine scaling and incorporates harmonic oscillations around the main power-law trend. The regular part B(y) contains contributions which should not be ignored by comparing theoretical values with experimental data. The law presented in (27), which relates the important model-parameters Δ, γ and H, is exactly valid for the scaling part D(y). We suggest that a similar form of the diameter-law may also be expected for analogeous situations (neuronal trees, arteries etc.) where dichotonous branching processes are important. For the human bronchial tree we demonstrated good agreement between theory and Weibel's experimental data for the parameter choice $\mu = 1.1$ and $\gamma = 11$.

This work has been supported in part by the Forschungsschwerpunkt (FS 24) of the state of Baden-Württemberg.

Autowave Mechanochemical Model for Physarum Shuttle Streaming

Yu. M. Romanovskii, V. A. Teplov

Introduction

Self-generated movement has been considered an essential attribute of a living organism. In many cases, it results from regulated interaction between actin and myosin and is often of an oscillatory character. Rhythmical contractile activity is observed in single cells with amoeboid motility, in myxomycetes and in cellular aggregates (e. g. smooth muscle). A wide range of cellular processes can be involved in generating sustained oscillations (Berridge, Rapp, 1979). If an instability responsible for autooscillations exists in metabolic or regulatory systems it can result in periodic activation of the actomyosin system. However, it is quite possible that the basic instability is in the mechanochemical system of the cell. In this case autooscillations arise from a local feedback between tension and contraction of the contractile apparatus. Oscillations in insect flight muscles illustrate the latter point of view (Deshcherevskii, 1977). Such feedbacks can be mediated through a series of other cellular processes. For example, stretching the arterial smooth muscles increases myosin light chain phosphorylation in a graded fashion, which in its turn results in the

Thermodynamics and Pattern Formation in Biology

development of active tension upon release of the stretch (Barany et al., 1985). Oscillators including mechanochemical systems have the advantage of mechanical and hydrodynamic interactions.

This paper attempts to explain the oscillatory contractile activity in plasmodia of the acellular slime mold *Physarum polycephalum*. We shall formulate and analyse a mathematical model for rhythmic contractions of plasmodial strand. Advances in our understanding of plasmodium motility have been made as a result of extensive experimental investigations.

We draw attention to comparatively recent review articles (Wohlfarth-Bottermann, 1979; Kamiya, 1981; Teplov et al., 1981 a; Kessler, 1982; Tyson, 1982; Ueda, Kobatake, 1982; Teplov, Romanovskii, 1986) in which one can find references to original papers. Mathematical models of plasmodium oscillatory motility have appeared quite recently (Romanovskii et al., 1981; Romanovskii, Khors, 1981; Beylina et al., 1984; Arkhangelskaya et al., 1984; Oster, Odell, 1984; Odell, 1984; Vasilev et al., 1986; Teplov, Romanovskii, 1986).

Autowave phenomena in contractile activity of Physarumplasmodium

The plasmodium of a myxomycete is a multinuclear mass of protoplasm characterised by its large size (up to a square metre). It represents a suitable object for study of the general phenomena of non-muscle cell motility, especially its mechanical and thermodynamic aspects. The migrating plasmodium has an anterior-posterior polarity. It is differentiated into a frontal zone presenting a compact sheet of protoplasm and, at the rear, into a network of interconnected protoplasmic strands (veins) (Fig. 1). The motility of the plasmodium is manifested as rhythmic pulsations of the plasmodial body and as fast shuttle streaming of endoplasm and these are superimposed on a slow net movement of the whole organism. The pulsations have a complex spatio-temporal organisation. Typically, their periods are in the range of minutes.

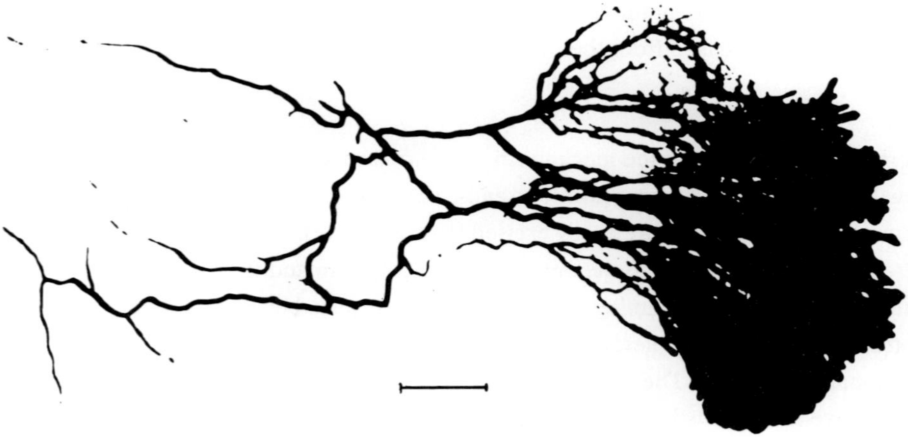

Fig. 1 The migrating plasmodium of Physarum polycephalum. Bar 1 = 1 cm.

Time lapse cinematographic analysis of migrating plasmodia reveals the following kinds of autowave phenomena:

(i) a wave-like propagation of changes in thickness in the frontal sheet with velocities of 5 mm/s to 2 cm /s (Stewart, 1964; Baranowskii, 1978; Teplov, Romanovskii, 1986). Sometimes circulation of such waves can be observed. Running waves can be induced artificially by gradients in physiological factors in the underlayer (Hejnowicz, Wohlfarth-Bottermann, 1980);

(ii) quasi-stochastic oscillations of sheet thickness with subsequent synchronisation (Baranowski, 1978);

(iii) nearly synphasic radial pulsations of the strands in moderately small plasmodia (5 cm or less in diameter) (Grebecki, Cieslawska, 1978);

(iv) radial pulsations of long strands in situ in the form of running or standing waves (Stewart, Stewart, 1961; Baranowski, Wohlfarth-Bottermann, 1982; Ermakov, Priezzhev, 1984);

(v) synphasic longitudinal homogeneous contractions of any subsegments in a strand segment excised from the plasmodium (Yoshimoto, Kamiya, 1978).

A great body of data on the mechanism of protoplasmic streaming in the plasmodium has been obtained on protoplasmic strands. Strand segments with diameters of 0.2 - 1.0 mm and 3 - 20 mm long are easily excised from a network region of plasmodium. The strand is differentiated into the outer gel-like ectoplasm and the more fluid inner endoplasm . Shuttle streaming of the endoplasm is caused by nonstationary gradients of intracellular pressure generated by quasi-periodic contraction of the relatively stationary ectoplasm. The contractile activity of the strands can be measured easily with a sensitive force transducer. One of the striking characteristics of these strands is their response to stretching. When a segment of plasmodial strand under isomeric conditions is quickly stretched by 10 - 20 % of its length, the tension and the amplitude of oscillation are conspicuously increased. Under isotonic conditions, the increase in the load gives rise to an increase in the amplitude of length oscillations (Kamiya, Yoshimoto, 1972; Teplov, Budnitskii, 1973).

Though ectoplasm and endoplasm are interconvertible, it is the ectoplasmic gel structure that is mainly responsible for the dynamic activities of the strand. There are numerous actomysin fibrils in ectoplasm, which are attached to invaginations in the plasma membrane (Wohlfarth-Bottermann, 1964; Fleischer, Wohlfarth-Bottermann, 1975). The fibrils consist mainly of numerous parallel actin filaments. Thick filaments composed of myosin are not normally seen in plasmodia. Myosin is present among actin filaments in the form of oligomers (Wolf et al., 1981) and its ATPase activity is similar to that of myosin from muscle (Nachmias, 1979).

There is a convincing evidence that Physarum actin-myosin interaction is controlled by calcium ions. A definite Ca^{2+} requirement for plasmodial strand contraction has been shown (Ueda et al., 1978; Yoshimoto, Kamiya, 1984). The plasmodium contains vesicles capable of sequestering and releasing CA^{2+} (Kato, 1979). There may be a shift of Ca^{2+} between the cytoplasm and vesicles during the contraction-relaxation cycle. Oscillations in the free calcium level with the shuttle streaming period have been demonstrated (Teplov et al., 1973; Ridgway, Durham, 1976; Yoshimoto et al., 1981), although the phase relation between oscillations of the force output and Ca^{2+} concentration within fibrillar space is not known exactly. The physical mechanism of these oscillations has not yet been elucidated. In this paper we shall adhere to the widespread view that mechanical strain may alter calcium homeostasis by affecting mechanosensitive calcium channels (Kessler, 1982).

Theoretical model

Events occuring in plasmodium strands can be described as follows. Let the protoplasmic strands be long cyclindrical tubes with viscous-elastic walls which are filled with an incompressible fluid. Actomyosin fibrils are homogeneously distributed in the walls. Under certain conditions myosin oligomers can interact with active centres of neighbouring, and oppositely polarised, actin filaments. It is known that oligomeric myosin can participate in smooth muscle contraction (Horio et al., 1985). If, in a given section of the strand all these movements occur, on the average synchronously, e. g. owing to an increase of the local Ca^{2+} concentration, then this leads to a decrease in strand radius. The endoplasm is evacuated from the contracting region, streams into the neighbouring regions and stretches the corresponding parts of the strand. This stretching can promote an increase in local Ca^{2+} concentration. In this way conditions for a new active contraction are created, but now in the sections adjacent to the original contraction. It is assumed that there is a sufficient local supply of ATP.

The above description can be expressed in mathematical terms. Let us consider, as a first approximation, nonstationary axi-symmetric streaming of incompressible viscous Newtonian fluid in a long cylindrical tube possessing an impermeable wall (Fig. 2). Let

R denote the mean radius of the tube, ϱ and η denote endoplasm density and viscosity, respectively. For endoplasm shuttle streaming with a frequency ω , the Reynold's number, Re = $\varrho R^2 \omega / \eta \approx 10^{-6}$. Thus, we can completely neglect the effects of inertial forces. In this case the motion of endoplasm may be described by the equation for Poiseulle flow in a tube of variable cross-section:

$$\partial P/\partial z = \frac{8\eta}{R^2} V, \qquad (1)$$

where V(z, t) is the velocity averaged over a cross-section of the strand lumen, and $\partial P / \partial z$ is the longitudinal pressure gradient.

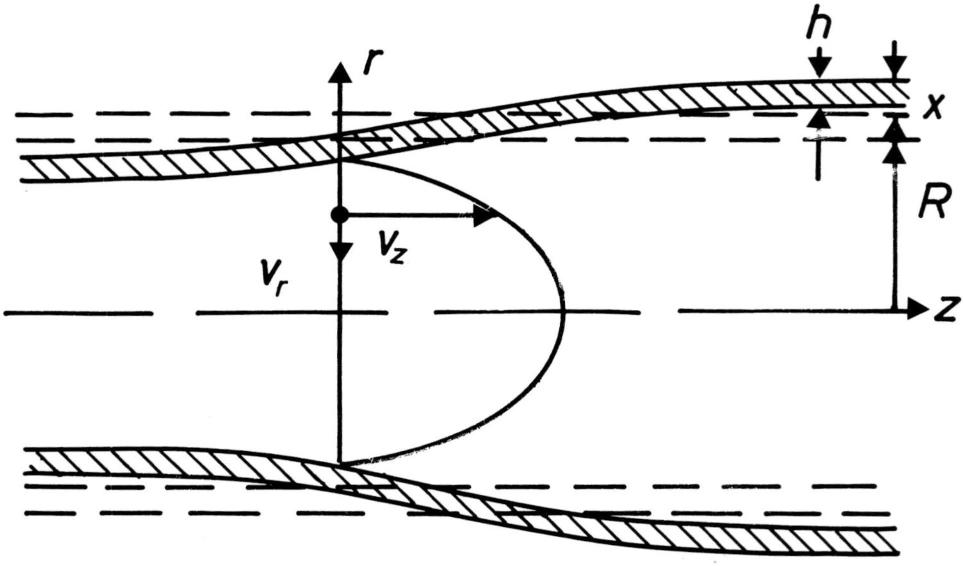

Fig. 2 Schematic diagram to show a longitudinal section of a strand and a profile of endoplasm velocity.

The problem to be solved is to find an expression for the pressure P(z,t) which depends on wall parameters and on properties of the contractile proteins. Let us assume that the passive viscoelasticity of the wall is described by a Kelvin model, that the ratio of the wall thickness to the strand radius is small and that longitudinal contraction of the wall is impossible, e. g. because of stronger adhesion of the strand to the underlying substrate. When the radial inertial forces affecting the wall element and when heterogeneous distribution of stresses and strains along the wall thickness are neglected, the following expression for the pressure can be obtained:

$$P = \frac{Eh}{R^2} x + \frac{\kappa h}{R^2} \frac{\partial x}{\partial t} + P_A + P_E + L_z\left(x, \frac{\partial x}{\partial t}\right). \qquad (2)$$

Here x is the small deviation from the mean radius R, h is the thickness of the wall, E and κ are the Young's modulus and viscosity coefficient of the wall for tangential deformations, respectively. The active pressure , P_A, is P_A = hG/R where G is the active tangential stress averaged over the wall thickness. P_e is the pressure of the surrounding medium on the strand. L_z is the linear differential operator depending on transmission of tension along the strand induced by the local changes in diameter. For small, radially symmetrical deformations of the strand one can use the following approximation:

$$L_z\left(x, \frac{\partial x}{\partial t}\right) = \frac{E'I}{2\pi R} \frac{\partial^4 x}{\partial z^4} + \frac{\kappa' I}{2\pi R} \frac{\partial^5 x}{\partial z^4 \partial t} + \frac{T}{2\pi R} \frac{\partial^2 x}{\partial z^2}, \qquad (3)$$

where E' and κ' are the Young's modulus and viscosity coefficient of the wall for the longitudinal deformations, respectively, I is the moment of inertia of the wall section ($I = \pi R h^3/6$), T is the longitudinal tension of the strand, which may be described approximately by the formula $T = 2\pi R h E' \varepsilon$, where ε is the relatively constant longitudinal deformation of the strand.

It is known (Ermakov, Priezzhev, 1984) that if there is a cosine-shaped standing wave along the strand:

$$x(z,t) = x_n(t) \cos \frac{\pi n}{L} z \qquad n = 1, 2, 3 \ldots \tag{4}$$

loops appear at the ends of the strand and the following inequalities:

$$\lambda_n = \frac{2L}{n} \gg R \gg x \simeq h \tag{5}$$

are nearly always fulfilled. Then it is easy to see that the quantity L_z can be neglected in comparison with the first terms of expression (2).

To find a relationship between the average velocity and the velocity of the small wall deformation a continuity equation may be used:

$$\frac{\partial v_z}{\partial z} = \frac{1}{r} \frac{\partial}{\partial r} (r v_r), \tag{6}$$

where v_z is the longitudinal component of endoplasm velocity and v_r is the radial component (Fig. 2). Integrating equation (6) over the cross-sectional area of the strand lumen with regard to (5) we obtain:

$$\frac{\partial V}{\partial z} = -\frac{2}{R} v_r \Big|_{r=R} = -\frac{2}{R} \frac{\partial x}{\partial t}, \tag{7}$$

where the average velocity $V(z,t) = \frac{1}{\pi R} \int_0^R 2\pi r v_z \, dr$.

Following differentiation of equation (1) with respect to z and introducing expressions (2) with P_e = const and (7), an equation governing the strand contraction can be obtained

$$\frac{16\eta}{R^3}\frac{\partial x}{\partial t} = \frac{Eh}{R^2}\frac{\partial^2 x}{\partial z^2} + \frac{\kappa h}{R^2}\frac{\partial^3 x}{\partial z^2 \partial t} + \frac{\partial^2 P_A}{\partial z^2} . \tag{8}$$

Let us derive the equations connecting active pressure P_A, Ca^{2+} concentration and deformation x for some section of the strand. A great deal of our knowledge of actin-myosin was originally obtained from the study of muscle proteins (for example see: Deshcherevskii, 1977; Bagshaw, 1982). Actin and myosin interact in a cycle which involves the attachment of a myosin head to actin and its subsequent detachement at a later stage in the cycle. During this cycle an ATP molecule is hydrolysed. Noting that Physarum actin and myosin are very similar to muscle actin and myosin, it can be assumed that the qualitative peculiarities of their mechanochemical behaviour are much alike.

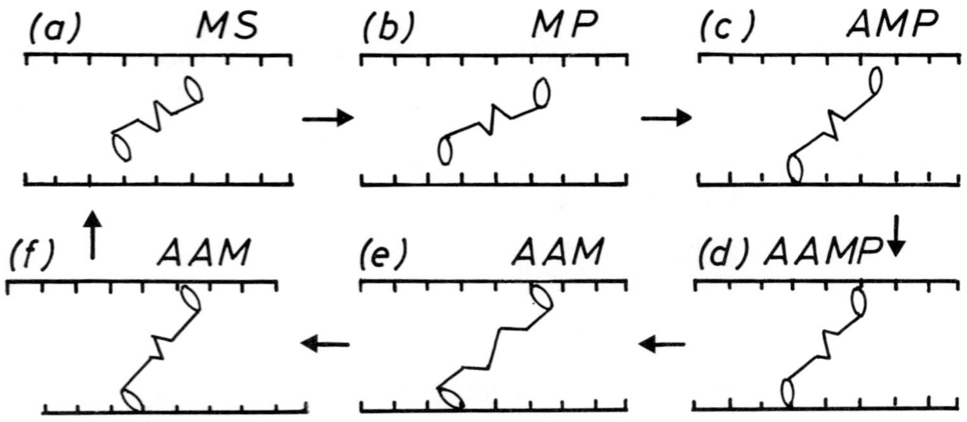

Fig. 3 Scheme suggested for mechanochemical cycle in Physarum. Explanations in text.

Fig. 3 shows a possible model for the interaction of myosin oligomers (dimers) with two oppositely polarised actin filaments. The following abbreviations are used: M is the myosin oligomer, A is the actin filament, S is the substrate (Mg ATP), P is the products (ADP and P_i). By analogy with the Lymn-Taylor scheme (Lymn, Taylor, 1971) ATP is hydrolysed by free myosin (MS → MP). The products of ATP hydrolysis remain attached to the enzymatic sites for a long time unless the myosin oligomer contacts two actin filaments. This myosin-product intermediate (MP stores the free energy of ATP hydrolysis in the form of some internal strain. After Ca^{2+} appearance and, probably, myosin light chain phosphorylation, the triple complex, actin-myosin-actin (AAMP), can be formed, and this produces some favourable conformational changes in actomyosin.

To simplify matters, we assume the myosin heads to move from an angle of 90 ^{o} to an angle of 45 ^{o} just as it occures in muscle. Let the step of product release (AAMP → AAM) be achieved only by stretching the spring (state e on Fig. 3) or by allowing actin filaments to move (state f) as in muscle (Bagshaw, 1982). Hence, equilibrium between the states d,e, and f will depend on the relative longitudinal position of the active centres of the two actin filaments with respect to the given oligomer bound. During steps e → f work can be extracted from the cycle by releasing the elastic energy of the spring. However, it must be assumed that direct transition e → a is prevented by a high activation energy, otherwise state f will be bypassed and little tension or work can be obtained.

Though no explanation for this valve-like property is known at present an explanation could shed light on the mechanisms controlling formation of Physarum actomyosin fibrils (Wohlfarth-Bottermann, 1964; Kamiya, 1968). Under isometric conditions tension is generated by myosin oligomers fluctuating between states d and e, so that their cycles are almost at a standstill. There may exist a little e → a transition, and in this case free energy is released as heat. It should be noted that the conditions for ectoplasm contraction are closer to isometric, inasmuch as there is a substantial tension level caused by the constant component of intracellular hydrostatic pressure (Layrand et al., 1972; Teplov et al., 1981 b; Teplov, Romanovskii, 1986). In this case myosin oligomers may take part in binding actin filaments without considerable energy consumption. This may account for fibrillogenesis: the higher the tension, the greater the number of myosin oligomers bound to actin filaments and the greater the number of actin bundles. It has been shown, that stretching of a strand by 50 % or more of its original length results in de novo formation of actomyosin fibrils within the endoplasm (Fleischer, Wohlfarth-Bottermann, 1975; Wohlfarth-Bottermann, Fleischer, 1976).

However, under isotonic conditions, and in the presence of ATP, there occurs tight binding of ATP to myosin heads in state f (Fig. 3). Such binding drives the dissociation of the AAM complex. Once formed the AAMS complex dissociates to MS before ATP hydrolysis takes place. All this can account for the disappearance of newly generated endoplasmic fibrils immediately after allowing the strand to contact isotonically without an external load (Fleischer, Wohlfarth-Bottermann, 1975). On a macroscopic scale, such a valve-like property of actomyosin can be illustrated by the principle of action of the drag used by skiers. It is only at a particular tension that the drags contact the wire rope and pull skiers up the mountain slope. When the tension decreases the drags detach from the rope.

Noting the cyclical behaviour of actomyosin complexes we shall consider, as a first approximation, only two mechanically significant states of myosin oligomers, namely: the free state M (detached oligomer; states a, b, c on Fig. 3) and the bound state N (oligomers attached to two actin filaments; states d, e, f). Thus, such an elementary mechanochemical cycle can be written as

$$A + M \underset{k_2}{\overset{k_1'}{\rightleftharpoons}} N$$

where k_1' and k_2 denote the rate constants. A kinetic equation for the number of bound oligomers $n(z,t)$ is therefore given by:

$$\frac{\partial n}{\partial t} = k_1' am - k_2 n . \tag{9}$$

Here a is the number of the active centres on actin filaments, which could, but do not, bind with myosin oligomers. It is obvious that $a = a_o - n$, $m = m_o - n$, where a_o and m_o are the total numbers of active centres on actin filaments and of myosin oligomers, respectively.[1)] It is known (Kessler et al., 1976) that for Physarum $a_o >> m_o$, hence one may assume $a \simeq a_o$. In addition, taking into account the fact that tangential active stress G is proportional to n, equation (9) can be rewritten as:

$$\frac{\partial G}{\partial t} = k_1 (G_o - G) - k_2 G, \tag{10}$$

where $k_1 = k_1 a_o$ and k_2 are the effective rate constants for formation and dissociation of actomyosin complexes, respectively, G_o is the maximal active stress which occurs when all myosin oligomers m_o participate simultaneously in force generation.

In contrast to muscle the plasmodium has essentially lower velocities for both contraction and relaxation. This difference may lie in the regulatory mechanisms of the contractile process. The rate of tension development probably reflects the kinetics of several processes occurring in plasmodium. The following factors may contribute

(i) the rate of Ca^{2+} entry into fibrillar spaces

(ii) the rate of activation of contractile proteins, that is the rate of phosphorylation of myosin light chains by the Ca^{2+}-activated light chain kinase as well as possible changes in the actin filament-linked regulatory system

(iii) the rate of interaction of activated myosin oligomers and actin filaments. Similarly, the slowness of plasmodium relaxation may be explained by the kinetics of the reverse processes.

---

1) It should be noted that we consider only myosin oligomers, which are attached to two actin filaments or stereometrically capable of this 'correct' attachment. In prinicple, their number can depend on the extent of the strech.

It is most likely that, along with calcium ions, cyclic nucleotides and phosphatidylinositol derivatives are involved in chemical control of the contraction cycle. Unfortunately, not much is known about these specific processes for Physarum and so we cannot derive detailed equations relating to the concentrations of the second messengers and defining the chemical states of the contractile proteins. For simplicity, let us suppose that, in the entire chemical chain, there is a rate-limiting step. Since it is well known that the rhythmic contractions in Physarum are accompanied by synchronous Ca^{2+} concentration oscillations, let Ca^{2+} be such a rate-controlling substance. Calcium control of actin-myosin interaction can be ensured by allowing the rate constant k_1 in the equation (10) be dependent on Ca^{2+} concentaration. Let $k_1(c) = k_1 f(c)$, where c denotes the Ca^{2+} concentration, $f(c)$ is a dimensionless function of activation.

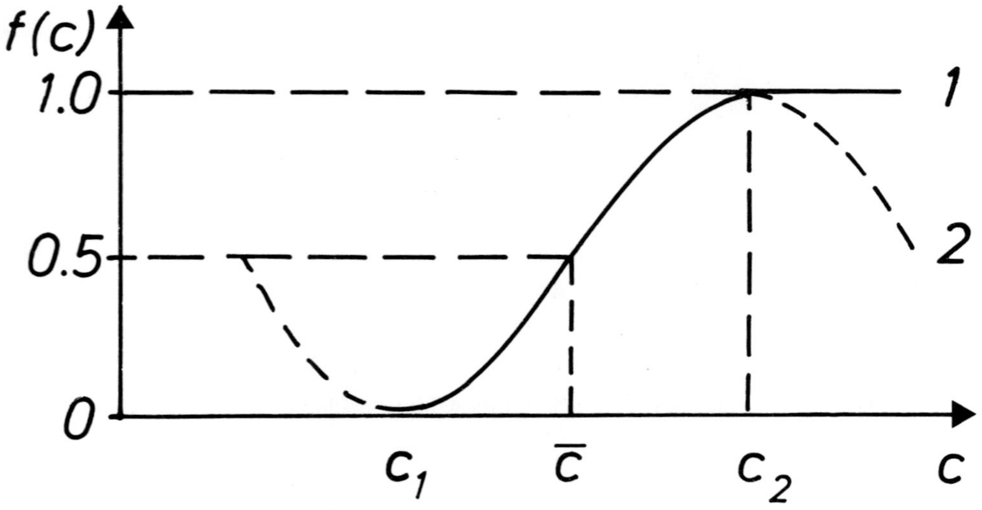

Fig. 4 Dependence of activation function (1) on the concentration of calcium ions, and its approximation for $c_1 \leqslant c \leqslant c_2$ by polynomial of third degree (2).

It is known that the dependence of contractile activity in plasmodial strands on Ca^{2+} concentration has a threshold character. From injection experiments threshold concentration for oscillatory activity was found to be $2 \cdot 10^{-7}$ M (Ueda et al., 1978). Recent experiments have shown that, in saponin models of plasmodial strands, tension starts to increase at $c_1 = 6.10^{-8}$ M and reaches its maximum at $c_2 = 10^{-7}$ M (Yoshimoto, Kamiya, 1984). To simplify let us suppose that the activation function has the same dependence on c and is approximated by a polynomial of the third degree (Fig. 4). Then taking into account that $P_A = h\,G/R$, the following equation for the active pressure can be written from equation (10):

$$\frac{\partial P_A}{\partial t} = k_1 f(c)(P_0 - P_A) - k_2 P_A\,, \tag{11}$$

where $P_0 = h\,G_0/R$ is the maximal active pressure.

The simplest equation coupling mechanical strain and calcium homeostasis is:

$$\frac{\partial c}{\partial t} = k_3 x - k_4 c\,. \tag{12}$$

Here it is assumed that the rate of Ca^{2+} inflow into fibrillar space increases linearly with radial deviation x and that Ca^{2+} sequestering into vesicles obeys first order kinetics. We have not considered constant calcium influx through the plasmalemma, since this seems to be well-balanced by a corresponding efflux associated with permanent exocytosis of vesicles. In addition in this paper, for simplicity, we restrict ourselves to elucidating the possibility of only hydrodynamic interaction between different sections of the strand and therefore terms describing diffusion and drift of trigger chemicals are omitted in equation (12).

Equations (8), (11) and (12) constitute an autonomous set of three differential equations of partial derivatives, which describes the protoplasmic strand as an essentially nonlinear active distributed system. These equations admit of a large variety of solutions and these can be evaluated by approximate methods. We shall obtain some of these approximate solutions in the following manner. First, it is necessary to find those solutions which are stationary and spatially uniform. Let $\overline{x}$, $\overline{P_A}$, $\overline{c}$ denote such solutions. Next, their stability with respect to small perturbations is to be investigated. Finally, we shall find exact solutions in the form of standing autowaves according to expression (4).

Assuming that $\partial x/\partial t$, $\partial P_A/\partial t$, and $\partial c/\partial t$ are equal to 0, the system of three algebraic equations that define the stationary state is

$$\frac{Eh}{R^2}\overline{x} + \overline{P_A} = P_H$$

$$k_1 f(\overline{c})(P_0 - \overline{P_A}) - k_2\overline{P_A} = 0 \qquad (13)$$

$$k_3\overline{x} - k_4\overline{c} = 0$$

Here P_H is the integration constant, an important parameter defining an intracellular hydrostatic pressure. The necessary condition for the existence of a hydrostatic pressure in the cell is an osmotic inequality between the surrounding fluid and the interior of the cell (Layrand et al., 1972; Teplov, Romanovskii, 1986).

The stationary state depends on the system parameters, and in particular, upon the value of P_H and this is not known exactly. Therefore we must solve the inverse. For example, let $\overline{c}$ be equal to $(c_1 + c_2)/2$, then

$$\overline{x} = \frac{k_4}{k_3}\overline{c}, \quad \overline{P_A} = \frac{k_1 f(\overline{c})}{k_1 f(\overline{c}) + k_2} P_0, \quad P_H = \overline{P_A} + \frac{Eh}{R^2}\overline{x}.$$

In this case the variables x, P_A and c may be replaced by $\alpha = x - \bar{x}$, $\beta = P_A - \bar{P}_A$, $\gamma = c - \bar{c}$. The activation function is then represented as $f(\gamma) = S_o + S_1\gamma - S_3\gamma^3$ and the set of equations (8), (11), and (12) takes the form

$$\frac{16\eta}{R^3}\frac{\partial\alpha}{\partial t} = \frac{Eh}{R^2}\frac{\partial^2\alpha}{\partial z^2} + \frac{\kappa h}{R^2}\frac{\partial^3\alpha}{\partial z^2\partial t} + \frac{\partial^2\beta}{\partial z^2} \tag{14}$$

$$\frac{\partial\beta}{\partial t} = k_1S_1(P_o - \bar{P}_A)\gamma - (k_1S_o + k_2)\beta - k_1S_3(P_o - \bar{P}_A)\gamma^3 - k_1S_1\gamma\beta + k_1S_3\gamma^3\beta$$

$$\frac{\partial\gamma}{\partial t} = k_3\alpha - k_4\gamma$$

In order to investigate the stability of stationary solutions let us assume that perturbations have the form:

$$\alpha = \tilde{\alpha}\exp(pt + iKz), \quad \beta = \tilde{\beta}\exp(pt + iKz), \quad \gamma = \tilde{\gamma}\exp(pt + iKz) \tag{15}$$

where $K = 2\pi/\lambda$ is the wave number and $\tilde{\alpha}/\bar{x}, \tilde{\beta}/\bar{P}_A, \tilde{\gamma}/\bar{c} \ll 1$.

Substituting expression (15) into (14) and disregarding small nonlinear terms we obtain the characteristic equation:

$$A(k^2)\cdot p^3 + B(K^2)\cdot p^2 + C(K^2)\cdot p + D(K^2) = 0 \tag{16}$$

where the functions A, B, C, and D are positive, linear with respect to K^2 and depend on the parameters of the system (14). In such a system, according to the Rauss-Gurvich criterion, only an oscillatory instability is possible and the condition for self- excitation is given by:

$$\Lambda(K^2) = B(K^2)\cdot C(K^2) - A(K^2)\cdot D(K^2)F \leqslant 0. \tag{17}$$

When the equality criterion holds the critical value of the wave number, K_{min}, can be found. Only waves with lengths of $\lambda \leqslant 2\pi/K_{min}$ may exist.[2)]

The next step is to solve the system (14) in the form of stationary standing autowaves:

$$\alpha = \alpha_0 \cos \omega t \cdot \cos Kz , \qquad \beta = \beta_0 \cos(\omega t + \psi) \cdot \cos Kz$$
$$\gamma = \gamma_0 \cos(\omega t + \varphi) \cdot \cos Kz . \tag{18}$$

For a strand of length L, K is given by $K = \pi n/L$, with $n = 1,2,3,...$ It is necessary to determine the following quantities $\omega, \psi, \varphi, \alpha_0, \beta_0, \gamma_0$. Equations defining them can be obtained in the following way. When expressions (18) are substituted into (14) then since the first and the third equations are linear the result is an algebraic relationship between the parameters. The second equation of system (14) is nonlinear and contains terms with the multipliers cos Kz, cos 2Kz, cos 3Kz, cos 4Kz. Application of the Galerkin procedure, and multiplication of the second equation by cos Kz and integration over z from 0 to L, allows equating of the coefficients of sin and cos to obtain a system of six algebraic equations from which the quantities in question can be found:

$$\omega^2 = \frac{(k_1 S_0 + k_2 + k_4) E h}{\kappa h + 16(\mu/R)(L/\pi n)^2} + (k_1 S_0 + k_2) k_4 ;$$

$$\operatorname{tg} \psi = \frac{\omega[\kappa h + 16(\eta/R)(L/\pi n)^2]}{E h} ; \qquad \operatorname{tg} \varphi = -\frac{\omega}{k_4} ;$$

$$\gamma_0 = \frac{4}{3\sqrt{S_3}} \sqrt{S_1 - \frac{(\omega^2 + k_4)^2 [E h + (k_1 S_0 + k_2)[\kappa h + 16(\eta/R)(L/\pi n)^2]]}{k_1 k_3 (P_0 - \bar{P}_A) R^2}} ; \tag{19}$$

$$\alpha_0 = \frac{\gamma_0}{k_3} (k_4 \cos \varphi - \omega \sin \varphi) ; \qquad \beta_0 = -\alpha_0 \frac{E h}{R^2 \cos \psi} .$$

---

2) Considering the operator L_z in expression (2) one can obtain the second critical value of the wave number, K_{max}.

Consideration of (7) and (19) yields a formula for the average velocity amplitude:

$$V(z) = \frac{2L\omega\alpha_o}{\pi R n} \sin \frac{\pi n}{L} z . \tag{20}$$

To compare the results of mathematical modelling with experimental data it is necessary to know values for the strand parameters. For lack of space we omit the description of our experimental study which gave us the principal parameters of plasmodial strands (Teplov, Romanovskii, 1986). Mechanical properties and their dependence on deformation at different phases of autooscillation have been determined on the basis of the transient tension responses of the strand to rapid length changes in the form of a sequence of short rectangular impulses with increasing, small, amplitude. To a first approximation we assume that the values of Young's modulus and viscosity coefficient for the ectoplasm obtained for longitudinal deformations are equal to those for radial deformations. Additionally we suppose that those values calculated for the phase of maximum relaxation corresponds to the passive properties of the strand.

In the following table we bring together the values of all parameters involved in the model (Teplov, Romanovskii, 1986):

$R = 10^{-3} - 10^{-1}$ cm	$P_H = 7.10^4$ dyn cm $^{-2}$	$c_1 = 6.10^{-8}$ M
$h = 10^{-4} - 10^{-2}$ cm	$k_1 = 2.10^{-2}$ s^{-1}	$c_2 = 10^{-7}$ M
$\eta = 20$ dyn s cm^{-2}	$k_2 = 5.10^{-3}$ s^{-1}	$S_o = 0.5$
$\kappa = 2.10^6$ dyn s cm^{-2}	$k_3 = 3.10^{-7}$ M cm^{-1}s^{-1}	$S_1 = 3.7.10^7$ M^{-1}
$E = 5.10^5$ dyn cm^{-2}	$k_4 = 5.10^{-3}$ s^{-1}	$S_2 = 3.1\ 10^{22}$ M^{-3}
$P_o = 10^5$ dyn cm^{-2}		

For strand fragments ($L = 1\text{-}5$ cm, $R = 0.2\text{-}0.5$ mm) only standing waves with $n = 1,2$ are observed (Ermakov, Priezzhev, 1984). From (19) and (20) when $L = 3.14$ cm, $R = 0.25$ mm it follows that
$\omega = 5.10^{-2}\ s^{-1}$, tg $\psi = 0.35$, tg $\varphi = -10$, i. e. the period $T = 100$ s, $\varphi = -85^{o}$, $\psi = 15^{o}$, $\alpha_o = 0.08$ mm, $\beta_o = 2.10^{4}$ dyn cm^{-2}, $\gamma_o = 2.10^{-8}$M, $V = 0.3$ mm s^{-1}.

It should be noted that when the ectoplasm viscosity exceeds that of the endoplasm ($\kappa >> \eta$) by many orders of magnitude the period and phases found from (19) are weakly dependent on n and strand size. In this case the frequency can be evaluated from the formula $\omega \simeq \sqrt{E/\kappa\tau}$, where the time constante $\tau = 1/(k_1S_o + k_2 + k_4)$ (cf. Beylina et al., 1984).

From the model it follows that the oscillations in the Ca^{2+} concentration in fibrillar space and the tension oscillations are 100 ^{o} out of phase. On the other hand, Yoshimoto et al. (1981) have shown that the tension oscillations in strand treated with calcium ionophore and the oscillations of aequorin luminescence in the ambient solution are 180 ^{o} out of phase. However, the finite value of the permeability coefficient suggests a phase shift between oscillations of internal and external Ca^{2+} concentrations. This can resolve the above contradiction.

Thus, the simplest solution of the mathematical model agrees with the available experimental data. It implies that the autowave character of plasmodium motility can truly be based on the existence of feedback between deformation and contraction in the contractile apparatus.

Concluding remarks

The mechanochemical model proposed is undoubtedly not final and will be improved with the accumulation of new experimental data on the force generation mechanism and on the feedback processes. In conclusion therefore let us outline some theoretical problems that remain unsolved.

1. The one-dimensional model presented is suitable for investigating more complicated events. For example, nonharmonic oscillations enriched by second and higher harmonics are often observed experimentally. Solutions should be sought for those cases in which the self-excitation condition (17) is fulfilled for several values of K such that $K > K_{min}$. In this case the question arises how do small initial perturbations with these wave numbers interact and how are autowaves in the shape of running impulses formed if $L >> 1/K_{min}$?
2. To understand complicated autowave processes such as reverberators, quasi-stochastic waves and others occurring in the fanlike sheet at the plasmodium front it is necessary to construct and investigate a two-dimensional model for an active sheet of protoplasm.
3. Finally, in addition to elucidating autowave processes in the sheet and strands a perfect mathematical model must account for plasmodium migration. It is important to formulate such a model because analogous processes are typical of many other cells and play a significant role in morphogenesis.

Statistical Geometry of Tissues

N. Rivier

Introduction

In this chapter, we shall discuss the geometry of undifferentiated biological tissues, and their evolution. It will be understood that tissues are space- or area-filling mosaics of polyhedral or polygonal cells, which can grow and divide. The time at which a given cell divides is random, and there is no correlation in the process and in the phase of the mitotic cycle between two cells far apart in the tissue. This implies that, at any given time, the shape and size of the cells constituting the tissue is not unique, and the mosaic is random. Our aim is to describe this random, space-filling geometry.

This chapter is therefore an attempt to use geometry to describe and classify the simplest, disordered, biological forms. Apart from elementary mathematics, a few concepts from other disciplines will be used, notably from crystallogryphy (the concept of disclinations and dislocations, which, in disordered structures, must be defined topologically), and from statistical thermodynamics (the idea of most probable distribution or maximal entropy). The book by K. J. Dormer (1980) constitutes an

Thermodynamics and Pattern Formation in Biology

excellent introduction to the geometry of cells. That by V. V. Smoljaninov (1980) is much more complete, but has not so far been translated from the Russian. Geometry of two dimensional random mosaics, and its universality, is presented in the recent review by Weaire and Rivier (1984). The best visual introduction on dislocations and disclinations is that of Harris (1977). As far as maximum entropy is concerned, the original papers by Jaynes (1957, 1979) are without equal. They have been collected and edited by Rosenkrantz (1984).

Partioning space at random

The architecture of composite structures is determined by their growth mechanism or by boundary conditions. Pennies packed in two dimensions make a triangular network, because the geometrical figure achieving best (closest) local packing, an equilateral triangle, also tiles space (like the floor of a kitchen). This is no longer the case in three dimensions, where the closest packing figure, the tetrahedron, does not fill the Euclidean space: some form of compromise is usually achieved, either by distorting tetrahedra (as in face-centered cubic and hexagonal closed packed crystallographic phases), or by introducing structural defects (dislocations, grain boundaries etc.), or by a combination of these, regularly or at random. Daisies or sunflowers have a structure with florets forming families of spirals and a Fibonacci number of such spirals per family, which is entirely determined by its growth from a stem (the geometrical centre of the flower) (Rivier et al., 1984) (Fig. 1). Alternatively, this can be regarded as the closest packing consistent with cylindrical symmetry.

The structure of living tissues and their evolution is governed by growth of individual cells and by cellular divisions. They fluctuate within the tissue because the minimal size for a given cell to divide, and its rate of growth depend, among others, on the number of its neighbours. The end product is a space-filling assembly of cells, a network with elements - edges, faces, cells - of fluctuating sizes, yet uniform and isotropic as a whole, without a recognisable centre, origin or orientation, like a forest in which one has got lost. Compare such a structure with the daisy of Fig. 1 or with that of a crystal.

Fig. 1 A daisy. Most florets are hexagons, except for those marked x = heptagon, o = pentagon. Pairs xo are dislocations through which additional spirals of florets are inserted. Note the cylindrical symmetry. The structure is generated by florets sprouting from the stem at regular time intervals and angles, the younger pushing out the older ones (Rivier et al., 1984).

Moreover, two undifferentiated tissues are indistinguishable from each other, apart from a characteristic length scale, not because they can be superposed exactly, but because they can be interchanged without anybody noticing the substitution. Identity of the two tissues is therefore a topological, rather than a geometrical equivalence. The two biological tissues on the frontispiece of Dormer's (1980) book (the wing of a fly and a crab apple) are indistinguishable in this fashion, after their respecitive cell size has been adjusted by relevant blowing up (scaling) of the photograph. Similarly, the picture of polycrystalline aluminium-tin alloy in C. S. Smith's (1962) essay looks exactly like that of a soap froth.

Undifferentiated tissues are also homogeneous and isotropic again not because a pattern and its translated counterpart are exactly superposable, as in crystals or in proving theorems in elementary geometry, but because any difference is not objectively (without using words like *I*, *this*, *here* or *now*) relevant or measurable.

Botanists from Hales (1727), through Thompson (1917), to Matzke (1950) have been using packings of peas, cannonballs or lead shot, or soap froths as geometrical models for botanical tissues, because they all have similar architecture, even though the forces shaping tissues and the two physical models (hard core repulsion versus surface tension) are very different.

What matters is filling space with 'random avoidance of the niceties of adjustment' (Lewis, 1943). Matzke (1946) even tried to quantify the structural differences between botanical tissues, soap froths and lead shot packings, without success. However, if the structures of soap froths, sintering metallurgical grains, and botanical tissues are similar, their evolution and long term fate are different. Small bubbles or grains shrink, whereas all cells keep growing and dividing. Botanical tissues are in a steady state.

Two observations form the basis of our understanding of the tissue structure, and its evolution:

- The structural elements of the tissue, edges, faces of cells and the cells themselves, have fluctuating sizes.
- Different tissues are indistinguishable from each other, and from other random, space-filling structures like soap froths, compressed lead shot, metallurgical grain aggregates, etc., at least to a first approximation.

These observations (beautifully illustrated in gels under stress (Tanaka et al., 1987)) suggest that the relevant geometry is *topology*, where precise values of angles and sizes do not matter, as if the tissue was imprinted on deformable rubber. Only adjacency and connectivity are properties independent of the deformation of rubber. Similarly, indistinguishability between tissues, is a statement of topological equivalence, which may have interesting legal implications illustrated by Denis Weaire's cartoon (Fig. 2). By contrast, superposability would be *geometrical* identity.

Fig. 2 The difference between topological equivalence and geometrical identity (Weaire, 1982).

Furthermore, the shape of a cell in an evolving tissue is not constant: cells divide (mitosis) when they grow too large, or exchange neighbours when an edge shrinks to nothing (Fig. 3). These topological transformations (discrete changes of cell shape), occurring at random, are ultimately responsible for the tissue's geometry. (Indeed, Wooten, Winer and Weaire (1985) have succeeded in generating three-dimensional random covalent networks by a succession of T1 transformations randomly in space, and structural relaxations, in a series of computer simulations. All the networks produced after a sufficient number of T1 transformations have the same structural characteristics, independent of the choice of initial, crystalline network).

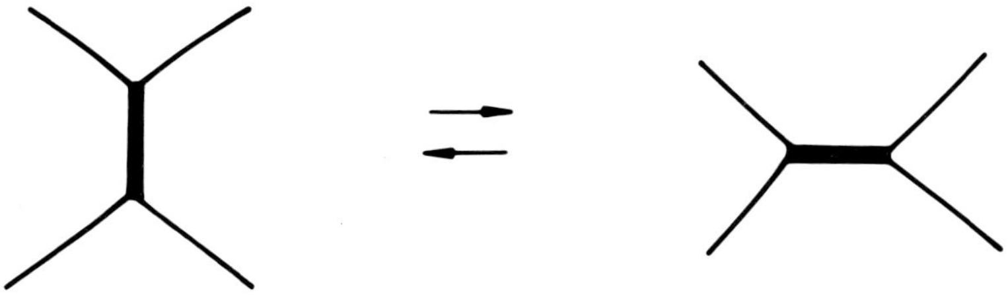

Fig. 3 Elementary topological transformation: T1 or neighbour exchange.

In our analysis, elementary topological transformations are the essential agents producing a structure which is homogeneous, isotropic and in a steady state. This structure can then be described very simply by a few relations between its observable parameters, like the cell sizes and shapes. These relations are (1) Aboav's law, a statement of detailed balance, (2) Lewis' law, expressing statistical equilibrium and (3) von Neumanns's law, governing evolution (see the following chapters).

The topological tissue

In two dimensions, cells are polygons, filling space in a random pattern. Three cells meet at vertices. An arrangement like the Four Corners boundary between New Mexico, Arizona, Utah and Colorado, where four cells meet at a vertex, is not structurally stable, as it splits into two stable vertices through an infinitesimal deformation of the boundaries. Probability that such an arrangement occurs in a random pattern is strictly nil.

In three dimensions, cells are space-filling polyhedra ('an aspect of tissue geometry so obvious that it can hardly escape the attention of any person who seriously considers the question at all' (Dormer, 1980, page 8)), with three cells meeting on a common edge, or four joining at a vertex constituting a structurally stable configuration. Because every cell is at a different phase of its mitotic cycle, the undifferentiated tissue is random, and remains random despite evolving in time, as observed. Randomness is fully characterised by distributions of shapes and sizes of the polyhedra, and description of their correlations in space and time. To obtain these distributions and correlations is the aim of statistical histology or crystallography (Rivier, 1985). In this chapter, we shall be less ambitious, and concentrate on the average properties of the random, evolving tissue, exactly as the laws of thermodynamics constitute an observable average of the results of statistical mechanics.

An elementary construction of random space-filling cellular structures is the *random Voronoi froth (RVF)*. Distribute seeds at random (Poisson distribution) in space. Cells are defined as containing all points in space which are closer to one seed than to any other, very much like a territorial partition between foxes. Boundaries are therefore perpendicular bissectors of the segments joining every pair of seeds; their existence implies that the two seeds are neighbours. This is a structure which can be easily generated on a computer (Crain, 1976; Boots, 1982), but it does not look like a natural tissue: its cells are much too anisotropic. However, it can be relaxed very rapidly and efficiently to a more natural-looking tissue, for example by moving the seed to the centre of mass of its cell, redrawing the boundaries and so on iteratively, or by adding a hard-core repulsion between the seeds. See Weaire and Rivier (1984) for a review.

Elementary structural transformations

Consider now a tissue in two dimensions, and assume that the cell boundaries (edges) can shrink or expand independently of each other. One distinguishes two local, elementary *topological transformations*, (i) T1, or neighbour-switching (Fig. 3), (ii) T2, or cell division (with one triangular daughter) (its inverse, bubble disappearance, is readily observable in soap froth) (Fig. 4). General mitosis, where the cell divides into two cells of arbitrary shape (Fig. 5), is simply an iterate of the T2 transformation, or a combination of one T2, followed by several T1 transformations.

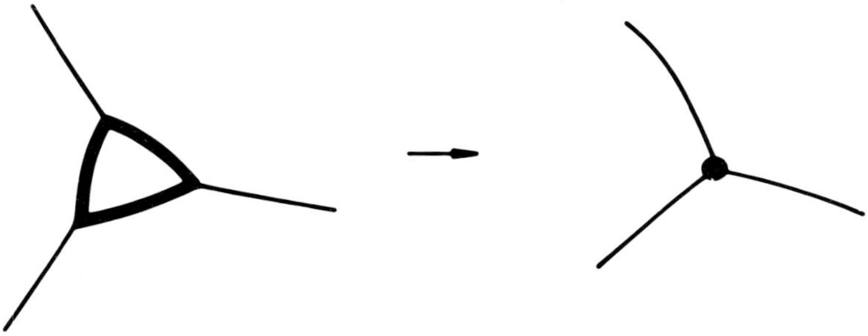

Fig. 4 Elementary topological transformation: T2 or cell disappearance.

These transformations change the number of sides of the cells involved. In 2D, this corresponds to the introduction of topological disclinations: hexagonal cells tile the plane, and one pentagon can be constructed by removing a wedge in the tiling and reglueing. One recovers a perfect tiling, apart from one pentagon, and the fact that the tiling is now warped into a cone. 12 pentagons, i. e. 12 topological disclinations make up a football, or close a sphere, and six pentagons, together with hexagons, would cover an appendix, a finger or a deep cavity. These results are elementary consequences of

Euler's relation. Conversely, heptagons are responsible for negative curvature of the tissue, like saddle points. The intestine, which is globally Euclidean (a cylinder), is locally full of appendices, and cavities, separated by saddle points, and the intestinal crypt is therefore covered with pentagons and heptagons, in addition to flat, hexagonal cells.

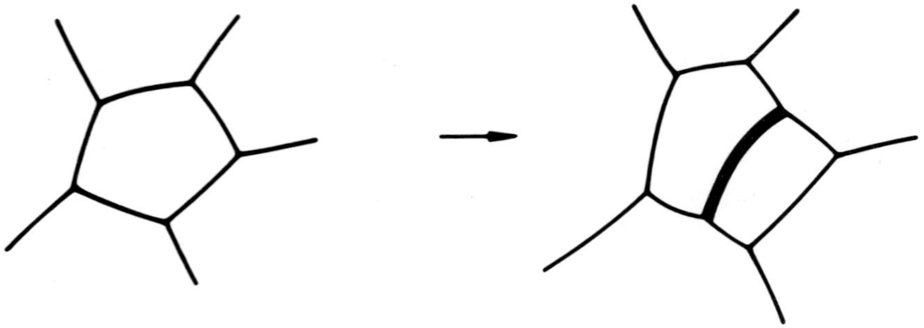

Fig. 5 Topological transformation: mitosis (cell division), a combination of T2 and T1 elementary transformations.

Two neighbouring 5- and 7-sided cells form a dislocation (Fig. 6). Adding new material is dislocation's main structural role. Accordingly, they are essential constituents of the structure of composite flowers (daisies, sunflowers), pinecones or pineapples (phyllotaxis), and the agents in their growth (Rivier et al., 1984) (Fig. 1). Dislocations and disclinations are defined topologically, and do not depend, as in crystals, on a symmetry or metric which the random tissue does not possess.

Of greater biological importance is the fact that dislocation can climb: Mitosis (or a T1 transformation) creates a pair of dislocations in an otherwise hexagonal tissue. Subsequent mitoses of the two, now heptagonal cells[1)] (or subsequent T1) make the two

1) The two heptagonal cells are then growing faster than the others (5.4), and will be the first to divide.

dislocations *climb* away from each other, leaving an additional layer of cells in between (Fig. 7). Through this climb mechanism, additional material is inserted or created locally, and the tissue grows by topological means, a succession of elementary structural transformations. This growth mechanism was first suggested by F. T. Lewis (1943 b) and applied to the growth of intestinal crypt by Pyshnov (1980). In the intestine, dislocation climb stops at positive disclinations, i. e. at the top of appendices and the bottom of cavities, so that the mechanism can also occur on a curved substrate. Note that, whereas an isolated dislocation inserts material from the boundary of the tissue, a pair of climbing dislocation does it locally.

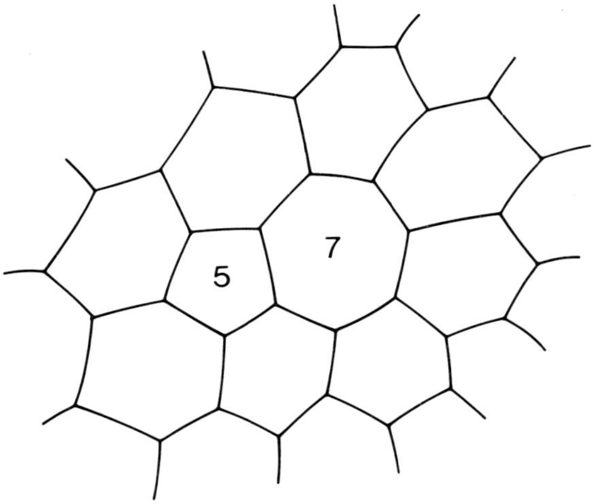

Fig. 6 A topological dislocation: a 5-7 pair of cells in an otherwise hexagonal tissue.

We shall see in the next section that elementary transformations leave the statistical properties of the random tissue invariant, so that they can occur locally anywhere in space, independently of each other and in any order, at any time, without affecting the statistical properties of the structure. They establish microreversibility, or detailed balance in the tissue.

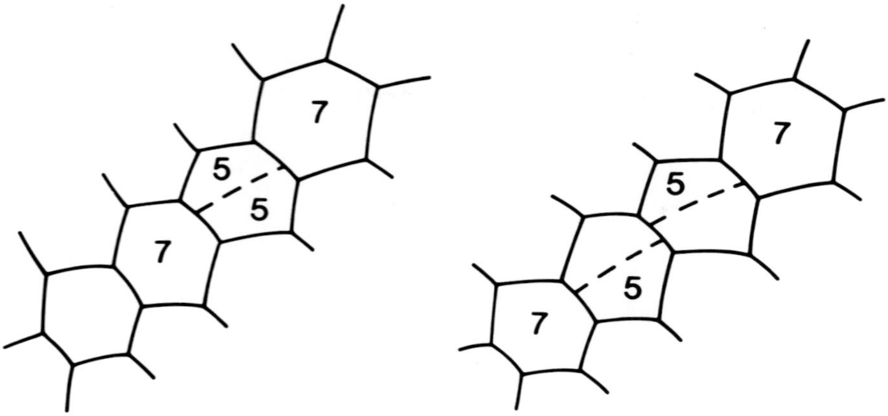

Fig. 7 Dislocation climb by successive mitoses, leaving behind an additional cellular layer by local operations. This is a mechanism for tissue growth.

In three dimensions, one should add (iii) face disappearance or its inverse, face division, to the set of elementary structural transformations. If edges are no longer able to shrink or expand independently of each other - they form interfaces with their own elasticity or rheology -, then more complicated, cooperative transformations can occur, which are sometimes observed in soap froth (Desselberger and Kooros, private communication). Also, T1 itself may not occur because it would warp the interfaces, as suggested by Fortes and Ferro (1985). There is, however, no evidence for cooperative shrinking of edges and interfaces in biological tissues, at least to first order of approximation.

In a regular tissue in three dimensions, disclination and dislocations form uninterrupted lines, and move accordingly. For example, additional material is inserted through the creation of a small dislocation loop around a divided cell, which stretches out through further mitoses. For a random structure, however, neither dislocations, nor disclinations are topologically or structurally stable, because there is no translational or rotational symmetry to define them (Rivier, 1979; Duffy, Rivier, 1982). The only topologically stable linear structures in 3D disordered cellular structures are the *odd lines* (Rivier, 1979), which thread through odd faces in an uninterrupted fashion. It is akin to a disclination, but characterised by its existence rather than by its intensity.

Topological conservation laws

Topologically stable constituents of a tissue are those which retain their identity and their existence under small, continuous deformations. For example, an arbitrary elastic distortion, or a 4-coordinated vertex in 2D, are not topologically stable, unlike a 3-coordinated vertex in 2D or an odd line in 3D. Topological stability is guaranteed by a conservation law. In random tissues, there are two conservation laws. The first is Euler's Relation

$$C - E + V = \chi \qquad \text{(2D)} \qquad (1)$$

$$-C + F - E + V = 0 \qquad \text{(3D)} \qquad (2)$$

for a tissue with C cells, F interfaces in 3D, E edges and V vertices. χ is a characteristic of the substrate of the tissue, $\chi = 2$ for a sphere, $\chi = 1$ for a plane or a cylinder, not counting the cells at infinity. At any rate, χ is an integer of order 1, and thus negligible in comparison with C, E, V in a tissue containing many cells. Euler's relations are easily proven by induction, adding elements to a given tissue. In particular, one can check that the left-hand sides of (1) and (2) are invariant under the elementary transformations of the last section.

The second conservation law is the continuity of odd lines in 3D (Rivier, 1979). These take over from conventional crystalline constituents like dislocations and disclinations, which are not topologically stable in 3D random tissues.

Euler's relations are combined with valence relations between edges and vertices, etc,

$$2E = 3V \qquad 3E = \langle n \rangle\, C \quad (2D) \tag{3}$$

The first relation states that every edge joins two vertices, whereas every vertex has z incident edges. Here $z = 3$, for topological stability. In the second relation, every edge separates two cells and cells have $\langle n \rangle$ sides on average. Here n is therefore the sole local, topological variable describing the tissue. (1) and (3) imply that

$$\langle n \rangle = 6 \qquad (2D). \tag{4}$$

On average, the topological variable of the 2D froth is constant. Its spatial correlations are described by Aboav's law (next section).

In three dimensions, there are two topological random variables, n and f, where f is the number of faces of a cell. Their expectation values are not fixed as in 2D, but only related by the 3D analogue of (4)

$$\langle f \rangle = 12/(6 - \langle n \rangle). \tag{5}$$

This is an identity, which holds both for an individual cell, and for the tissue as a whole. At the cell level, n can still fluctuate, and does, as demonstrated in the considerable variety of polyhedral shapes observed in cells or soap bubbles (Matzke, 1946).

Equations (4) and (5) are the only *topological* constraints on the structure of the tissue.

It is found empirically that $<f> \approx 14$ in 3D, but, unlike $<n> = 6$ in 2D, this is not an exact result. A random tissue with isotropic cells of equal volume has $<f> = 13.40$. Fluctuation in cell volume decreases $<f>$, whereas cell anisotropy increases $<f>$ (Rivier, 1982). Indeed, for a random Voronoi froth, $<f> = 15.54$ (Meijering, 1953) and the cells are highly anisotropic. These fractional averaged shapes - and others to be encountered later, like the space of a daughter cell after mitosis - indicate that tissue geometry is random and that there is no single elementary cell which can be repeated to fill the whole space. This observation has been emphasised by Matzke (1950). It is an expression of geometrical *frustration*, the fact that local perfection cannot be achieved consistently everywhere, or even within one single cell. Some compromise must be struck locally, which produces a random structure globally in the absence of conspiracy ('avoidance of the niceties of adjustment').

Topological correlations, Aboav relation and detailed balance

In this section, we shall investigate shape correlations between neighbouring cells in the random tissue. Recall that $<n> = 6$ in two dimensions (4), and that $<f>$ and $<n>$ are related by (5) in three dimensions, with $<f> \approx 13.40$ for isotropic cells of equal volumes. What is, on average, the number of sides or faults of a cell adjacent to a n- or f-sided cell? This is given by Aboav's relation (11, 12).

Aboav's relation works on two levels. First of all, it describes the correlations between shapes of neighbouring cells. One notices immediately that cells with many sides are surrounded with cells with few sides. But also, it guarantees detailed balance between shapes of cells which fluctuate, through the T1, T2 transformations and mitosis. In this respect, it is a prerequisite for statistical equilibrium of the tissue, which can then be described by an equation of state and a most probable distribution of cell sizes and shapes.

Detailed balance not only guarantees the existence of an ideal random cellular network, architypical of undifferentiated biological tissues or soap bubble froths, metallurgical aggregates, etc., but also suggests methods of generating such a network on the computer, for example by alternating random T1 transformations, with relaxation (Wooten et al., 1985).

Aboav's relation had been known only by metallurgists, at least until a few years ago, and only in its direct interpretation, as describing spatial correlations. (By contrast, Lewis' law (next section) has been the preservee of a few botanists, but can only be found in rare textbooks on tissue geometry (Smoljaninov, 1980)). Discovered empirically by Aboav (1970), its first derivation was given by Bland and Mocellin (1979), and some later emphasis and generalisation by the present author (Rivier, 1985, 1986) which established its central position as the result of detailed balance between cells with fluctuating shapes. Let us now derive Aboav's relations and the equations of detailed balance.

In view of its biological importance, we shall discuss the two-dimensional (2D) tissue first, and defer the - less-known - three-dimensional tissue geometry for later. By two-dimensional is meant either a single layer of cells, or a multilayer of cells so anisotropic that all topological transformations take place within the layers (columnar tissue, for example).

Recall that the average cell shape in 2D tissues, $\langle n \rangle = 6$, remains invariant at all times in a fluctuating and evolving tissue (except at the critical instant of a T1 transformation, where the vertex is tetravalent). What about the cell's neighbours? Denote by $m(a)$ the average number of sides of the neighbours of a given a-sided cell (so that $a\, m(a)$ is the total number of sides of all its neighbours). Consider first a T2 transformation, where cell $d = 3$ disappears, and its inverse. Neighbours to cell d are cells a, b and c, and neighbours to a are therefore cells b, c and d - when it exists -, and other cells s_i, whose shape remains unaffected by the transformation. Thus, before the transformation, we have

$$a\, m(a) = b + c + d + \Sigma s_i \tag{6}$$

and after,

$$(a - 1)\, m(a - 1) = (b - 1) + (c - 1) + \Sigma s_i \tag{7}$$

so that, with $d = 3$, eliminating $b + c + \Sigma s_i$, we obtain

$$a\, m(a) = (a - 1)\, m(a - 1) + 5. \tag{8}$$

Similarly, for a T1 transformation where cells a and b are pulled apart, we have

$$a\, m(a) = (a - 1)\, m(a - 1) + (b - 1) - 1. \tag{9}$$

But, after the transformation, cell b - which has now b-1 sides - is no longer adjacent to cell a, so that it has $\langle(b\text{-}1)\rangle = 6$ sides on average (4), if one assumes no correlation beyond nearest neighbours. Then, (9) is exactly the same recursion relations as (8) or (10) below and transformations T1, T2, mitosis and their inverses can occur independently of each other in space and in time, without offsetting the structural properties of the tissue, which are in detailed balance under these transformations. This is a necessary, but not a sufficient condition for statistical equilibrium and for steady state of the tissue.

Under the same assumption of no correlation beyond adjacent cells, the recursion relation under mitosis is simply the rth iterate of (8)

$$a\, m(a) = r\, m(r) + 5(a - r) \qquad (10)$$

when cell a splits into daughters with r and $a - r + 4$ sides.

Equations (8) to (10) are easily solved,

$$m(n) = 5 + B/n \qquad (11)$$

which is the Aboav relation (1970). The parameter B is related to the second moment of the shape distribution $\mu_2 = < (n - < n >)^2 >$ where $< >$ denotes average, by $B = \mu_2 + 6$, through a sum rule $<nm(n)> = <n^2>$ (Lambert, Weaire, 1983). Aboav's relation is almost universally obeyed (Weaire, Rivier, 1984), except in the *random Voronoi froth* (RVF) which will be discussed below. One checks easily that (11) holds for a regular hexagonal tissue, and also if $n = \infty$. An infinite-sided cell, like the tissue's boundary, has indeed 5-sided cells on average as its neighbours.

In three dimensions, there are two topological variables n and f, and, although one would expect a similar correlation as (11), but involving f only instead of n, none has been found which is stable under all topological transformations. It is the relation between a n-sided *face* of a f-faceted cell and its neighbours which gives the same results as in two dimensions (unique recursion relations and detailed balance without correlation beyond nearest neighbours), the 3D Aboav relation,

$$n\, m_f(n) = 5f - 11 - K(f-1-n) \qquad (12)$$

(Rivier, 1985), where $m_f(n)$ denotes the average number of sides of faces neighbour to face n, all belonging to the same, f-faceted cell. K is a constant of the froth which can again be related to $\mu_{2f} = (n-<n>_f)^2>_f$ (average restricted to f-faceted cells) by

$$\mu_{2f} = (f-4)\,(f-3)\,[\,5 - K - 12/f\,]\,\,(1/f)\,. \tag{13}$$

Typically, $-1 \leqslant K \leqslant 2$.

Absence of correlation beyond nearest neigbour cells, ($<b-1> = 6$ in equation (9)) may look merely like a simplifying assumption. It is not the case, and freedom through detailed balance, to undergo any topological transformation locally and in any order is an overriding advantage which destroys shape correlation beyond direct neighbours. This point is demonstrated in the mathematical model of *random Voronoi froth* (RVF). Boots (1982) has computed Aboav's relation, and finds a m(a) linear in a, but with a slope larger than 5. This is because, by definition, only T1 transformations do occur in RVF, and detailed balance is trivially achieved, with arbitrary $<(b-1)>$ in (9). Consequently, correlations do extend beyond immediate neighbours in this case.

Lewis' law as the equation of state of an ideal tissue

We have seen that specific, short-ranged biological, physical or chemical forces are less important in framing the tissue's architecture than the mathematical constraints that cells are filling space.

It turns out that these mathematical constraints are sufficient to determine the average characteristics of the random structures generated under the overriding influence. This is due to the fact, well-known in statistical thermodynamics, that the most probable distribution of cells is overwhelmingly more probable than any other. The structure is then in statistical equilibrium. This is the case of biological tissues.

The most probable distribution of cells is obtained by maximising the arbitrariness, or entropy of the structure, subject to the mathematical constraints of space-filling (Jaynes, 1957).

In detailed balance under topological transformations, the tissue can also be, with overwhelming probability, in statistical equilibrium. This is a central concept of statistical mechanics familiar to physicists, which has been established at the turn of the century by Boltzmann and Gibbs, and whose characteristics (equation of state and distribution of microscopic configurations) can be obtained simply by maximum entropy inference, an application of information theory and probability theory, due to Jaynes (1957, 1979). Details can be seen in two papers by the author (Rivier, 1985, 1986).

Statistical equilibrium describes the structure of the tissue as a whole (its macroscopic configuration). The tissue consists of large numbers of cells, each with several possible microscopic configurations (their shape and, possibly, their size), related by elementary structural transformations and occurring at random. The macroscopic configuration of the tissue is characterised by averaged shapes, sizes of cells, and by correlations thereof, that is, by the probability distribution of microscopic parameters. Because of detailed balance (Aboav's law), it is invariant under elementary transformations.

Statistical equilibrium can be obtained by maximising the arbitrariness or entropy of the tissue, subject of a few inescapable and obvious constraints (volume or area to fill, topology). This expresses the fact that the tissue is large and loose enough (it has undergone enough elementary transformations) so that the least restricted macroscopic configurations can be realised by many more microscopic cell configurations than any other and is therefore overwhelmingly the most probable one. The tissue can grow and evolve while remaining in statistical equilibrium.

Statistical equilibrium also leaves an observable signature, the equation of state, which is a relationship between averaged microscopic parameters. In (ideal) random tissues, the equation of state is Lewis' (1928) relation between averaged shape and size of a cell, which we shall now derive.

The complete statistical problem involves two random variables (in 2D), the number n of sides per cell - the topological variable - and the area A of a cell - the metric variable. To obtain an equation of state, it is sufficient to concentrate on the topological variable n only.

One looks for the most probable distribution, of shapes of the cells, where $p_n = \int dA\, p(n, A)$ is the probability of finding a n-sided cell. It is that distribution which maximises the entropy or information

$$S = - \Sigma p_n \ln p_n \qquad (14)$$

(Jaynes, 1957), subject to constraints corresponding to our prior knowledge of the system. For *any* random tissue, the constraints are

$$\Sigma p_n \overline{A}_n = A_o/C \qquad \text{(space filling)} \qquad (15)$$

$$\Sigma p_n n = 6 \qquad \text{(topology)} \qquad (4)$$

in two dimensions, and similarly, mutatis mutandis ((4) replaced by (5)) in 3D. Here $\overline{A}_n = \int dA\, p(n, A)\, A$ is the average area of a n-sided cell, and A_o, the total area available to the C cells in the tissue. The problem, and the constraints in particular, are entirely mathematical. Biology, or physics are absent at this level, so that the resulting structures are universal.

The 'equation of state' of the tissue can be obtained without evaluating the entropy or the most probable distribution $\{ p_n \}$, by the following argument (Rivier, Lissowski, 1982. See Rivier, 1985 for a full calculation): The constraints are a linear system of equations between the p_n's, so that the smaller the dimensionality of the space of contraints, the larger that of the (vector) space of possible solutions $p = (p_3, p_4, p_5, ...)$, and, in the absence of any other information, the more probable one such solution will be. The most probable distribution is obtained therefore by reducing as much as possible the dimensionality of the space of constraints, by making them linearly dependent, that is

$$A_n = (A_o/C)\lambda[n - (6 - 1/\lambda)] = (A_o/C)[\lambda(n - 6) + 1]. \tag{16}$$

The averaged area of a n-sided cell, A_n, is linearly related to the number n of it sides. There is a finite intercept $n_o = 6 - 1/\lambda$. This relation was suggested empirically by Lewis (1928), on the basis of observations on cucumber epidermis, human amnion, and the pigmented epithelium of the retina. As far as I know, all undifferentiated biological tissues obey it, with intercept $n_o \approx 1 - 2$ (Smoljaninov, 1980). It is also obeyed by random Voronoi froths (as can be read from the tables of Crain, 1978), with a smaller intercept $n_o \approx 0$, but not, usually, by metallurgical grain aggregates. It is a relation between averaged, observable structural parameters of the tissue, i. e. the equation of state of statistical histology describing an ideal tissue (in the same sense as 'ideal gas' in thermodynamics).

The parameter λ in (16) is the Lagrange multiplier imposing the constraints (15) and (4). (The other multiplier is eliminated by the linear dependence of the constraints). It measures the slope of Lewis' relation and is therefore an important descriptive parameter of the structure. Indeed, Lagrange multipliers have an habit in thermodynamics and in mechanics of possessing a physical meaning of their own. They are not mere arbitrary factors. We shall see in the next section that λ is a constant of the steady state in biological tissues. In soap froths and other networks coarsening or ripening in time, it measures the ageing of the structure (it is simply the time).

An ideal tissue has the smallest possible number of constraints. Additional constraints cause (observable) departures from ideality and deviations from Lewis' equation of state. The equation of state classifies therefore the tissues. Undifferentiated biological tissues are ideal, and satisfy Lewis' law. Metallurgical grains, whose interfaces cost energy, obey a different equation of state characteristic of mosaics with additional energy constraint. It is grain's perimeter, rather than its area, which is proportional to n (Rivier, 1985). (The terminology (equation of state, ideal, ...) is borrowed from physics, where a gas of non-interacting particles is called ideal and satisfies the equation of state $pV = NkT$ between pressure p, volume V and temperature T. Here N is the number of particles and k Boltzmann's constant. A gas with finite potential energy between its particles is classified as non-ideal, and its equation of state deviates from the ideal law.)

This example of metallurgical aggregates emphasises the *diagnostic* power of the Maximum Entropy formalism. A relevant physical constraint has been uncovered by noticing the discrepancy between observed and ideal equations of state. It also classifies random tissues and mosaics. The class is defined by the physical and biological constraints imposed on the structure, or, equivalently, by its equation of state. Lewis' law corresponds to minimal number of constraints (2, both mathematical and unavoidable), or to the simplest tissue architecture, cheapest in information. Departure from Lewis' law betrays additional constraints, and is diagnostic of some pathology.

The effective elimination of one constraint, expressed in Lewis' law (16), increases the entropy further. This is because, as in statistical thermodynamics, the most probable entropy is a convex function of its variables, the constraints (the right-hand sides of the constraint equations (15) and (4)).

Lewis' law is therefore a result which goes beyond the standard applications of the Maximum Entropy formalism (Jaynes, 1959; Rosenkrantz, 1984), in physics or in crystallography, in which the functional form of the constraints or of the correlations is known precisely and only the most probable distribution remains to be found. Here, even the functional form of the constraints is not known a priori, and we have taken advantage of the freedom to adjust it in order to maximise the entropy further.

It is elementary to derive, by the same method, Lewis' law for three-dimensional tissues (Rivier, 1982)

$$\overline{V_f} = (V_o/C)\lambda[f - (<f> - 1/\lambda)] \tag{17}$$

but this relation has not yet been confirmed experimentally.

Evolution of a random tissue. Von Neumann's law. Steady State

Many random cellular structures evolve slowly in time. In soap froths, gas diffuses across the interfaces between bubbles. In metallurgical aggregates, large grains grow at the expense of small ones during the controlled process of sintering. In biological tissues, cells grow and divide, so that there is local activity even if the tissue is, over all, in a steady state. The time scale for this evolution is much longer than that associated with the mechanical response of the structure, which remains therefore in statistical equilibrium at all times. Consequently, Lewis' law (16) should hold throughout the evolution of the structure. The parameters A_o, C, n, and the hitherto undetermined multiplier λ vary with time in their own way, A_o and λ continuously, C and n by discrete jumps.

The simplest case is the evolution of a two-dimensional soap froth with fixed total area A_o and constant number of cells C. This applies to time intervals not long enough for bubbles to disappear (no T2 process). Differentiating Lewis' law with respect to time, we obtain the rate of growth of an averaged n-sided bubble

$$d\bar{A}_n/dt = (A_o/C)\ (d\lambda/dt)\ (n - 6). \qquad (18)$$

This law has been obtained first by von Neumann (1952) in a different derivation, valid for 2D soap froths only. Yet, the result (18) and our derivation (Rivier, 1983) are topological: A pentagonal cell loses area at the same rate as an heptagonal cell gains it, and at half the rate of an octagonal cell, etc.

For soap froths,

$$\mu = (A_o/C)\ (d\lambda/dt) = (2\pi/3)\,\delta\sigma \quad > 0 \qquad (19)$$

(von Neumann, 1952), where σ is the surface tension of the interfacial liquid and δ, the air diffusivity through it. The coefficient μ in von Neumann's law is a constant of the froth, independent of time as long as the interfacial liquid does not thin or drain out appreciably. This provides a physical interpretation of the Lagrange multiplier λ in soap froths:

$$\lambda = \mu(C/A_o)\ t + \lambda_o$$

so that λ measures the ageing of the froth, or simply the time t.

Like Lewis', von Neumann's law can be generalised to three dimensions. In particular,

$$dV_f/dt = (V_o/C)\,(d\lambda/dt)\,[\,f-<f>\,] \qquad (20)$$

(Rivier, 1983), over short times (constant V_o and C), exactly as in two dimensions.

Biological tissues do not age like froths. They are in a steady state, at least to a first approximation (Dormer, 1980). This means that Lewis' law coefficients A_o/C and λ are invariant over times longer than the period τ of the mitotic cycle. Steady state is achieved through interplay between continuous cellular growth $(dA_o/dt > o)$ and discrete mitoses $(\Delta C/\Delta t > 0)$.

Over times Δt shorter than τ, only A_o varies (cells grow). The dimensionless slope λ of Lewis' law is a constant of the steady state. Differentiating Lewis' law under these conditions, we obtain

$$d\bar{A}_n/dt = (\lambda/C)\,(dA_o/dt)\,(n-n_o), \qquad \Delta t < \tau \qquad (21)$$

with the same intercept $n_o = 6 - 1/\lambda$ as in Lewis' law, so that *all* cells grow in the biological tissue, in contrast to soap froths or metallurgical aggregates where small $(n < 6)$ cells shrink.

Let us now investigate the mitotic cycle within the steady state.

Mitosis and the topological steady state

In two dimensions, let a m-sided cell (mother cell) divide into two daughter cells with d_1 and d_2 sides respectively. Clearly,

$$m + 4 = d_1 + d_2. \tag{22}$$

The average number of sides per cell is $\langle n \rangle = 6$ (4), and this result should hold throughout the mitotic cycle in the steady state. Hence,

$$\langle m \rangle + 2\langle d \rangle = 3\langle n \rangle = 18 \tag{23}$$

since $\langle d_1 \rangle = \langle d_2 \rangle = \langle d \rangle$. Comparing (22) and (23), we obtain

$$\langle m \rangle = 7, \qquad \langle d \rangle = 5.5 \qquad \text{(2D)} \tag{24}$$

(Lewis, 1943 b; Dormer, 1980, p. 41). Incidentally, the same argument applies for cells neighbouring both daughter cells, which acquire one side through nearby mitosis, $a \rightarrow a + 1$. Since

$$\langle a \rangle + \langle a+1 \rangle = 2\langle n \rangle, \tag{25}$$

$\langle a \rangle = 5.5$. Steady state holds also for (metric) areas of cells, as a direct consequence of Lewis' law, as we shall see presently.

In three dimensions, the influence of topological steady state upon mitosis is determined in the same fashion. m, d and a refer now to the number of facets of mother, daughter and neighbours to both daughter cells, respectively. Let the face separating the two daughter cells have c sides. Then,

$$m + 2 + c = d_1 + d_2. \quad (22')$$

On average, cells have $<f>$ facets each, and facets have $<n>$ sides, related by (5). Consequently,

$$2 <a> + 1 = 2<f> \quad (25')$$

or $<a> = <f> - 1/2$, for neighbour cells. For the cells directly involved in the mitosis,

$$<m> + 2 <d> = 3 <f> . \quad (23')$$

Moreover, the interface between the two daughter cells should be a typical facet of any $<d>$-faceted cell if the tissue is in a steady state, $<d> = 12/(6-<c>)$, so that (22') becomes

$$<m> = 2 <d> - 8 + 12/<d> \quad (26)$$

(using relation (5)). Combining the last two equations, we obtain,

$$<m> = (9/4)<f> - 2 - \phi$$
$$<d> = [(3/4)<f> + 2 + \phi]/2 \quad (27)$$

with $\phi^2 = [(3/4)\langle f\rangle + 2]^2 - 12$. For isochoric and isotropic cells, $\langle f\rangle = 13.4$,

$$\langle m\rangle = 16.61, \quad \langle d\rangle = 11.80 \qquad (3D). \qquad (24')$$

Both values increase with anisotropy and decrease with volume fluctuations, following $\langle f\rangle$ itself. Dormer (1980, p. 64) obtains slightly larger values in a derivation which contains an additional assumption (as to the value of $\langle c\rangle$), besides steady state. Matzke and Duffy (1956) observe $\langle m\rangle = 16.84$ and $d = 12.61$, which suggests, if confirmed, cellular anisotropy and $\langle c\rangle$ larger than typical for daughter cells ($\langle c\rangle > 6$!).

We can now obtain the period τ of the mitotic cycle in the steady state. It is approximatively the time necessary for an average daughter cell to grow to the size of a mother cell (neglecting the time taken by the mitotic process itself), thus $\bar{A}_{\langle d\rangle}(\tau) = \bar{A}_{\langle m\rangle}(0)$. Integrating (21), we obtain

$$\bar{A}_{\langle d\rangle}(\tau) = \bar{A}_{\langle d\rangle}(0) + (\lambda/C)(dA_0/dt)\,[\langle d\rangle - n_0]\,\tau$$

assuming as constant rate of growth over τ, and

$$\tau = A_0\,[\langle m\rangle - \langle d\rangle]\,/\,\{(dA_0/dt)\,[\langle d\rangle - n_0]\}. \qquad (28)$$

Typically, $1/3 \leqslant \tau(dA_0/dt)/A_0 \leqslant 3/7$ for biological tissues, $1 \leqslant n_0 \leqslant 2$, respectively.

Finally, it is possible to obtain a bound for the intercept n_0 of Lewis' and von Neumann's laws, i. e. for the smallest cell. If mitosis did occur instantaneously, the ratio between sizes of mother and daughter cells would be 2. Because mitosis takes a finite time, during which both mother and daughter are still growing, $A_{<m>}/A_{<d>} < 2$, so that, from Lewis,

$$n_0 < 4, \qquad \text{or} \quad \lambda < 1/2$$

an inequalitty verified empirically ($n_0 \approx 1 - 2$) (Smoljaninov, 1980). An important consequence of Lewis' law is that steady state holds not only for the shapes, but also for the sizes of cells in the tissue, that is

$$\bar{A}_{<m>} + 2\,\bar{A}_{<d>} = 3\,A_0/C \tag{29}$$

is the metric equivalent of the topological steady state condition (21) (and similarly, in three dimensions). Only a linear relation between shapes and sizes guarantees steady state for both.

Conclusions

We have seen that statistical histology, a method to describe and classify the structure of random cellular networks, can usefully be set up along the lines of statistical thermodynamics (see Table I). It actually goes one step beyond standard statistical mechanics (Jaynes, 1957), as one takes advantage of the arbitrariness in the functional dependence of one of the constraints to maximise further the entropy. The simplest random structure has minimal numbers of constraints. It is called *ideal*, by analogy with ideal gas in thermodynamics. Its equation of state (Lewis' law) is experimentally observable, as is the ideal gas law. Deviations from Lewis' law betray the presence and structural relevance of additional (biological, physical) forces, as departures from the ideal gas law yield information on interatomic forces and, ultimately, on other phases of matter. As in statistical thermodynamics, statistical equilibrium of the tissue is a balance between entropy (most probable distribution) and the lowest level of organisation (territorial partition).

Table I.

Comparison between methodologies of statistical thermodynamics and histology

Thermodynamics		Histology
	Random variables:	
velocity		cell shape n, size A
	Detailed balance under:	
collisions		elementary transformations - Aboav's law - shape correlations
	▼	
	Statistical equilibrium (maximum entropy) - equation of state (ideal systems):	
$pV = NkT$		Lewis' law
	- most probable distributions:	
Maxwell-Boltzmann		Universality of cellular structures
	▼	
	Slow evolution	
quasi-static transformations		(growth, mitosis) - von Neumann's law - steady state

The biological content of this chapter is small, but not nil. Whereas all the constraints are unavoidable in ideal tissues, and Aboav's and Lewis' are general relations of statistical equilibrium, there are two structural features specific to living tissues:

- Steady state, achieved by interplay of cellular growth and mitosis. (By contrast, physical froths coarsen and age). It implies a linear relationship between cells shapes and sizes (Lewis' law).
- Topological transformations enable the tissue to grow locally without requiring additional material to be brought in from the outside. They create a cellular partition of space which is fluid and capable of growth.

We recognise therefore the central position of Lewis' law (16, 17), which is altogether the equation of state of an ideal random tissue, a concept of statistical mechanics, and a sufficient condition for steady state evolution, a biological property.

Dormer (1980, p. 12) has accused the biological literature on tissue structure of lacking the courage to try any hypothesis which is not immediately manifest. In this chapter, we have derived a set of observable laws, based on the abstract and far from manifest hypothesis of statistical equilibrium, and consistent with steady state evolution of the tissue. These laws sort out normal, undifferentiated tissues from any pathological structure or evolution. They may also indicate the cause of the pathology. We have also justified the hypothesis, and given criteria for its validity.

Acknowledgements

I have been introduced to Lewis' law by A. Lissowski. I am also grateful to V. V. Smoljaninov for sending me a copy of his book, to T. Elsdale for recommending Dormer's book, and to R. de Almeida, R. Englman, M. Fortes, A. Lawrence, R. Occelli, D. Srolowitz and D. Weaire for helpful discussions.

Surface Changes and Shape Formation During Organism Growth

G. V. Timchenko, E. A. Prokofiev, A. I. Zotin

In the most general terms the development and growth of an organism has three components, change of the mass (growth), appearance of differences in various parts of the system (differentiation), and changes in geometric shape (shape formation). All these phenomena can be essentially analysed in quantitative terms. Changes of mass, or growth phenomena, have been most thoroughly explored and represented in numerous quantitative models and equations (Hofman, 1938; Medawar, 1945; Parks, 1970, 1982; Zotin, 1972, 1985; Zaike, 1975; Walter, Lamprecht, 1978; etc.). Differentiation and formation have proved harder to model.

This paper will concentrate on shape formation because no phenomenological approach to differentiation phenomena has been developed thus far. Thoroughly developed methods are not available, either, for analysis of shape formation because, firstly, it is difficult to characterise the shape of most organisms in quantitative terms and, secondly, a shape change measure is difficult to introduce. We try to express the shape formation in terms of body surface area changes. Naturally, in a growing organism the surface is principally a function of the size or volume of the animal. However it also depends on the body shape itself. Both possibilities will be discussed.

Thermodynamics and Pattern Formation in Biology

Surface changes

The available experimental data on surface area changes suggest that in ontogenesis it is S-shaped during the growth period, flat at the stationary phase, and declines in ageing. Fig. 1 shows the experimental data on body surface changes in ontogenesis of humans together with the growth curve (Meeh, 1879; Lewis, 1938). The mass and surface obviously correlate over the whole ontogenesis. Numerous formulae and nomograms have therefore been developed for determining the organism surface from the body weight. In particular, for white rats it is (Lee, 1929)

$$S = 12.54\ M^{0.6} \tag{1}$$

and for man it is (Brody, 1945)

$$S = 240\ H^{0.4}\ M^{0.53} \tag{2}$$

where S is the surface area; M is the body mass; and H is the body weight.

An allometric dependence such as (1) leads to equations describing the kinetics of the animal body surface growth as a growth equation where M is expressed in terms of S. This is more difficult to do using equation (2) which also involves the factor of body height.

Equation (1) and (2) have led to relationships for body surface change in rats as a function of weight (Kleiber, 1947) and in man (Quetelet, 1869) as a function of age. We tried to describe these curves using equations similar to the growth equations on the assumptions that:

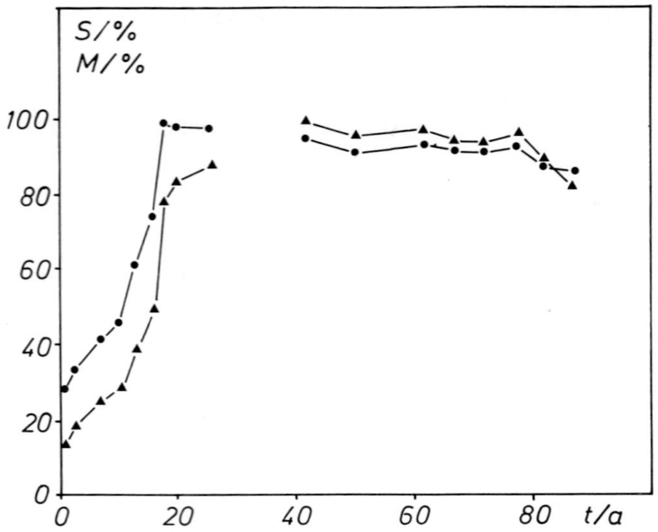

Fig. 1 Body mass and surface area changes in humans (Meeh, 1879; Lewis, 1938), during lifetime.

1. the greater the surface area change the bigger the initial organism surface;
2. the surface change is proportional to some gradient between its current and final values. Consequently, two kinds of equations are obtained

$$1/S \cdot dS/dt = k\{(S_m - S)/S\} \quad (3)$$

$$1/S \cdot dS/dt = k(S_m - S), \quad (4)$$

whose solutions are, respectively,

$$S = S_m - C \exp(-kt); \; C = S_m - S_0 \quad (5)$$

$$S = S_m / \{C \exp(-S_m k t) + 1\}, \quad C = (S_m - S_o)/S_o \tag{6}$$

where S is the current surface; S_m is the maximal surface; t is time; S_o is the body surface at birth; and k is a factor.

The check of fit to experimental data has revealed, Figs. 2 and 3, that these equations provide a satisfactory description of surface changes both in the period of mammalian growth and at the stationary phase but not during ageing and during weight decrease. Equation (3) is a better fit for the empirical curve for rats (Fig. 2) and equation (4), for men (Fig. 3).

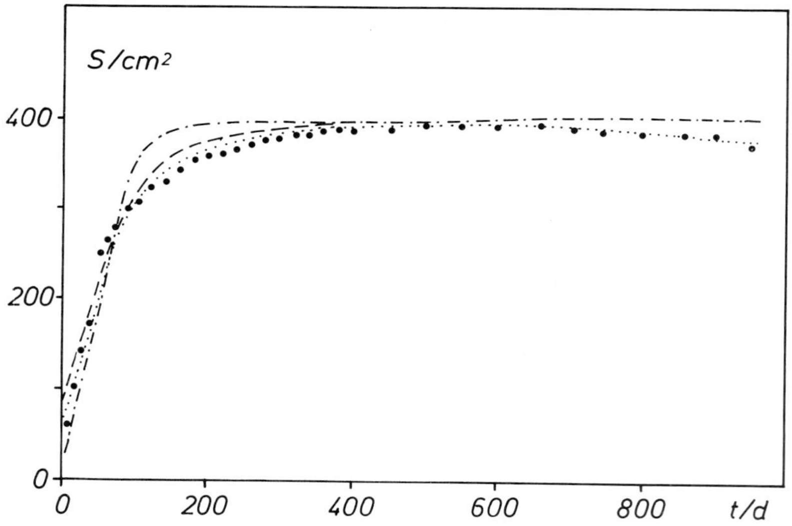

Fig. 2 Comparison of experimental and theoretical curves for surface area changes in rats, the factors being

in equation (3): $k = 0.014$

in equation (4): $k = 0.000109$

in equation (10): $k_{fg} = -0.05$ $k_{fd} = 1 \times 10^{-6}$ $k_{ff} = 0.024$ $M_m = 310$ g

$T_m = 650$ d $S_m = 391.8$ cm^2 $S_o = 38.5$ cm^2 $M_o = 6.5$ g $b = 0.25$

The failure of equations (3) and (4) to be valid over the entire lifetime of mammals comes as no surprise because conventional growth equations do not fit the complete mass variation curve, either (Zotin, 1972). As noted above, three components, growth, differentiation, and shape formation, interact during ontogenesis. Recognising these three phenomena, an equation describing mass changes at the stationary phase and describing weight decrease in ageing as well as during the growth period can be derived (Zotin, 1978; Prokofiev et al., 1982). The following discussion will concentrate on this approach.

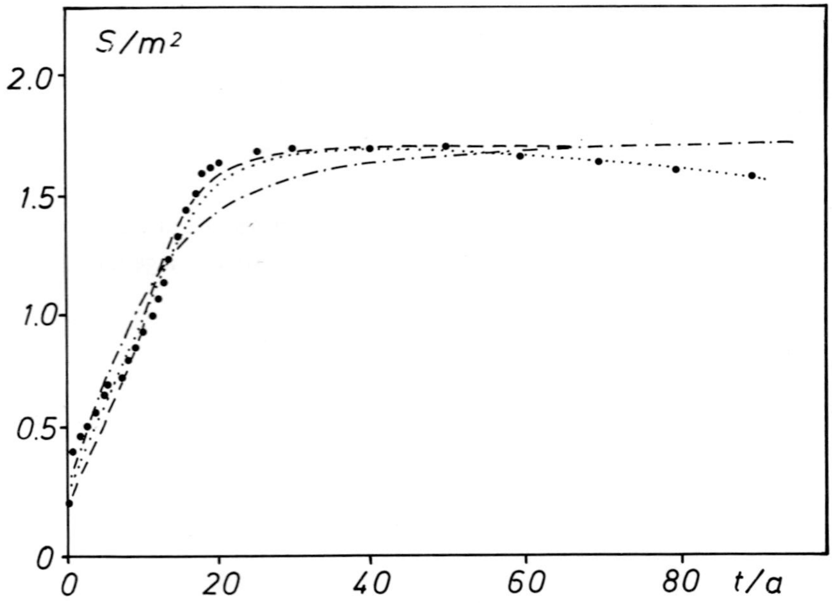

Fig. 3 Comparison of experimental and theoretical curves for surface area changes in humans, the factors being

in equation (3): $k = 0.0856$

in equation (4): $k = 0.000133$

in equation (11): $k_{fg} = -0.1 \times 10^{-5}$ $k_{fd} = 0.00027$ $k_{ff} = 0.123 \times 10^{-4}$

$M_m = 63.7$ kg $T_m = 40$ years $S_m = 16871$ cm^2 $S_o = 2125$ cm^2

$M_o = 3.2$ kg

Phenomenological equations describing the interaction of growth, differentiation and shape formation

In terms of the thermodynamics of nonequilibrium processes, the interaction of any irreversible phenomena in a system is represented as

$$\bar{J}_i = \sum_{j=1}^{n} L_{ij}(x) X_j \qquad (7)$$

where $\bar{J}_i$ are specific thermodynamic flows; X_j are thermodynamic forces and L_{ij} are phenomenological coefficients.

Assuming that relation (7) is applicable to animal ontogenesis, the interaction of growth, differentiation, and shape formation can be represented in a general form as:

$$\begin{aligned} 1/M \cdot dM/dt &= k_{gg} X_g + k_{gd} X_d + k_{gf} X_f \\ 1/D \cdot dD/dt &= k_{dg} X_g + k_{dd} X_d + k_{df} X_f \\ 1/F \cdot dF/dt &= k_{fg} X_g + k_{fd} X_d + k_{ff} X_f \end{aligned} \qquad (8)$$

where

$1/MdM/dt$	is the specific flow of animal mass
$1/DdD/dt$	is the specific flow of differentiation
$1/F \cdot dF/dt$	is the specific flow of shape formation
X_g	is the force responsible for growth
X_d	is the force responsible for differentiation
X_f	is the force responsible for shape formation
M	is the organism mass
D	is differentiation
F	is shape.

The forces represented in equation (8) are somewhat difficult to express. In earlier papers (Zotina, Zotin, 1978; Zotin, 1985) they are described by the equations

$$X_g = r_g\left(\frac{M_m^b}{M^b} - 1\right); \quad X_d = r_d\,(t_m - t); \quad X_f = r_f\,(M_m^b - M^b), \tag{9}$$

where M_m is the maximal weight of the organism; t_m is the time required to achieve maximal differentiation; r_f, r_g, r_d are dimension constants; b is a constant from the relation

$$q_{O_2} = a\,M^{-b}.$$

We applied the third equation of system (8) to describe surface area changes in rats and humans. Equations (3) and (4) were used as the expression for the force which characterised organism surface changes in rats

$$1/S \cdot dS/dt = L_{fg}\left(\frac{M_m^b}{M^b} - 1\right) + L_{fd}\,(t_m - t) + L_{ff}\,((S_m - S)/S) \tag{10}$$

and in humans

$$1/S \cdot dS/dt = L_{fg}\,(M_m^b - M^b) + L_{fd}\,(t_m - t) + L_{ff}\,(S_m - S). \tag{11}$$

These equations provide a more accurate description of the surface growth curve in rats and humans (Figs. 2, 3) including the stationary phase and decline in old age.

A phenomenological shape-formation equation

As the organism and its size grow, so does the body surface not only as a result of volume changes but also as a result of shape changes. These two phenomena can be conceptually separated as follows. Let us imagine that the organism is a sphere and its growth results in a mass increase in proportion to the volume of the sphere (Fig. 4). Let us also introduce body shape changes during growth as a difference between the actual surface area S of an organism whose mass is M and the surface area S_{sp} of a spherical organism having the same mass M. In the former case (Fig. 4), where the organism maintains the spherical form, no shaping occurs and $F = S - S_{sp} = 0$. If, however, shaping does occur (Figs. 4, II) then $F = S - S_{sp} \neq 0$ and this difference varies as the shape. Consequently, $F = S - S_{sp}$ may be viewed as the phenomenological

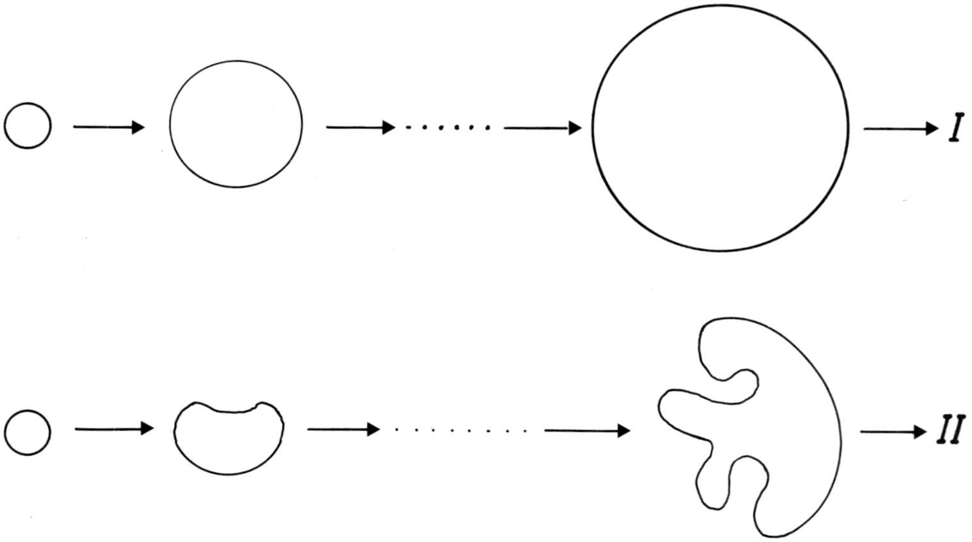

Fig. 4 Schematic respresentation of development without and with changes of shape.

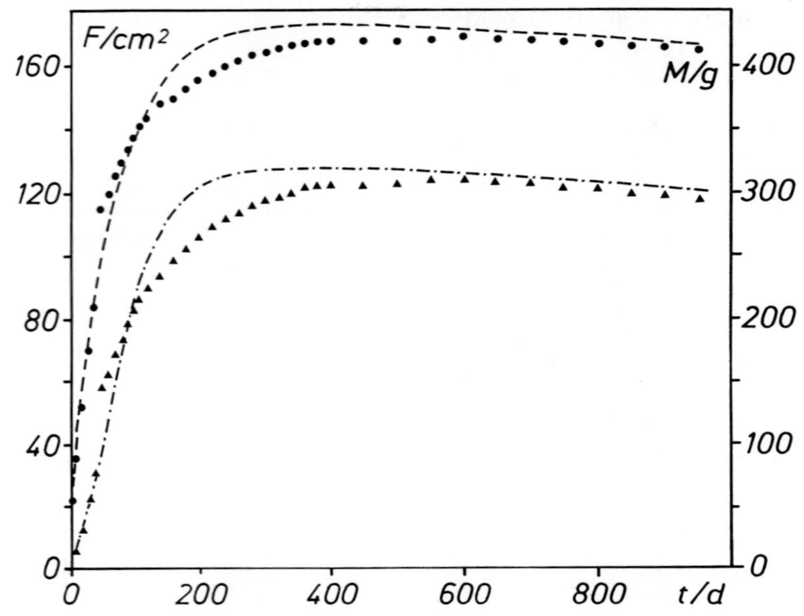

Fig. 5 Comparison of mass and shape changes in rats and the calculated curves, the factors being in equation (13):

k_{gg} = 0.097	k_{gd} = 0.2×10^{-5}	k_{gf} = - 0.01
k_{fg} = - 0.01	k_{fd} = 0.1×10^{-5}	k_{ff} = 0.018
T_m = 650 d	F_m = 170.4 cm^2	F_o = 21.7 cm^2

measure of shaping. If an organism of mass M were spherically shaped, its surface area would be $S_{sp} = 3.84\ M^{2/3}$ and

$$F = S - 3.84\ M^{2/3} \quad (12)$$

where F is the shape-formation measure.

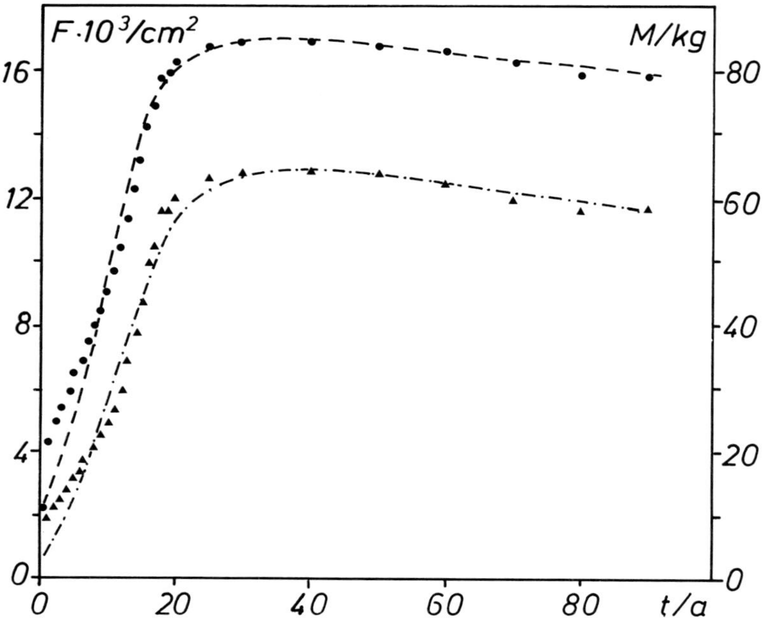

Fig. 6 Comparison of mass and shape changes in humans and the calculated curves, the factors being in equation (14):

k_{gg} = 0.227	k_{gd} = 0.00035	k_{gf} = - 0.1 10^{-5}
k_{fg} = - 0.1 x 10^{-5}	k_{fd} = 0.00027	k_{ff} = 0.123 x 10^{-4}
F_m = 16 794 cm^2	T_m = 40 years	F_o = 2115 cm^2

This relation (12) may be helpful in describing the shape formation curve for postnatal development in rats and humans in the knowledge of the data cited above (Fig. 2, 3)

The shape formation curve may be described by an equation which recognises the interaction of growth and differentiation; for rats

$$1/F \cdot dF/dt = k_{fg}\left(\frac{M_m^{b}}{M^{b}} - 1\right) + k_{fd}(t_m - t) + k_{ff}((F_m - F)/F) \qquad (13)$$

and for humans

$$1/F \cdot dF/dt = k_{fg}(M_m^{b} - M^{b}) + k_{fd}(t_m - t) + k_{ff}(F_m - F). \qquad (14)$$

The computer simultaneously yields theoretical weight variation and shape formation curves. Equations (13) and (14) are found (Fig. 5, 6) to provide a good insight into shape formation.

These models are most general representations of growth and shaping, basic ontogenetic phenomena, and as such seem to be useful.

List of Contributors

Akhromeyeva, T. S.
Institute of Applied Mathematics, Academy of Sciences of USSR, Miusskaja Street 4, Moscow 125 047, USSR

Beloussov, L. V.
Department of Embryology, Faculty of Biology, Moscow State University, Moscow 119 899, USSR

Betz, A.
Botanisches Institut, Universität Bonn, Kirschallee 1, D-5300 Bonn 1, FRG

Busse, H.-G.
Biochemisches Institut, Medizinische Fakultät, Christian-Albrecht-Universität Kiel, Olshausenstraße, D-2300 Kiel 1, FRG

Davydov, A. S.
Institute for Theoretical Physics, Academy of Sciences, Ukrainian SSR, Metrologischeskaja Street 14, Kiev 252 130, USSR

Ebeling, W.
Sektion Physik, Humboldt-Universität zu Berlin, Invalidenstraße 42, DDR-1040 Berlin, GDR

Epstein, I. R.
Department of Chemistry, Brandeis University, Edison-Lecks 211, Waltham, MA 02254, USA

Engel, A.
Sektion Physik, Humboldt-Universität zu Berlin, Invalidenstraße 42, DDR-1040 Berlin, GDR

Eschrich, K.,
Institut für Biochemie, Karl-Marx-Universität, Liebigstraße 16,
DDR-7010 Leipzig, GDR

Haken, H.
Institut für Theoretische Physik, Universität Stuttgart, Pfaffenwaldring 57/IV,
D-7000 Stuttgart 80, FRG

Hofmann, E.
Institut für Biochemie, Karl-Marx-Universität, Liebigstraße 16,
DDR-7010 Leipzig, GDR

Isaeva, V. V.
Institute of Developmental Biology, Academy of Sciences of USSR, Vavilov
Street 26, Moscow 117 334, USSR

Jean, R. V.
University of Québec, 300 avenue des Ursulines Rimouski, Québec, G5L 3Al,
Canada

Keller, J. U.
Institut für Fluid- und Thermodynamik, Universität Siegen,
Paul-Bonatz-Straße 9-11, D-5900 Siegen, FRG

Kerner, B. S.
Institute of Biological Physics, Academy of Sciences of USSR , Pushchino,
Moscow Region, 142 292 USSR

Kretschmer, M.
Institut für Biochemie, Karl-Marx-University, Liebigstraße 16,
DDR-7010 Leipzig, GDR

Kreuzberg, K.
Botanisches Institut, Universität Bonn, Kirschallee 1, D-5300 Bonn 1, FRG

Krinskii, V. I.
Institute of Biological Physics, Academy of Sciences of USSR, Pushchino,
Moscow Region, 142 292 USSR

Kurdjumov, S. P.
Institute of Applied Mathematics, Academy of Sciences of USSR, Miusskaja
Street 4, Moscow 125 047, USSR

Lakirev, A. V.
Department of Embryology, Faculty of Biology, Moscow State University, Moscow 119 899, USSR

Lamprecht, I.
Institut für Biophysik, Freie Universität Berlin, Thielallee 63, D-1000 Berlin 33, FRG

Malinetski, G. G.
Institute of Applied Mathematics, Academy of Sciences of USSR, Miusskaja Street 4, Moscow 125 047, USSR

Malyghin, S. N.
Institute of Developmental Biology, Academy of Sciences of USSR, Vavilov Street 26, Moscow 117 334, USSR

Mazenko, V. G.
Faculty of Mathematics, Chernovtsy State University, Chernovtsy 374 001, USSR

Mikhailov, A. S.
Department of Physics, Moscow State University, Moscow 117 234, USSR

Mineev, L. D.
L. D. Landau Institute for Theoretical Physics, Academy of Sciences of USSR, Moscow 117 334, USSR

Müller, S. C.
Max-Planck-Institut für Ernährungsphysiologie, Rheinlanddamm 201, D-4600 Dortmund 1, FRG

Nonnenmacher, Th. F.
Abteilung für Mathematische Physik, Universität Ulm, Oberer Eselsberg, D-7900 Ulm, FRG

Osipov, V. V.
Institute of Biological Physics, Academy of Sciences of USSR, Pushchino, Moscow Region, 142 292 USSR

Plesser, Th.
Max-Planck-Institut für Ernährungsphysiologie, Rheinlanddamm 201, D-4600 Dortmund 1, FRG

Potapov, A. B.
Institute of Applied Mathematics, Academy of Sciences of USSR, Miusskaja Street 4, Moscow, USSR

Presnov, E. V.
Institute of Developmental Biology, Academy of Sciences of USSR , Moscow 117 334, USSR

Prokofief, E. A.
Institute of Developmental Biology, Academy of Sciences of USSR, Vavilov Street 26, Moscow 117 334, USSR

Rivier, N.
Blackett Laboratory, Imperial College, Prince Consort Road, London, SW 7 2BZ, GB

Romanovskii, Yu. M.
Faculty of Physics, Lomonosov State University, Moscow, USSR

Samarskii, A. A.
Institute of Applied Mathematics, Academy of Sciences of USSR, Miusskaja Street 4, Moscow 125 047, USSR

Schellenberger, W.
Institut für Biochemie, Karl-Marx-Universität, Liebigstraße 16, DDR-7010 Leipzig, GDR

Teplov, V. A.
Institute of Biological Physics, Academy of Sciences of USSR, Pushchino, Moscow Region, 142 292, USSR

Timchenko, G. V.
Institute of Developmental Biology, Adademy of Sciences of USSR, Vavilov Street 26, Moscow 117 334, USSR

Volovik, G. E.
D. D. Landau Institute for Theoretical Physics, Academy of Sciences of USSR, Moscow 117 334, USSR

Zeldovich, Ya. B.
Institute for Physical Problems, Kosygin Street 2, Moscow 117 234, USSR

Zotin, A. I.
Institute of Developmental Biology, Academy of Sciences of USSR,
Vavilov Street 26, Moscow 117 334, USSR

References

Aboav, D. 1970. The arrangement of grains in a polycrystal. Metallogr. 3, 383-390

Adler, J. 1966. Chemotaxis in bacteria. Science 173, 708-716

Agladze, K. I., Krinskii, V. I. 1982. Multi-armed vortices in an active chemical medium. Nature 296, 424-426

Agladze, K. I., Tzyganov, M. A., Krinskii, V. I., Budrene, E. O., Skulachev, V. P., Ivanitsky, G. R. 1987. On autowave mechanism of pattern formation in a growing bacterial population. Nature (submitted)

Akhromeyeva, T. S., Kurdjumov, S. P., Malinetskii, G. G. 1985. Paradoxes in the world of nonstationary structures. Moscow, Znanie

Akhromeyeva, T. S., Kurdjumov, S. P., Malinetskii, G. G., Samarskii, A. A. 1984. On the classification of two-component systems in the neighbourhood of a bifurcation point. Dokl. Akad. Nauk SSSR 279, 591-595 (Russian)

Akhromeyeva, T. S., Kurdjumov, S. P., Malinetskii, G. G., Samarskii, A. A. 1986. Two-component systems in the neighbourhood of bifurcation points. In: Mathematical modelling. Processes in Nonlinear Media. (A. A. Samarskii, S. P. Kurdjumov, V. A. Galaktionov, eds.). Nauka, Moscow, 7-59 (Russian)

Akhromeyeva, T. S., Malinetskii, G. G., 1985. On strange attractors in one synergetic problem. Preprint N89, Keldysh Inst. Appl. Math. Acad. Sci, USSR, Moscow (Russian)

Akkerman, J. W. W. 1984. More on metabolic control. TIBS 10, 187-188

Alad'ev, V. Z. 1981. Cybernetic modelling of developmental biology. In: Parallel Information Processing and Parallel Algorithms. Tallin, Valgus, 211-280 (Russian)

Andronov, A. A., Vitt, A. A., Khaikin, S. E. 1981. Theory of Oscillations. Nauka, Moscow, 568 p. (Russian)

Arkhangelskaya, T. A., Barch, G., Kolin'ko, V. G., Romanovskii, Yu. M. 1984. Models of autowave motility in the myxomycete plasmodium. In: Dynamics of Biological Population. Univ. Gorky Press, Gorky, 98-108 (Russian)

Arnold, R., Showalter, K., Tyson, J. J. 1987. Luther's 1906 discovery and analysis of chemical waves. J. Chem. Ed. (submitted)

Arnold, V. I. 1980. Chapitres Supplémentaires de la Théorie des Equations Différentielles Ordinaires. Mir, Moscow. 323 p.

Avigad, G. 1981. Stimulation of yeast phosphofructokinase activity by fructose 2,6-bisphosphate. Biochem. Biophys. Res. Commun. 102, 985-991

Avnir, D., Kagan, M. 1983. Chemical reactions of gas-liquid interfaces. A novel source of spatial dissipative structures. Naturwiss. 70, 361-363

Avnir, D., Kagan, M. 1984. Spatial structures generated by chemical reactions at interfaces. Nature 307, 717-720

Avnir, D., Kagan, M., Levi, A. 1983. Structure formation in chemical systems driven far from equilibrium by light. Naturwiss. 70, 144-146

Bagshaw, C. R. 1982. Muscle Contraction. Chapman and Hall, Londen, New York. 79 p.

Balkarei, Yu. I., Evtikhov, M. G., Elinson, M. I. 1980. Localized stationary states in two-dimensional diffusive autowave medium. Mikroelektronika 9, 141-149 (Russian)

Balkarei, Yu. I., Evtikhov, M. G., Elinson, M. I. 1983. The formation of contrast dissipative structures with noises. Mikroelektronika 12, 65-69 (Russian)

Baranowski, Z. 1978. Contraction-relaxation waves in *Physarum polycephalum* plasmodia. Acta protozool. 17, 387-397

Baranowski, Z., Wohlfarth-Bottermann, K. E. 1982. Endoplasmic veins from plasmodia of *Physarum polycephalum*: a new strand model with defined age, structure, and behaviour. Eur. J. Cell Biol. 27, 1-9

Barany, K., Ledvora, R. F., Maugios, V., Barany, M. 1985. Stretch-induced myosin light chain phosphorylation and stretch-release-induced tension development in arterial smooth muscle. J. Biochem. 260, 7126-7130

Barenblatt, G. I., Zeldovich, Ya. B. 1957. On stability of flame propagation. Prikl. Math. Mekh. 2, 856-964

Barenblatt, G. I., Zeldovich, Ya. B. 1959. Theory of flame propagation. Combust. and Flame 3, 61-75

Bartrons, R., van Schaftingen, E., Vissers, S., Hers, H.-G. 1982. The stimulation of yeast phosphofructokinase by fructose 2,6-bisphosphate. FEBS letters 143, 137-140

Bataille, Edelen, D. G. B., Kestin, J. 1982. On the structuring of thermodynamic fluxes: A direct implementation of the dissipation inequality. In: Thermodynamics and Kinetics of Biological Processes. (I. Lamprecht, A. I. Zotin, eds.). Walter de Gruyter, Berlin, New York, 239-253

Baumann, G., Nonnenmacher, T. F. 1988. The fractal dimension of human granulocyte membranes: experiments and interpretations. Proceed. Locarno Confer. on Fractal Structures, April 22-23, 1988 (in press)

Bazsa, G., Epstein, I. R. 1985. Travelling waves in the citric acid-iron (II) reaction. J. Phys. Chem. 89, 3050-3053

Becker, P. K., Field, R. J. 1985. Stationary concentration patterns in the Oregonator model of the Belousov-Zhabotinskii reaction. J. Phys. Chem. 89, 118-128

Belintsev, B. N. 1983. Dissipative structures and the problem of biological pattern formation. Uspekhi Fiz. Nauk 111, N1, 54-101 (Russian)

Belintsev, B. N. 1983. Dissipative structures and the problem of the biological pattern formation. Sov. Phys. Usp. 26, 775-816

Belintzev, B. N., Beloussov, L. V., Zaraiskii, A. G. 1985 a. The model of epithelial morphogenesis basing on the elastic forces and contact polarization of cells. Ontogenez (Sov. J. Devel. Biol.) 16, 5-14 (Russian)

Belintzev, B. N., Beloussov, L. V., Zaraiskii, A. G. 1985 b. The model of epithelial morphogenesis basing on the elastic forces and contact polarization of cells. Biological consequences. Ontogenez (Sov. J. Devel. Biol.) 16, 437-449 (Russian)

Belousov, B. P. 1959. A Periodic reaction and its mechanism. Sb. Ref. Radiats. Med., Medgiz, Moscow, 145

Belousov, B. P. 1985. A periodic reaction and its mechanism. In: Oscillations and Traveling Waves in Chemical Systems. (R. J. Field, M. Burger, eds.). New York, Wiley, 605-613

Beloussov, L. V. 1973. Growth and morphogenesis of some marine Hydrozoa according to histological data and time-lapse studies. Publ. Seto Marine Biol. Lab. 20, 315-386

Beloussov, L. V. 1979. Experiments with changes of tension fields of axial rudiments in amphibian embryos. Ontogenez (Sov. J. Devel. Biol.) 10, 120-129 (Russian)

Beloussov, L. V. 1980. The role of the tensile fields and contact cell polarization in the morphogenesis of amphibian axial rudiments. Roux'Arch. Devel. Biol. 188, 1-7

Beloussov, L. V. 1984. Synergetics and biological morphogenesis. In: Self-Organization. Autowaves and Structures Far from Equilibrium. (V. J. Krinskii, ed.). Springer Series in Synergetics, Vol. 28. Springer, Berlin, Heidelberg. 204-208

Beloussov, L. V., Chernavskii, D. S., Solyanik, G. I. 1985. Applications of synergetics to ontogenesis (on parametrical control of development). Ontogenez (Sov. J. Devel. Biol. 16, 213-228 (Russian)

Beloussov, L. V., Dorfman, J. G. 1974. On the mechanics of growth and development in hydroid polypes. Amer. Zool. 14, 719-734

Beloussov, L. V., Dorfman, J. G., Cherdantzev, V. G. 1975. Mechanical stresses and morphological patterns in amphibian embryos. J. Embryol. Exp. Morph. 34, 559-574

Beloussov, L. V., Labas, Ju. A., Badenko, L. A. 1984. Growth pulsations and rudiment shapes in hydroid polypes. J. Obsch. Biol. 45, 796-806 (Russian)

Berlekamp. E. R., Conway, J. H., Guy, R. K. 1982. Winning Ways for Your Mathematical Plays. Academic Press, New York

Berridge, M. J., Rapp, P. E. 1979. A comparative survey of the function, mechanism and control of cellular oscillations. J. Exp. Biol. 81, 217-279

Berry, M. V., Lewis, Z. V. 1980. On the Weierstrass-Mandelbrot fractal function. Proc. R. Soc. London A370, 459-484

Betz, A. 1966. Metabolite flux in yeast cells with oscillatory controlled glycolysis. Physiol. Plantarum 19, 1049-1054

Betz, A. 1967. Oscillatory control of glycolysis as a model for biological timing processes. In: Quantitative Biology of Metabolism. (L. Locker, ed.). Springer, Berlin, 205-216

Betz, A., Becker, J. U., Kreuzberg, K., Jonnalagadda, S. 1981. Control interactions and flux rates in yeast glycolysis, revealed by metabolite oscillations. In: Current Developments in Yeast Research. (G. G. Stewart, J. Russell, eds.). Pergamon Press, Toronto, 363-368

Betz, A., Chance, B. 1965. Influence of inhibitors and temperature on the oscillation of reduced pyridine nucleotides in yeast cells. Arch. Biochem. Biophys. 109, 575-584

Betz, A., Giersch, C. 1978. The oscillatory network of glycolysis: a model for multiple phase output derived from a single master oscillator. Chronobiologia 5, 45-55

Betz, A., Hinrichs, R. 1968. Incorporation of glucose into an insoluble polysaccharide during oscillatory controlled glycolysis in yeast cells. Europ. J. Biochem. 5, 154-157

Betz, A., Röttger, U., Kreuzberg, K. 1975. Allosteric activation and competitive inhibition of yeast phosphofructokinase by D-fructose. Arch. Mikrobiol. 103, 123-126

Betz, A., Sel'kov, E. 1969. Control of phosphofructokinase (PFK) activity in conditions simulating those of glycolysing yeast extract. FEBS letters 3, 5-9

Bewerdorf, A. 1986. Transport durch chemische Wellen. Naturwiss. 73, 363-365

Beylina, S. I., Matveeva, N. B., Priezzhev, A. V., Romanovskii, Yu. M., Sukhorukov, A. P., Teplov, V. A. 1984. Plasmodium of the myxomycete *Physarum polycephalum* as an autowave self-organizing system. In: Self-Organization. Autowaves and Structures far from Equilibrium. Springer Series in Synergetics, Vol. 28 (V. I. Krinsky, ed.).Springer, Berlin, Heidelberg, New York, Tokyo, 218-221

Blanc, M., Mocellin, A. 1979. Grain coordination in plane sections of polycrystals. Acta Metall 27, 1231-1237

Bloxham, D. P., Lardy, H. A. 1973. Phosphofructokinase. In: The Enzymes. (P. D. Boyer, ed.). Vol. 8, 3rd ed. Academic Press, New York, 239-287

Boiteux, A., Hess, B. 1980. Spatial dissipative structures in yeast extracts. Ber. Bunsenges. Phys. Chem. 84, 392-398

Boiteux, A., Hess, B. 1981. Design of glycolysis. Phil. Trans. R. Soc. Lond. B293, 5-22

Boots, B. N. 1982. The arrangement of cells in 'random' networks. Metallogr. 15, 53-62

Bosca, L., Corredor, C. 1984. Is phosphofructokinase the rate limiting step of glycolysis? TIBS 9, 372-373

Bourbaki, N. 1948. L'Architecture des mathématiques. In: Les Grands Courants de la Pensée Mathématiques, presenté par F. Le Lionnai. Cahiers du Sud, 35-47

Bowers, P. G., Noyes, R. M. 1985. Gas evolution oscillators. In: Oscillations and Traveling Waves in Chemical Systems. (R. J. Field, M. Burger, eds.). Wiley, New York, 473-492

Bracewell, R. 1965. The Fourier Transform and its Application. McGraw Hill, New York. 381 p.

Bray, W. C. 1921. A periodic reaction in homogeneous solution and its relation to catalysis. J. Am. Chem. Soc. 43, 1262

Brody, S. 1945. Bioenergetics and Growth. Reinhold, New York. 1023 p.

Budrene, E. O. 1985. Space-ordered structure formation in motile bacteria colonies on agar. Dokl. Akad. Nauk SSSR 283, 470-473 (Russian)

Burns, J. A., Cornish-Bowden, A., Groen, A. K., Heinrich, R., Kacser, H., Porteous, W., Rapoport, S. M., Rapoport, T. A., Stucki, J. W., Tager, J. M., Wanders, R. J. A., Westerhoff, H. V. 1985. Control analysis in metabolic systems. Trends Biochem. Sci. 10, 16

Burton, K. 1974. The enthalpy change for the reduction of nicotinamide-adenine dinucleotide. Biochem. J. 143, 365-368

Busenberg, S., Ianelli, M. 1985. Separable models in age-dependent population dynamics. J. Math. Biol. 22, 145-173

Busse, H. G. 1968. Personal communication of an experiment

Busse, H. G. 1971. Temperature variations in oscillating chemical reactions. Nature 233, 137-138

Busse, H. G. 1976. Personal communication of a movie about nonlinear phenomena in thermodynamics

Buvet, R., Ponnamperuma, C. (eds.). 1971. Molecular Evolution. Vol. 1, 2. North-Holland, Amsterdam

Calvet, E., Prat, H. 1963. Recent Progress in Microcalorimetry. (H. A. Skinner, ed., transl.). Pergamon Press, Oxford. 177 p.

Carley, A. F., Joyner, R. W. 1979. The application of deconvolution methods in electron spectroscopy - A review. J. Electron. Spectrosc. Relat. Phenom. 16, 1-23

Cesari, E., Tora, V., Maqaueron, J. L., Probst, R., Dubes, J. P., Tachoire, H. 1982. Thermogenese: Application comperative de l'analyse harmonique et du filtrage inverse. Thermochim. Acta 53, 17-27

Cesari, E., Tora, V., Maqueron, J. L., Probst, R., Dubes, J. P., Tachoire, H. 1982. Therrmogenese: Application comperative de l'analyse harmonique et du filtrage inverse. Thermochim. Acta 53, 17-27

Cesari, E., Vinals, J., Torra, V., Ortin, J., Maqueron, J. L., Dubes, J. P., Tachoire, H. 1984. Microccalorimetrie et thermogenese: Identification des dispositifs experimentaux permettant des mesurer directement les enthalpies d'exces. Thermochim. Acta 79, 23-24

Chance, B., Hess, B., Betz, A. 1964. DPNH oscillations in a cellfree extract of *S. carlsbergensis*. Biochem. Biophys. Res. Commun. 16, 182-187

Chandrasekhar, S. 1961. Hydrodynamic and Hydromagnetic Stability. Oxford, Clarendon Press

Chernavskii, D. S., Ruijgrok, Th. W. 1978. Dissipative structures in morphogenetic models of the Turing type. J. Theor. Biol. 73, 585-607

Chernavskii, D. S., Ruijgrok, Th. W. 1982. On the formation of unique dissipative structures. Biosystem 15, 75-81

Child, D. M. 1941. Patterns and Problems of Development. Univ. of Chicago Press, Chicago. 811 p.

Church, A. H. 1920. On the Interpretation of Phenomena of Phyllotaxis. Hafner. Pub., New York (1968 facsimile of the 1920 edition)

Conrad, M. 1983. Adaptability. Plenum Press, New York. 383 p.

Cooley, J. W., Tukey, J. W. 1965. An algorithm for the machine calculation of complex Fourier series. Math. Computation 19, 297-301

Crain, I. K. 1978. The Monte-Carlo generation of random polygons. Comput. and Geosci. 4, 131-141

Davydov, A. S. 1974. Molecular mechanism of contraction of striated muscles. studia biophysica 47, 221-228

Davydov, A. S. 1979 a. Solitons, bioenergetics, and the mechanism of muscle contraction. Int. J. Quant. 16(1), 5-17

Davydov, A. S. 1979 b. Influence of electron phonon interaction on the motion of an electron in a 1-dimensional molecular system. Teor. Math. 40(3), 831-840

Davydov, A. S. 1982 a. Sov. Phys. Usp. 25, no. 12, 908-918

Davydov, A. S. 1982 b. Biology and Quantum Mechanics. Pergamon Press, 229 p.

Davydov, A. S. 1984. Solitons in Molecular Systems. Reidel Publishing Comp., 288 p.

Davydov, A. S., Kislukha, N. I. 1973. Solitary excitons in one-dimensional molecular chains. Phys. Stat. Solidi (b) 59, 465-470

de Boor, C. 1978. A Practical Guide to Splines. Applied Sciences, Vol.27, Springer, Berlin, Heidelberg. 392 p.

de Gennes, P. G. 1974. The Physics of Liquid Crystals. Clarendon Press, Oxford

De Kepper, P., Epstein, I. R., Orban, M. 1982. Batch oscillations and spatial wave patterns in chlorite oscillating systems. J. Phys. Chem. 86, 170-171

Degn, H. 1972. Oscillating chemical reactions in homogeneous phase. J. Chem. Educ. 49, 302-307

Degn, H. 1972. Oscillating chemical reactions in homogeneous phase. J. Chem. Educ. 49, 302-307

Deshcherevskii, V. I. 1977. Mathematical Models of Muscle Contraction. Nauka, Moscow. 160 p. (Russian)

Dobjansky, T. 1970. Genetics of Evolutionary Process. Columbia Univ. Press, New York. 505 p.

Dormer, K. J. 1980. Fundamental Tissue Geometry for Biologists. Cambridge Univ. Press. 149 p.

Doubrovine, B. A., Novikov, S. P., Fomenko, A. T. 1985. Geométrie contemporaire. Mir, Moscow. Vol. 1: 438 p., Vol. 2: 371 p.

Driesch, H. 1892. Entwicklungsmechanische Studien. Zeitschr. wiss. Zool. 53, 160-184

Dubitskii, A. L., Kerner, B. S., Osipov, V. V. 1986 a. The kinetics of autosoliton formation in the stable heated plasma of intrinsic semiconductors. Fiz. Tverd. Tela (Leningrad) 28, 1290-1295

Dubitskii, A. L., Kerner, B. S., Osipov, V. V. 1986 b. Spontaneous formation of collisional-ionization localized regions in homogeneous semiconductors in weak electric fields. Fiz. Tekh. Poluprovodn. (Leningrad) 20, 1195-1198 (Russian)

Duffy, D. M., Rivier, N. 1982. The glass as an elastic continuum. J. de Phys. 43, C9-475-479

Durston, A. J. 1974. Pacemaker mutants of *Dictyostelium discoideum*. Develop. Biol. 38, 308-319

Ebeling, W. 1976. Strukturbildung bei Irreversiblen Prozessen. Teubner, Leipzig. 277 p.

Ebeling, W., Engel, A., Feistel, R. Mazenko, V. G. 1986. Models of evolution processes including age-structure. Proc. Conf. Dynamical Systems and Environmental Models. March, 16-21, Wartburg, Eisenach

Ebeling, W., Engel, A., Mazenko, V. G. 1986. Modelling of selection processes with age-dependent birth and death rates. BioSystems 19, 213-221

Ebeling, W., Feistel, R. 1982. Physik der Selbstorganisation und Evolution. Akademie-Verlag, Berlin. 451 p.

Eigen, M. 1971. Selforganization of matter and the evolution of biological macromolecules. Naturwiss. 58, 216-528

Eigen, M., Schuster, P. 1977. The Hypercycle. Naturwiss. 64, 541-565

Elenin, G. G., Kurdjumov, S. P., Samarskii, A. A. 1983. Nonstationary thermal structures in nonlinear heat-conducting medium. Zh. Vychisl. Mat. i Mat. Fiz. 23, 380-390 (Russian)

Elliot, D. F., Rao, K. R. 1982. Fast Transforms, Algorithms, Analysis, Applications. Academic Press, New York. 488 p.

Epstein, I. 1984. Complex dynamical behavior in 'simple' chemical systems. J. Phys. Chem. 88, 187-198

Epstein, I. R., Kustin, K., De Kepper, P., Orban, M. 1983. Oscillating chemical reactions. Scient. Amer 248(3), 112-123

Ermakov, V. B., Priezzhev, A. V. 1984. Wave regimes of contractile activity of plasmodium myxomycete Physarum strands and their relationship with protoplasm transfer. Biophysics 24, 100-104 (Russian)

Eschrich, K., Schellenberger, W., Hofmann, E. 1983. Sustained oscillations in a reconstituted enzyme system containing phosphofructokinase and fructose 1,6-bisphosphatase. Arch. Biochem. Biophys. 222, 657-660

Feeney, R., Schmidt, S., Ortoleva, P. 1981. Experiments on electric field - chemical wave interactions: Annihilation and the crescent wave. Physica D 2, 536-544

Feinn, D., Ortoleva, P. Scalf, W., Schmidt, S., Wolff, M. 1978. Spontaneous pattern formation in precipitating systems. J. Chem. Phys. 69, 27-39

Field, R. J., Burger, M. (eds.). 1985. Oscillations and Traveling Waves in Chemical Systems. Wiley, New York. 681 p.

Field, R. J., Körös, E., Noyes, R. M. 1972. Oscillations in chemical systems. II: Thorough analysis of temporal oscillations in the Ce-BrO_3-malonic acid system. J. Am. Chem. Soc. 94, 8649-8644

Field, R. J., Noyes, R. M. 1974 a. Oscillations in chemical systems. V: Quantitative explanation of band migration in the Belousov-Zhabotinskii reaction. J. Am. Chem. Soc. 96, 2001-2006

Field, R. J., Noyes, R. M. 1974 b. Oscillations in chemical systems. IV: Limit cycle behavior in a model of a real chemical reaction. J. Chem. Phys. 60, 1877-1884

Fleischer, M., Wohlfarth-Bottermann, K. E. 1975. Correlation between tension force generation, fibrillogenesis, and ultrastructure of cytoplasmic actomyosin during isometric and isotonic contractions of protoplasmic strands. Cytobiology 10, 339-365

Flicker, M., Ross, J. 1974. Mechanism of chemical instability for periodic precipitation phenomena. J. Chem. Phys. 60, 3458-3465

Fliege, R., Flügge, U.-I., Werdan, K., Heldt, H. W. 1978. Specific transport of inorganic phosphate, 3-phosphoglycerate and triosephosphates across the inner membrane of the envelope in spinach chloroplast. Biochim. Biophys. Acta 502, 232-247

Fomenko, A. T. 1984. Topological Variational Problems. Moscow State Univ. Press, Moscow. 216 p.

Fortes, M. A., Ferro, A. C. 1985. Topology and transformations in cellular structures. Acta Metall. 33, 1697-1708

Francois, J., van Schaftingen, E., Hers, H.-G. 1983. On the mechanism of inhibition of neutral fructose 1,6-bisphosphatase by fructose 2,6-bisphosphate. Eur. J. Biochem. 134, 269-273

Frank, F. C. 1953. On spontaneous asymmetric synthesis. Biochim. Biophys. Acta 11, 459-465

Frank, U., Geiseler, W. 1971. Zur periodischen Reaktion von Malonsäure mit Kaliumbromat in Gegenwart von Cer-Ionen. Naturwiss. 58, 52-53

Fraser, S., Kapral, R. 1986. Ring dynamics and percolation in an excitable medium. J. Chem. Phys. 85, 5682-5688

Fricke, R., Suwelack, O. 1926. Untersuchungen an einem Liesegangschen 'rhythmischen' Fällungssystem. Z. Phys. Chem. 124, 359-393

Fuks, D. B., Fomenko, A. T., Gutenmaher, W. L. 1969. Homotopy Topology. Moscow State Univ. Press, Moscow. 460 p.

Furuya, E., Kotaniguchi, H., Hagikara, B. 1982. The effect of oxygen on fructose-2,6-bisphosphate concentration and fructose-6-phosphate, 2-kinase activity in yeast cells. Biochem. Biophys. Res. Commun. 105, 1519-1523

Furuya, E., Yokoyoma, M., Uyeda, K. 1982. Regulation of fructose 6-phosphate-2-kinase by phosphorylation and dephosphorylation; possible mechanism for coordinated control of glycolysis and gluconeogenesis. Proc. Acad. Sci. USA 79, 325-329

Gafüchuk, V. V., Kerner, B. S., Lazurchak, I. I., Osipov, V. V. 1986. Pulsating autosolitons in active distributed media. Mikroelektronika 15, 180-183

Galaktionov, V. A., Kurdjumov, S. P., Samarskii, A. A. 1984. On approximate self-similar solutions of a class of quasilinear heat equations with a source. Mat. Sbornik 124(166), 163-188 (Russian)

Galitskii, V. V. 1985. Horizontal structures and the dynamics of a plant community of the same age: numerical simulation. In: Mathematical Modelling of Biogeocenotical Processes. Moscow, Nauka, 59-69 (Russian)

Gardner, M. 1971. On cellular automata, self-reproduction, the Garden of Eden and the game 'life'. Scient. Amer. 224, N2, 112-117

Gasaryan, K. G., Belousov, L. V. 1983. Developmental Biology of Animals. V. Shkola, Moscow. 287 p. (Russian)

Gerisch, G. 1968. Cell aggregation and differentiation in *Dictyostelium discoideum*. Curr. Top. Dev. Biol. 3,157

Gerisch, G. 1971. Periodische Signale steuern die Musterbildung in Zellverbänden. Naturwiss. 58, 430-438

Gershuni, G. Z., Juhovitskii, E. M. 1978. Convective stabilitty of incompressible liquids. Nauka, Moscow (Russian)

Gierer, A. 1981. Physik der biologischen Gestaltbildung. Naturwiss. 68, 245-251

Gierer, A., Meinhardt, M. 1972. A theory of biological pattern formation. Kybernetik 12, 30-39

Gierer, A., Meinhardt, M. 1974. Application of a theory of biological pattern formation based on lateral inhibition. J. Cell. Sci. 15, 321-349

Giersch, C., Betz, A., Richter, O. 1975. A new method for determining the individual reaction rates of the glycolytic system. Biosystem 7, 147-153

Glansdorff, P., Prigogine, I. 1971. Thermodynamic Theory of Structure, Stability and Fluctuations. Wiley Interscience, New York, 306 p.

Goldberger, A. L., West, B. J., Dresselhaus, T., Bhargava, V. 1985. Bronchial asymmetry and Fibonacci scaling. Experientia 41, 1537-1538

Goldbeter, R., Caplan, S. R. 1976. Oscillatory enzymes. Annu. Rev. Biophys. Bioeng. 5, 449-476

Goldblatt, R. 1979. Topoi, the Categorial Analysis of Logic. North-Holland, Amsterdam. 486 p.

Gosper, R. 1984. Exploiting regularity in large cellular spaces. Physica 10D, 75-80

Gowland, R. J., Stedman, G. 1983. A novel moving boundary reaction involving hydroxylamine and nitric acid. J. Chem. Soc. Chem. Commun. 1038-1039

Gradsteyn, J. S., Ryzhik, J. M. 1980. Tables of Integrals, Series and Products. Academic Press, New York. 1160 p.

Grebecki, A., Cieslawsky, M. 1978. Plasmodium of *Physarum polycephalum* as a synchronous contractile system. Cytobiologie 17, 335-342

Green, D. E. 1973. Mechanism of energy transduction in biological systems. Science 181, 583-587

Green, P. B. 1985. Surface of the shoot apes: A reinforcement-field theory for phyllotaxis. J. Cell. Sci. (Suppl. 2), 181-201

Greenberg, J. M. 1980. Spiral waves for lambda-omega systems. SIAM J. Appl. Math. 39, 301-310

Greenberg, J. M., Hastings, S. P. 1978. Spatial patterns for discrete models of diffusion in excitable media. SIAM J. Appl. Math. 34, 515-526

Gribschaw, T. A., Showalter, K., Banville, D. L., Epstein, I. R. 1981. chemical waves in the acidic iodate oxidation of Arsenite. J. Phys. Chem. 85, 2152-2155

Grillner, S. 1982. Possible analogies in the control of innate motor acts and the production of sound in speech. In: Speech Motor Control. (S. Grillner, B. Lindblom, J. Lubker, A. Persson, eds.). Pergamon, Oxford, 217-229

Gul'ko, F. B., Petrov, A. A. 1972. Mechanisms of the formation of closed pathways of conduction in excitable media. Biophysics 17, 271-282

Hadeler, K. P. 1985. Models for parasitic diseases. Math. Research (Berlin) 23, 162-169

Hafke, C., Gilles, E. D. 1968. Experimentelle Unersuchungen des dynamischen Verhaltens von Rührkesselraktoren. Mess. Steuern Regeln 11, 204-208

Haken, H. 1978. Synergetics. Springer, Berlin, Heidelberg, New York. 404 p.

Haken, H. 1980. Synergetics. Mir, Moscow (Russian)

Haken, H. 1981. Chaos and Order in Nature. In: Springer Series in Synergetics, Vol. 11. Berlin, Springer, 218 p.

Haken, H. 1983. Advanced Synergetics. Springer Series in Synergetics, Vol. 20. Springer, Berlin, Heidelberg, New York, 356 p.

Haken, H. 1983. Synergetics. An Introduction 3rd ed. Springer Series Synergetics, Vol. 1, Springer, Berlin, Heidelberg, New York, 355 p.

Haken, H., Kelso, J. A. S., Bunz, H. 1985. A theoretical model of phase transitions in human hand movements. Biol. Cybern. 51, 347-356

Hales, S. 1727. Vegetable Staticks. London, Oldbourne, reprinted 1961

Hanna, A., Saul, A., Showalter, K. 1982. Detailed studies of propagating fronts in the iodate oxidation of arsenous acid. J. Am. Chem. Soc. 104, 3838-3844

Harris, W. F. 1977. Disclinations. Scient. Amer. 237, 130-145

Harsch, G., Bursenas, H. H. 1985. Bilder, die sich selber malen. Der Chemiker Runge und seine Musterbilder für Freunde des Schönen. Studio DuMont, Köln. 135 p.

Harwell Subroutine Library 1985. AERE Harwell, Didcot Oxfordshire, England

Haynes, K. E., Phillips, F. Y., Mohrfald, J. W. 1980. The entropies: Some roots of ambiguity. Socio-Econ. Plan. Sci. 14, 137-145

Hedges, E. S. 1932. Liesegang Rings and Other Periodic Structures. Chapman and Hall, London), 122 p.

Heinrich, R., Rapoport, T. A. 1974. A linear steady-state treatment of enzymatic chains. Eur. J. Biochem. 42, 89-95

Heinrich, R., Rapoport, T. A., Rapoport, S. M. 1977. Metabolic regulation and mathematical models. Prog. Biophys. Molec. Biol. 32, 1-82

Hejnowicz, Z., Wohlfarth-Bottermann, K. E. 1980. Propagated waves induced by gradients of physiological factors within plasmodia of Physarum polycephalum. Planta 150, 144-152

Hemminger, W., Höhne, G. 1984. Calorimetry. Verlag Chemie, Weinheim, 99-109

Hers, H.-G., van Schaftingen, E. 1982. Fructose 2,6-bisphosphate 2 years after its discovery. Biochem. J. 206, 1-12

Hess, B., Boiteux, A. 1968. Mechanism of glycolytic oscillation in yeast (I). Hoppe-Seyler's Zeitschr. Phys. Chem. 349, 1567-1574

Hess, B., Boiteux, A. 1969. Cooperation of glycolytic enzymes. In: Advances in Enzyme Regulation, Vol. 7. Pergamon Press, Oxford, 149-167

Hess, B., Boiteux, A. 1971. Oscillatory phenomena in biochemistry. Annu. Rev. Biochem. 40, 237-258

Hess, B., Chance, B., Betz, A. 1964. Isolierung eines oszillierenden Systems aus *S. carlsbergensis*. Ber. Bunsenges. Phys. Chem. 68, 768-773

Hoffman, W. C. 1973. A system of axioms for mathematical biology. Math. Biosciences 16, N1/2, 11-29

Hofman, P. B. 1938. Quantitative laws of animal growth. Usp. Sovr. Biol. 9, No. 1, 39-67 (Russian)

Hofmann, E. 1976. The significance of phosphofructokinase to the regulation of carbohydrate metabolism. Rev. Physiol. Biochem. Pharmacol. 75, 1-68

Hofmann, E., Eschrich, K., Schellenberger, W. 1985. Temporal organization of the phosphofructokinase/fructose 1,6-bisphosphate cycle. Adv. Enzyme Regul. 23, 331-362

Horio, S., Yamada, T., Shimizu, H., Ishikawa, H. 1985. Can oligomeric myosin participate in smooth muscle contraction? J. Muscle Research and Cell Motility 6, 709-723

Horsfield, K., Cumming, G. 1968. Morphology of the bronchial tree in man. J. Appl. Physiol. 24, 373-383

Hue, L. 1982. The role of futile cycles in the regulation of carbohydrate metabolism. In: Metabolic Compartimentation. (M. Sies et al., eds.). Academic Press, London, 71-97

Hugo, P., Wirges, H. P. 1980. Approximation to calculate self-sustained oscillations of a stirred tank reactor. Ind. Eng. Chem. Fund. 19, 428-435

Hyman, J. M., McLaughlin, D. W., Scott, A. C. 1979. Los Alamos Sci. Lab. Los Alamos NM 87545

Hyman, J. M., McLaughlin, D. W., Scott, A. C. Davydovs alpha-helic soliton. Physica D 3(1-2) 23-44

Ivanitskii, G. R., Krinskii, V. I., Sel'kov, E. E. 1978. Mathematical Biophysics of Cell. Nauka, Moscow. 310 p.

Ivanitskii, G. R., Krinskii, V. I., Tzyganov, M. A., Agladze, K. I. 1987. Changes in the autowave modality in a growing bacterial population. Biophysics (submitted)

Ivanitskii, G. R., Krinskii, V. I., Zaikin, A. N., Zhabotinskii, A. M. 1981. Autowave processes and their role in disturbing the stability of distributed excitable systems. Sov. Sci. Rev. 2D, 280-324

Jaynes, E. T. 1957. Information theory and statistical mechanics. Phys. Rev. 106, 620-630; 108, 171-190

Jaynes. E. T. 1978. Where do we stand on maximum entropy? In: The Maximum Entropy Formalism (R. D. Levine, M. Tribus (eds.). MIT press, Boston. 15-118

Jean, R. V. 1980. A systemic model of growth in botanometry. J. Theor. Biol. 87, 569-584

Jean, R. V. 1981. An L-system approach to non-negative matrices for the spectral analysis of discrete growth functions of populations. Math. Biosci. 55, 155-168

Jean, R. V. 1982. The hierarchical control of phyllotaxis. Ann. Bot. 49, 747-760

Jean, R. V. 1983 a. Mathematical modeling in phyllotaxis: The state of the art. Math. Biosci. 64, 1-27

Jean, R. V. 1983 b. Allometric relations in plant growth. J. Math. Biol. 18, 189-200

Jean, R. V. 1984. Mathematical Approach to Pattern and Form in Plant Growth. Wiley-Interscience, New York

Jean, R. V. 1986 a. A basic theorem on, and a fundamental approach to pattern formation on plant. Math. Biosci. 79(2), 127-154

Jean, R. V. 1986 b. Predictions confronted to observations in phyllotaxis. J. Inf. Deduct. Biol. (to appear)

Jessen, W. 1978. Chemische Oszillationen und Strukturen als Grundlage einer zeitlichen und räumlichen Organisation. Naturwiss. 65, 449-455

Jorne, J. 1980. Role of diffusion in trigger wave propagation in the Belousov-Zhabotinskii reaction. J. Am. Chem. Soc. 102, 6196-6198

Joseph. D. 1976. Stability of Fluid Motions. Springer, Berlin

Kai, S., Müller, S. C. 1985. Spatial and temporal macroscopic structures in chemical reaction systems - precipitation patterns and interfacial motion. Sci. Form 1, 9-39

Kai, S., Müller, S. C., Ross, J. 1982. Measurements of temporal and spatial sequences of events in periodic precipitation processes. J. Chem. Phys. 76, 1392-1406

Kai, S., Müller, S. C., Ross, J. 1983. Periodic precipitation patterns in the presence of concentration gradients. II. Spatial bifurcation of precipitation bands and stochastic pattern formation. J. Phys. Chem. 87, 806-813

Kamiya, N. 1968. The mechanism of cytoplasmic movement in a myxomycete plasmodium. In: Aspects of Cell Motility. Symp. Soc. Exp. Biol. 22. Cambridge Univ. Press, Cambridge, 199-214

Kamiya, N. 1981. Physical and chemical basis of cytoplasmic streaming. Ann. Rev. Plant Physiol. 32, 205-236

Kamiya, N., Yoshimoto, Y. 1972. Dynamic characteristics of the cytoplasm. A study on the plasmodial strand of a myxomycete. In: Aspects of Cellular and Molecular Physiology. Univ. Tokyo Press, Tokyo, 167-189

Kato, T. 1979. Ca^{2+} uptake of Physarum microsomal vesicles. In: Cell Motility: Molecules and Organization. Tokyo Univ. Press, Tokyo, 211-223

Keener, J. P., Tyson, J. 1986. Spiral waves in the Belousov-Zhabotinskii reaction. Physica 21D, 307-324

Keizer, J., Rock, P. A., Lin, S. W.. 1979. Analysis of the oscillations in 'beating mercury heart' systems. J. Am. Chem. Soc. 101, 5637-5649

Keller, J. B., Rubinow, S. I. 1981. Recurrent precipitation and Liesegang rings. J. Chem. Phys. 74, 5000-5007

Keller, J. U. 1984. Evolution of biothermodynamic systems. In: Synergetics - From Microscopic to Macroscopic Order. Springer Series in Synergetics, Vol. 22, Springer, Berlin, (E. Frehland, ed.). 140-149

Keller, J. U. 1984. Evolution processes of populations of simple biothermodynamic systems. J. Non-Equilib. Thermodyn. 9, 293-308

Keller, J. U. 1984. Thermodynamics of processes and the evolution of simple biothermodynamic populations. In: Thermodynamics and Regulation of Biological Processes. (I. Lamprecht, A. I. Zotin, eds.). Walter de Gruyter, Berlin, New York. 305-329

Keller, J. U. 1987. On the evolution of populations of biothermodynamic systems with limited lifetime (in preparation)

Kelso, J. A. S. 1981. Contrasting perspectives on order and regulation in movement. In: Attention and Performance. (J. Long, A. Badenley, eds.). Erlbaum, Hillsdale, NJ, 437-457

Kelso, J. A. S. 1981. On the oscillatory basis of movement. Bull. Psychon. Soc. 18, 63

Kelso, J. A. S. 1982. Transitions in human bimanual coordination. Invited paper presented at Kroc Foundation Conference on Nonlinear Mechanics in Brain Function. Santa Barbara

Kelso, J. A. S. 1984. Phase transitions and critical behavior in human bimanual coordination. Am. J. Physiol. Reg. Integ. Comp. 15, R1000-R1004

Kelso, J. A. S., Scholz, J. 1985. Cooperative phenomena in biological motion. In: Complex Systems - Operational Approaches. (H. Haken, ed.). Springer Series in Synergetics, Vol. 31. Springer, Berlin, Heidelberg, New York, 124-149

Kerner, B. S., Kuznetsova, E. M., Osipov, V. V. 1984 a. Analytical investigation of the model of the media with diffusion. Mikroelektronika 13, 456-461 (Russian)

Kerner, B. S., Kuznetsova, E. M., Osipov, V. V. 1984 b. On variety of inhomogeneous states of active distributed media. Mikroelektronika 13, 407-426 (Russian)

Kerner, B. S., Kuznetsova, E. M., Osipov, V. V. 1984 c. On the stochastisity of inhomogeneous states of active media with diffusion. Sov. Phys.-Dokl. 29, 644-646

Kerner, B. S., Osipov, V. V. 1977. Nonlinear theory of nonisothermal current spiking in transistor structures. Mikroelektronika 6, 337-353 (Russian)

Kerner, B. S., Osipov, V. V. 1978. Nonlinear theory of stationary strata in dissipative systems. Sov. Phys. JETP 47, 874-885

Kerner, B. S., Osipov, V. V. 1979. Strata in a heated electron-hole plasma. Sov. Phys. Semicond. 13, 424-435 (Russian)

Kerner, B. S., Osipov, V. V. 1980. Stochastic inhomogeneous structures in nonequilibrium systems. Sov. Phys. JETP 52, 1122-1132

Kerner, B. S., Osipov, V. V. 1981. Stationary and moving dissipative structures in active kinetic media. Microelektronika 10, 407-432 (Russian)

Kerner, B. S., Osipov, V. V. 1982 a. Dynamic rearrangement of dissipative structures. Sov. Phys.-Dokl. 27, 484-486

Kerner, B. S., Osipov, V. V. 1982 b. Pulsating 'heterophase' regions in nonequilibrium systems. Sov. Phys. JETP 56, 1275-1282

Kerner, B. S., Osipov, V. V. 1982 c. Properties of stationary dissipative structures in mathematical models of morphogenesis. Biofisika 27, 137-143 (Russian)

Kerner, B. S., Osipov, V. V. 1983 a. On spontaneous appearance of irregular or pulsating structures in the stratification of the uniform state of nonequilibrium systems. Sov. Phys.-Dokl. 28, 485-487

Kerner, B. S., Osipov, V. V. 1983 b. Running 'heterophase' regions in nonequilibrium systems. Mikroelektronika 12, 512-529 (Russian)

Kerner, B. S., Osipov, V. V. 1985 a. Effects in active distributed systems. Mikroelektronika 14, 389-407

Kerner, B. S., Osipov, V. V. 1985 b. Autosolitons in a hot semiconductor plasma. Sov. Phys. JETP 62, 337-348

Kerner, B. S., Osipov, V. V. 1985 c. Autosolitons in active systems with diffusion. Proceedings of the Third International Conference: Self-Organization by Nonlinear Irreversible Processes. DDR. Springer, Berlin 33, 118 p.

Kerner, B. S., Osipov, V. V. 1985 d. Highly nonequilibrium localized states in a system slightly away from thermodynamic equilibrium. JETP Lett. 41, 386-388

Kerner, B. S., Osipov, V. V. 1987. 'Ball lighting' in the mixture of neutral gases. Dokl. Akad. Nauk SSSR 292, 82-85

Kessler, D. 1982. Plasmodial structure and motility. In: Cell Biology of Physarum and Didymium. Vol 1. Academic Press, New York, 145-208

Kessler, D., Nachmias, V. T. Loewy, A. G. 1976. Actomyosin content of *Physarum polycephalum* plasmodia and detection of immunological crossreactions with myosins from related species. J. Cell Biol. 69, 393-406

Khriplovich, I. B. 1981. Nonconservation of Parity in Atomic Phenomena. Nauka, Moscow (Russian)

Kitabatake, A., Suga, H. 1978. Diastolic stress-strain relations of nonexcised blood-perfused canine papillary muscle. Amer. J. Physiol. 234, 416-420

Kizel, V. A. 1985. The Physical Causes of Dissymmetry of Living Systems. Nauka, Moscow (Russian)

Kleiber, M. 1947. Body size and metabolic rate. Physiol. Rev. 27, No. 3, 511-544

Klöck, G. Kreuzberg, K. 1987 a. Sn-glycerol-3-phosphate is a product of starch degradation in isolated chloroplasts from *Chlamydomonas reinhardii*. Z. Naturforschg. 42c, 567-569

Klöck, G., Kreuzberg, K. 1988. Determination of compartmented glycolytic metabolite pools during starch catabolism in *Chlamydomonas reinhardii* protoplasts by a rapid fractionation. Biochim. Biophys. Acta (in press)

Koga, S., Kuramoto, Y. 1980. Localized patterns in reaction-diffusion systems. Prog. Theor. Phys. 63, 106-121

Kolmogorov, A. N., Petrovskii, I. G., Piskunov, N. S. 1937. The study of equations of diffusion, combined with the increase of quantity of substance and its application to one biological problem. Bull. MGU, sect. A, 1, N6 1-27 (Russian)

Kondepudi, D. K., Nelson, G. W. 1983. Chiral symmetry breaking in nonequilibrium systems. Phys. Rev. Lett. 50, 1023-1027

Kondepudi, D. K., Prigogine, I. 1981. Sensitivity of nonequlibrium systems. Physica 107A, 1-14

Kopell, N., Howard, L. N. 1981. Target pattern and spiral solutions to reaction-diffusion equations with more than one space dimension. Adv. Appl. Math. 2, 417-449

Kowalczy, K. S., Januszewska, B., Cymerska, E., Malskowski, P. 1984. The occurrence of inorganic pyrophosphate: D-fructose-6-phosphate 1-phosphotransferase in higher plants. I. Initial characterization of partially purified enzyme from *Sanseveria trifasciata* leaves. Physiol. Plant. 60, 31-37

Körös, E., Orbán, M., Nagy, Zs. 1973. Periodicity in the rate of heat evolution during the temporal oscillation in the 2,4-pentanedione-bromate-catalyst system. J. Phys. Chem. 77, 3122-3123

Körös, E., Orbán, M. 1978. Uncatalyzed oscillatory chemical reactions. Nature 273, 371-372

Körös, E., Orbán, M., Nagy, Zs. 1973. Periodic heat evolution during temporal chemical oscillations. Nature Phys. Sci. 242, 30-31

Körös, E., Orbán, M., Nagy, Zs. 1979. Calorimetric studies on the Belousov-Zhabotinskii oscillatory chemical reaction. Acta Chim. Acad. Hung. 100, 449-461

Kretschmer, M., Schellenberger, W., Hofmann, E. 1985. Quasistationary states in the phosphofructokinase-2/fructose 2,6-bisphosphatase cycle. Biochem. Biophys. Res. Commun. 131, 869-875

Kretschmer, M., Schellenberger, W., Hofmann, E. 1987. The appearance of a critical input concentration of fructose 6-phosphate in the phosphofructokinase-2/ fructose 2,6-bisphosphatase cycle. J. theor. Biol. (in press)

Kreuzberg, K. 1977. Substratabhängige Regulation der oszillierenden Glykolyse bei *Saccharomyces carlsbergensis*. Thesis, Bonn

Kreuzberg, K. 1978. Interaction of D-fructose and fructose-1-phosphate with yeast phosphofructokinase and its influence on glycolytic oscillations. Biochim. Biophys. Acta 527, 229-238

Kreuzberg, K. 1984. Starch fermentation via a formate producing pathway in *Chlamydomonas reinhardii*, *Chlorogonium elongatum* and *Chlorella fusca*. Physiol. Plant. 61, 87-94

Kreuzberg, K. 1985. Pyruvate degradation via pyruvate formate-lyase (EC 2.3.1.54) and the enzymes of formate fermentation in the green alga *Chlamydomonas reinhardii*. Planta 163, 60-67

Kreuzberg, K., Betz, A. 1979. Amplitude and period length of yeast NADH oscillations fermenting on different sugars in dependence of growth phase, starvation and hexose concentration. J. Interdiscpl. Cycle Res. 10, 41-50

Kreuzberg, K., Klöck, G., Grobheiser, D. 1987. Subcellular distribution of pyruvate-degrading enzymes in *Chlamydomonas reinhardii* studied by an improved protoplast fractionation. Physiol. Plantarum 69, 481-488

Kreuzberg, K., Klöck, G., Schlipf, B. 1986. The role of phosphofructokinase (PFK) in algal chloroplast glycolysis. Abstract: 5th Congress of FESPP, Hamburg, 227

Kreuzberg, K., Martin, W. 1984. Oscillatory starch degradation and fermentation in the green alga *Chlamydomonas reinhardii*. Biochim. Biophys. Acta 799, 291-297

Kreuzberg, K., Richter, O., Martin, W., Betz, A. 1977. Statistical analysis of NADH oscillations in the yeast *Saccharomyces carlsbergensis* fermenting on different sugars. J. Interdiscipl. Cycle Res. 8, 135-146

Krinskii, V. I. (ed.). 1984. Self-Organization. Autowaves and Structures Far from Equilibrium. Springer Series in Synergetics, Vol. 28. Springer, Berlin 28, 263 p.

Krinskii, V. I., Zhabotinskii, A. M. 1981. Autowave structures and perspective of their investigations. In: Autowave Processes in the Systems with Diffusion. Gorky: IPPh. AN SSSR, 6-32 (Russian)

Krompholz, H., Herzinger, G. 1981. Phenomena of self organization in dense plasma. In: H. Haken (ed.): Springer Series in Synergetics. Berlin, Springer, pp. 131-143

Krumhansl, J. A. 1986. Vibrational anomalies are not generally due to fractal geometry. Phys. Rev. Lett. 56, 2696-2699

Kuramoto, Y., Tsuzuki, T. 1975. On the formation of dissipative structures in reaction-diffusion systems. Progr. Theor. Phys 54, 687-699

Kurdjumov, S. P. 1982. Eigenfunctions of combustion of a nonlinear medium and constructive laws for constructing its organization. In: Current Problems in Math. Phys. and Numer. Math. (A. N. Tikhonov, ed.). Moscow, Nauka, 217-243 (Russian)

Kurdjumov, S. P., Kurkina, E. S., Potapov, A. B., Samarskii, A. A. 1986. Complex multidimensional structures of nonlinear medium combustion. Zh. Vychisl. Mat. i. Mat. Fiz. 26, 1189-1205 (Russian)

Kurdjumov, S. P., Malinetskii, G. G., Poveschenko, Yu. A., Popov, Yu. P., Samarskii, A. A. 1981. Dissipative structures in trigger media. Differentsial'nye Uravneniya 17, 1875-1885 (Russian)

Kurik, M. V., Lavrentovich, O. D. 1982. Topological defects of cholesteric liquid crystals for volumes with spherical shape. Mol. Cryst. Liq. Cryst. 72 (Letters), 239-246

Lakirev, A. V., Beloussov, L. V. 1986. Computer modelling of the gastrulation and neurulation in amphibian embryos on the base of tensile field concept. Ontogenez (Sov. J. Devel. Biol.) 17 (in press) (Russian)

Lakirev, A. V., Beloussov, L. V., Petuchov, S. V. 1987. A modelling of epithelial morphogenesis in hydroid polypes. Ontogenez (Sov. J. Devel. Biol.) (in press) (Russian)

Lambert, C. J., Weaire, D. 1983. Order and disorder in two-dimensional random networks. Phil. Mag. B47, 445-450

Lamprecht, I., Schaarschmidt, B. 1978. Microcalorimetric investigations of chemical oscillations: The Zhabotinskii reaction. Thermochim. Acta 22, 257-266

Lamprecht, I., Schaarschmidt, B., Plesser, Th. 1987. Heat production in oscillating chemical reactions: Three examples. Thermochim. Acta 112, 95-100

Layne, S. P. 1984. A possible mechanism for general anesthesia. Related topics. Los Alamos Science 10, 23-26

Layrand, D. B., Matveeva, N. B., Teplov, V. A., Beylina, S. I. 1972. The role of elastoosmotic parameters in locomotion of myxomycete plasmodia. Acta protozoologica 11, 339-354

Lee, M. O. 1929. Determination of the surface area of the white rat with its application to the expression of metabolic results. Amer. J. Physiol. 89, (1), 24-33

LeVan, M. E., Ross, J. Measurements and a hypothesis on periodic precipitation processes (preprint)

Lewis, F. T. 1928. The correlation between cell division and the shapes and sizes of prismatic cells. Anat. Records 38, 341-376

Lewis, F. T. 1943 a. A geometric accounting for diverse shapes of 14-hedral cells. Amer. J. Botany 30, 74-81

Lewis, F. T. 1943 b. The geometry of growth and cell division in epithelial mosaics. Amer. J. Botany 30, 766-776

Lewis, W. H. 1938. Changes with age in the basal metabolic rate in adult men. Amer. J. Physiol. 121, (2), 502-516

Liebhafsky, H. A., Roe, G. M. 1979. The detailed mechanism of the Dushman reaction explored by computer. Int. J. Chem. Kinet. 11, 693-703

Liesegang, E. S.. 1896. 'A-Linien'. Phot. Archiv 21, 321-326

Lindenmayer, A. 1975. Developmental algorithms for multicellular organisms: A survey of L-systems. J. Theor. Biol. 54, 3-22

Lippmann, G. 1873. Beziehungen zwischen den capillaren und elektrischen Erscheinungen. Ann. Phys. 2nd Series 149, 546-561

Lomdahl, P. S., Lane, S. P., Bigio, I. J. 1984. Los Alamos Science, Number 10, pp. 2-21

Lovett, R., Ortoleva, P., Ross, J. 1978. Kinetic instabilities in first order phase transitions. J. Chem. Phys. 69, 947-955

Luther, R. 1906. Räumliche Fortpflanzung chemischer Reaktionen. Z. Elektrochem. 12, 596-600 (German)

Lymn, R. W., Taylor, E. W. 1971. Mechanism of andenosine triphosphate hydrolysis by actomyosin. Biochemistry 10, 4617-4624

Maitra, P. K. 1966. Pulsating glucose flux in yeast. Biochem. Biophys. Res. Commun. 25, 462-467

Malinetskii, G. G. 1981. Stationary localized dissipative structures in trigger two-component media. Preprint N78, Keldysh Inst. Appl. Math. Acad. Sci., USSR, Moscow (Russian)

Mandelbrot, B. B. 1983. The Fractal Geometry of Nature. Freeman, and Company, New York. 468 p.

Maresin, V. M., Presnov, E. V. 1985. Topological approach to embryogenesis. J. Theor. Biol. 114, N3, 387-398

Marison, I. W., von Stockar, U. 1986. The application of a novel heat flux calorimeter for studying growth of *Escherichia coli W* in aerobic batch culture. Biotechn. Bioengin. 28, 1780-1793

Markus, M., Müller, S. C., Plesser, Th., Hess, B. 1987. On the recognition of order and disorder. Biol. Cybern. 57, 187-195

Marsden, J. E., McCracken, M. 1976. The Hopf Bifurcation and its Applications. Springer, Berlin. 408 p.

Marzek, C. 1986. Phyllotaxis as a dissipative structure. In: Mathematical modelling in Science and Technology. Proc. Vth ICMM, Symposium on Phyllotaxis, July 1985, Univ. of California at Berkeley. (X. J. R. Avula, G. Leitmann, C. D. Mote, E. Y. Rodin, eds.). Pergamon Press, New York (in press)

Mason, S. F., Tranter, G. E. 1985. The electroweak origin of biomolecular handedness. Proc. Roy. Soc. A397, 45-53

Masters, C. 1985. Phosphofructokinase, compartmentation and the regulation of glycolysis. TIBS 10, 189

Matzke, E. B. 1946. The three-dimensional shape of bubbles in foam. Amer. J. Botany 33, 58-80

Matzke, E. B. 1950. In the twinkling of an eye. Bull. Torrey Botan. Club 77, 222-227

Matzke, E. B., Duffy, R. M. 1956. Progressive three-dimensional shape changes of dividing cells. Amer. J. Botany 43, 205-225

Mazenko, V. G., Rubanovsky, V. N. 1983. Application of the direct Ljapunov method to the analysis of the age-structure of biopopulations. Zh. vych. matem. i metem. fiz. 23, 326-337 (Russian)

McGrane, M. M., El-Maghrabi, M. R., Pilkis, S. J. 1983. The interaction of fructose 2,6-bisphosphate and AMP with hepatic fructose 1,6-bisphosphatase. J. Biol. Chem. 258, 10445-10454

Medawar, P. B. 1945. Size, shape and age. In: Essays on Growth and Form. (W. E. Le Grog Clark, P. B. Medawar, eds.). Oxford Univ. Press, London, Milford, 157-187

Meeh, K. 1879. Oberflächenmessungen des menschlichen Körpers. Zeitschr. Biol. 15, 425-458

Meijering, J. L. 1953. Interface area, edge length, and number of vertices in crystal aggregates with random nucleation. Philips Res. Rep. 8, 270-290

Meinhardt, H. 1982. Models of Biological Pattern Formation. Acad. Press, London, 230 p.

Meinhardt, H., Gierer, A. 1974. Applications of a theory of biological pattern formation based on lateral inhibition. J. Cell. Sci. 15, 321-346

Meinhardt, H., Gierer, A. 1980. Generation and regeneration of sequence of structures during morphogenesis. J. Theor. Biol. 85, 429-450

Merkin, D. R. 1975. On the problem of the structure of forces. Appl. Math. Mech. 39, N5, 929-932 (Russian)

Mermin, N. D. 1979. The topological theory of defects in ordered media. Rev. Mod. Phys. 51, 591-648

Micheau, J. C., Gimenez, M., Borckmans, P., Dewel, G. 1983. Hydrodynamic instabilities and photochemical reactions. Nature 305, 43-45

Middleton, C. A. 1977. The effects of cell-cell contact on the spreading of pigmented retina epithelial cells in culture. Exptl. Cell Res. 109, 349-359

Miernyk, J. A., Dennis, D. T. 1982. Activation of the plastid isoenzyme of phosphofructokinase from developing endosperm of *Ricinus communis* by fructose 2,6-bisphosphate. Biochem. Biophys. Res. Commun. 105, 193-198

Milnor, J. W. 1965. Topology from the Differentiable Viewpoint. Univ. of Virginia Press, Virginia. 64 p

Minakami, S., de Verdier, C. H. 1976. Calorimetric study on human erythrocyte glycolysis. Heat production in various metabolic conditions. Eur. J. Biochem. 65, 451-460

Mineev, V. P. 1980. Topologically stable inhomogeneous states in ordered media. Sov. Sci. Rev. A2, 173-234

Mizani, S. M., Betz, A., Eilers-König, C. 1974. Zur Regulation des anaeroben Intermediärstoffwechsels in ausgehungerten Hefezellen bei Zugabe von Glucose und NH_4^+ Ionen. Planta 117, 11-27

Morozov, L. L. 1979. Mirror symmetry breaking in biochemical evolution. Orig. Life 9, 187-196

Morozov, L. L., Kuzmin, V. V., Goldanskii, V. I. 1983. Comparative analysis of the role of statistical fluctuations and factor of advantage (parity nonconservation) in the origins of optical activity. Orig. Life 13, 119-127

Mou, D.-G., Cooney, C. L. 1976. Application of dynamic calorimetry for monitoring fermentation processes. Biotechn. Bioengin. 18, 1371-1392

Möckel, R. 1977. Photochemisch induzierte dissipative Strukturen. Naturwiss. 64, 224 (German)

Müller, K. H., Plesser, Th. 1987. Deconvolution of periodic calorimetric data by Fast Fourier Transform. Thermochim. Acta 119, 189-201

Müller, S. C., Kai, S., Ross, J. 1982. Periodic precipitation in the presence of concentration gradients. I. Dependence on ion product and concentration difference. J. Phys. Chem. 86, 4078-4087

Müller, S. C., Plesser, T., Hess, B. 1986. Two-dimensional spectrophotometry and pseudo-color representation of chemical reaction patterns. Naturwiss. 73, 165-179

Müller, S. C., Plesser, Th. 1984. Spatial pattern formation in thin layers of NADH-solutions. In: Modelling of Patterns in Space and Time (Lecture Notes in Biomathematics, Vol. 55. (W. Jäger, J. Murray, eds.). Springer, Berlin, Heidelberg, 246-253

Müller, S. C., Plesser, Th. Boiteux, A., Hess, B. 1985. Pattern formation and Marangoni convection during oscillating glycolysis. Z. Naturforsch. 40c, 588-590

Müller, S. C., Plesser, Th., Hess, B. 1984. Hydrodynamic instabilities and pattern formation in a cytoplasmic medium from yeast. Naturwiss. 71, 637-638

Müller, S. C., Plesser, Th., Hess, B. 1985. Coupling of glycolytic oscillations and convective patterns. In: Temporal Order. (L. Rensing, N. I. Jaeger, eds.). Springer Series in Synergetics, Vol. 29. Springer, Berlin, 194-196

Müller, S. C., Plesser, Th., Hess, B. 1985. Surface tension driven convection in chemical and biochemical solution layers. Ber. Bunsenges. 89, 654-658

Müller, S. C., Plesser, Th., Hess, B. 1985. The structure of the core off the spiral wave in the Belousov-Zhabotinskii reaction. Science 230, 661-663

Müller, S. C., Plesser, Th., Hess, B. 1985. Two-dimensional spectrophotometry with high spatial and temporal resolution by digital video techniques and powerful computers. Anal. Biochem. 146, 125-133

Müller, S. C., Plesser, Th., Hess, B. 1986. Two-dimensional spectrophotometry and pseudo-color representation of chemical reaction patterns. Naturwiss. 73, 165-179

Müller, S. C., Plesser, Th., Hess, B. 1987. Chemical waves and natural convection. In: Physiochemical Hydrodynamics: Interfacial Phenomena. NATO ASI Series. (M. G. Velarde, B. Nichols, eds.). Plenum Press, New York (in press)

Müller, S. C., Plesser, Th., Hess, B. 1987. Two-dimensional spectrophotometry of spiral wave propagation in the Belousov-Zhabotinskii reaction. I. Experiments and digital data representation. Physica 24D, 71-86

Müller, S. C., Venzl, G. 1984. Pattern formation in recipitation processes. In: Modelling of patterns in space and time. Lecture Notes in Biomathematics, Vol. 55. (W. Jäger, J. D. Murray, eds.). Springer, Berlin, Heidelberg, 254-278

Nachmias, V. T. 1979. The contractile proteins of *Physarum polycephalum* and actin polymerization in plasmodial extracts. In: Cell Motility: Molecules and Organization. Tokyo Univ. Press, Tokyo, 33-57

Nagy-Ungvarai, Zs., Körös, E. 1985. Calorimetric studies on the bromate-malic acid-catalyst oscillating system. React. Kinet. Catal. Lett. 27, 83-89

Nagypal, I., Bazsa, G., Epstein, I. R. 1986. Gravity-induced anisotropies in chemical waves. J. Am. Chem. Soc. 108, 3635-3640

Nakache, E., Dupeyrat, M., Vignes-Adler, M. 1983. Experimental and theoretical study of interfacial instability at some oil-water interfaces involving a surface-active agent. J. Colloid. Interface Sci. 94, 187-200

Neumann, J., von 1952. Discussion - Shape of metal grains. Metal interfaces. Cleveland Amer. Soc. Metals, 108-110

Nicolis, G., Prigogine, I. 1977. Self-Organization in Nonequilibrium Systems. From Dissipative Structures to Order through Fluctuations. Wiley, New York. 490 p.

Nissler, K., Otto, A., Schellenberger, W., Hofmann, E. 1983. Similarity of activation of yeast phosphofructokinase by AMP and fructose 2,6-bisphosphate. Biochem. Biophys. Res. Commun. 222, 657-660

Nissler, K., Otto, A., Schellenberger, W., Hofmann, E. 1984. Cooperation of fructose 2,6-bisphosphate and AMP in the activation of yeast phosphofructokinase. Biomed. Biochim. Acta 43, 535-540

Nitzan, A., Ortoleva, P. 1981. Scaling and Ginzburg criteria for critical bifurcations in nonequilibrium reacting systems. Phys. Rev. Ser. A21, 1735-1755

Nitzan, A., Ortoleva, P., Ross, J. 1974. Symmetry breaking instabilities in illuminated systems. J. Chem. Phys. 60, 3134-3143

Nonnenmacher, J. D. F., Nonnenmacher, T. F. 1987. On the Weierstrass function and some self-affine modifications. Biol. Cybern. (in press)

Nonnenmacher, T. F. 1985 a. Correlation functions for the chemokinetic movement of granulocytes and tumor cells. J. Non-Equil. Thermodyn. 10, 119-130

Nonnenmacher, T. F. 1985 b. Stochastische Modelle für die Analyse von Zellbewegungen. Verh. Dtsch. Ges. Path. 69, 589-590

Nonnenmacher, T. F. 1987. A scaling model for dichotomous branching processes. Biol. Cybern. 56, 155-157

Odell, G. M. 1984. A mathematically modelled cytogel cortex exhibiting periodic Ca^{++}-modulated contraction cycles seen in Physarum shuttle streaming. J. Embryol. Exp. Morph. 83, Suppl., 261-287

Odell, G. M., Oster, G., Alberch, P., Burnside, B. 1981. The mechanical basis of morphogenesis. I. Epithelial folding and invagination. Devl. Biol. 85, 446-462

Oparin, A. I. 1968. Genesis and Development of Life. Academic Press, New York

Orbán, M. 1980. Stationary and moving structures in uncatalyzed oscillatory chemical reactions. J. Am. Chem. Soc. 102, 4311-4314

Ortoleva, P. J., Schmidt, S. L. 1985. The structure and variety of chemical waves. In: Oscillations and Traveling Waves in Chemical Systems (R. J. Field, M. Burger, eds.). New York, Wiley, 333-418

Osipov, V. V., Lazurchak, I. I., Kerner, B. S., Gafüchuk, V. V. 1987. Pulsating autosoliton and leading center in two-component systems with diffusion. Mikroelektronika 16, 22-32 (Russian)

Oster, G. F., Murray, J. D., Harris, A. K. 1983. Mechanical aspects of mesenchymal morphogenesis. J. Embryol. Expl. Morph. 78, 83-125

Oster, G. F., Odell, G. M. 1984. Mechanics of cytogels. I: Oscillations in Physarum. Cell Motility 4, 469-503

Packter, A. 1955. Inverted rhythmic precipitation. Nature 175, 556-557

Parks, J. R. 1970. Growth curves and the physiology of growth. I. Animals. Amer. J. Physiol. 219, (3), 833-836

Parks, J. R. 1982. A Theory of Feeding and Growth of Animals. Springer, Berlin. 322 p.

Passonneu, J. V., Lowry, O. H. 1962. Phosphofructokinase and the Pasteureffect. Biochem. Biophys. Res. Commun. 7, 10-15

Paumgartner, D., Losa, G., Weibel, E. R. 1980. Resolution effect on the stereological estimation of surface and volume and its interpretation in terms of fractal dimensions. J. Microscopy 121, 51-63

Perelson, A. S., Oster, G. F. 1979. Chemical reaction dynamics. Part II: Reaction networks. Arch. Rational. Mech. Anal. 57, 31-98

Pilkis, S. J., Regen, D. M., Stewart, B. H., Chrisman, T., Pilkis, J., Kountz, P., Pate, T., McGrane, M., El-Maghrabi, M. R., Claus, T. H. 1984. Rat liver 6-phosphofructo-2-kinase/fructose 2,6-bisphosphatase: A bifunctional enzyme regulated by cAMP-dependent phosphorylation. In: Enzyme Regulation by Reversible Phosphorylation. (P. Cohen, ed.). Elsevier, Amsterdam, New York, Oxford, 95-122

Pilkis, S. J., Walderhaug, M., Murray, K., Beth, A., Venkataramu, S. D., Pilkis, J., El-Maghrabi, M. R. 1983. 6-phosphofructo-2-kinase/fructose 2,6-bisphosphatase from rat liver. J. Biol. Chem. 258, 6135-6141

Platten, J. K., Legros, J. C. 1984. Convection in Liquids. Springer, Heidelberg, 679 p.

Plesser, Th., Müller, S. C., Hess, B., Lamprecht, I., Schaarschmidt, B. 1985. Periodic heat production by oscillating glycolysis in a cytoplasmic medium extracted from yeast. FEBS Lett. 189, 42-44

Poenaru, V., Toulouse, G. 1977. The crossing of defects in ordered media and the topology of 3-manifolds. J. de Phys. 38, 887-895

Polak, L. S., Mikhailov, A. S. 1983. Self-Organization in Equilibrium Physico-Chemical Systems. Nauka, Moscow, 276 p.

Polezhaev, A. A. 1984. Examination of stability of contrasting dissipative structures. Biofisika 29, 654-659

Poluektov, B. A. (ed.). 1974. Dynamical Theory of Biological Populations. Nauka, Moscow. pp. 350-386

Presnov, E. V. 1973. Potential character of the evolution criterium of thermodynamics of irreversible processes. J. Phys. Chem. 47, N11, 2902-2904 (Russian)

Presnov, E. V. 1978. Formalism of non-equilibrium phenomenological thermodynamics. In: Thermodynamics of Biological Processes. (I. Lamprecht, A. I. Zotin, eds.). Walter de Gruyter, Berlin, New York, 31-52

Presnov, E. V. 1984. Stochastic criterion in dynamical systems. In: Thermodynamics and Regulation of Biological Processes. (I. Lamprecht, A. I. Zotin, eds.). Walter de Gruyter, Berlin, New York, 73-76

Presnov, E. V., Isaeva, V. V. 1985. Modifications of Topology at Morphogenesis. Nauka, Moscow, 191 p. (Russian)

Presnov, E. V., Malyghin, S. N. 1982. Axiomatic models of phenomenological thermodynamics. In: Thermodynamics and Kinetics of Biological Processes. (I. Lamprecht, A. I. Zotin, eds.). Walter de Gruyter, Berlin, New York, 255-264

Preston, B. N., Laurent, T. C., Comper, W. D., Checkley, G. J. 1980. Rapid polymer transport in concentrated solutions through the formation of ordered structure. Nature 287, 449-503

Prigogine, I. 1973. Can thermodynamics explain biological order? Impact of Science on Sociology. Vol. XXIII, No. 3, 159-179

Prigogine, I. 1980. From Being to Becoming. Freeman, New York. 272 p.

Prigogine, I., Lefever, R. 1968. Symmetry breaking instabilities in dissipative systems. II. J. Chem. Phys. 48, 1695-1700

Prokofiev, E. A., Zotina, R. S., Zotin, A. I. 1982. Phenomenological growth equations and their utilization in determining the maximal life span. Matematicheskaya biologiya razvitiya. (A. I. Zotin, E. V. Presnov, eds.). Nauka, Moscow, 56-66 (Russian)

Pyshnov, M. B. 1980. Topological solution for cell proliferation in intestinal crypt. I - Elastic growth without cell loss. J. Theor. Biol. 87, 189-200

Quetelet, A. 1869. Physique Sociale. Un Essay sur le Development des Faculties de L'Homme. T. 1-2 Bruxelles, C. Muquardt. 384 p.

Rastogi, R. P., Singh, K., Rastogi, P., Rai, R. B. 1977. Nucleation phenomenon in oscillatory reactions. Ind. J. Chem. 15A, 295-297

Reich, J. G., Sel'kov, E. E. 1981. Energy Metabolism of the Cell. Academic Press, London,

Reusser, E. J., Field, R. J. 1979. The transition from phase waves to trigger waves in a model of the Zhabotinskii reaction. J. Am. Chem. Soc. 101, 1063-1071

Ridgway, E. B., Durham, A. C. H. 1976. Oscillations of calcium ion concentrations in *Physarum polycephalum*. J. Cell Biol. 69, 223-226

Rivier, N. 1979. Disclination lines in glasses. Phil. Mag. A40, 859-868

Rivier, N. 1982. Recent results on the ideal structure of glasses. J. Phys. Colloques 43, C9-91-95

Rivier, N. 1983. On the structure of random tissues or froths, and their evolution. Phil. Mag. B47, L45-49

Rivier, N. 1985. Statistical crystallography: structure of random cellular networks. Phil. Mag. B52, 795-819

Rivier, N. 1986. Structure of random cellular networks and their evolution. Physica 23, 129-137

Rivier, N., Lissowski, A. 1982. On the correlation between sizes and shapes of cells in epithelial mosaics. J. Phys. A15, 143-148

Rivier, N., Occelli, R., Pantaloni, J., Lissowski, A. 1984. Structure of Benard convection cells, phyllotaxis and crystallography in cylindrical symmetry. J. de Phys. 45, 49-63

Rodrigues, J. R., Perez Villar, V., Rey, C., Garcia, M. 1986. The application of deconvolution calorimetry to the Belousov-Zhabotinskii reaction. Thermochim. Acta 106, 27-34

Romanovskii, Yu. M., Chernyaeva, E. B., Kolin'ko, V. G., Khors, N. P. 1981. Mathematical models of protoplasm streaming. In: Autowaves Processes in Systems with Diffusion. Inst. Appl. Phys. USSR, Gorky, Acad. Sci. Press, 202-219 (Russian)

Romanovskii, Yu. M., Khors, N. P. 1982. The model of mechanical autooscillations in the myxomycete plasmodium. Biophysics 27, 707-710 (Russian)

Romanovskii, Yu. M., Stepanova, N. V., Chernavskii, D. S. 1984. Mathematical Biophysics. Nauka, Moscow, 304 p. (Russian)

Rosen, R. 1967. Optimality Principles in Biology. Butterworth, London

Rosenkrantz, R. D. (ed.) 1983. E. T. Jaynes, Papers on probability, statistics and statistical physics. D. Reidel Publ. Comp., Dordrecht, Holland. 434 p.

Roux, J. C., Vidal, C. 1979. On an experimental method of study of periodic reactions. Nouv. J. Chim. 3, 247-253

Runge, F. F. 1855. Der Bildungstrieb der Stoffe veranschaulicht in selbständig gewachsenen Bildern. Oranienburg, Selbstverlag (original not available)

Sabularse, D. C., Anderson, R. L. 1981. D-fructose 2,6 bisphosphate: a naturally occurring activator for inorganic pyrophosphate: D-fructose-6-phosphate-1-phosphotransferase. Biochem. Biophys. Res. Commun. 103, 848-855

Sackmann, E. 1985. Chemomechanical interfacial instabilities and waves. In: Temporal Order. Springer Series in Synergetics, Vol. 29. (L. Rensing, N. I. Jaeger, eds.). Springer, Berlin, Heidelberg, 157-162

Sakakibara, R., Kitajima, S., Uyeda, K. 1984. Differences in kinetic properties of phospho and dephospho forms of fructose-6-phosphate,2-kinase and fructose 2,6-bisphosphatase. J. Biol. Chem. 259, 41-46

Salomaa, M. M., Volovik, G. E. 1985. Half-quantum vortices in superfluid ^{3}He-A. Phys. Rev. Lett. 55, 1184-1187

Samarskii, A. A. 1984. Numerical simulation and nonlinear processes in dissipative media. In: Self-Organization. Autowaves and Structures far from Equilibrium. (V. I. Krinsky, ed.). Springer, Berlin, 119-129

Samarskii, A. A., Zmitrenko, N. V., Kurdjumov, S. P., Mikhailov, A. P. 1976. Thermal structures and fundamental length in a medium with nonlinear heat conduction and volume heat source. Dokl. Akad. Nauka SSSR, 227, 321-324 (Russian)

Saul, A., Showalter, K. 1985. Propagating reaction-diffusing fronts. In: Oscillations and Traveling Waves in Chemical Systems. (R. J. Field, M. Burger, eds.). New York, Wiley, 419-439

Savitzky, A., Golay, M. J. E. 1964. Smoothing and differentiation of data by simplified least squares procedures. Anal. Chem. 36, 1627-1639

Schaarschmidt, B., Lamprecht, I., Plesser, Th., Müller, S. C. 1986. Simultaneous measurements of the heat production and optical density in oscillating reactions. Thermochim. Acta 105, 205-213

Schellenberger, W., Eschrich, K., Hofmann, E. 1981. Self-stabilization of the energy charge in a reconstituted enzyme system containing phosphofructokinase. Eur. J. Biochem. 118, 309-314

Schellenberger, W., Eschrich, K., Hofmann, E. 1981. Selforganization of a glycolytic reconstituted enzyme system: alternate stable stationary states, hysteresis, and stabilization of the energy charge. Adv. Enzyme. Regul. 18, 257-283

Schellenberger, W., Eschrich, K., Hofmann, E. 1985. Influence of fructose 2,6-bisphosphate on the fructose 6-phosphate/ fructose 1,6-bisphosphate cycle. Biochem. Biophys. Res. Commun. 126, 571-577

Schmidt, R. A. 1982. Motor Control and Learning: A Behavioral Emphasis. Champaign. Il. Human Kinetics Publ.

Schmidt, S., Ortoleva, P. 1977. A new chemical wave equation for ionic systems. J. Chem. Phys. 67, 3771-3776

Schoener, G., Haken, H., Kelso, J. A. S. 1986. A stochastic theory of phase transitions in human hand movement. Biolog. Cybern. 53, 247-25

Schottky, W., Ulrich, H., Wagner, C. 1973. Thermodynamik. Springer, Berlin (reprint of the 1929 edition)

Schrödinger, E. 1944. What is life? Cambridge University Press, Cambridge

Sel'kov, E. E. 1979. The oscillatory basis of cell energy metabolism. In: Pattern Formation by Dynamic Systems and Pattern Recognition. Springer Series in Synergetics, Vol. 5 (H. Haken, ed.). Springer, Berlin, Heidelberg, New York, 166-174

Sevcikova, H., Marek, M. 1983. Chemical waves in electric field. Physica 9D, 140-156

Sevcikova, H., Marek, M. 1984. Chemical front waves in electric field. Physica 13D, 379-386

Shnol, S. E. 1979. Factors determining direction and rate of biological evolution. In: Mathematical Study of Biological Processes. Moscow, Nauka, 5-26 (Russian)

Showalter, K. 1980. Pattern formation in a ferroin-bromate system. J. Chem. Phys. 73, 3735-3742

Showalter, K. 1981. Trigger waves in the acidic bromate oxidation of ferroin. J. Phys. Chem. 85, 440-447

Showalter, K. 1985. Chemical waves. In: Kinetics of Nonhomogeneous Processes (G. R. Freeman, ed.). New York, Wiley

Showalter, K., Noyes, R. M., Turner, H. 1979. Detailed studies of trigger wave initiation and detection. J. Am. Chem. Soc. 101, 7463-7469

Shubnikov, A. V., Koptsik, V. A. 1972. Symmetry in Science and Art. Nauka, Moscow (Russian)

Smith, C. S. 1954. The shape of things. Scient. Amer. 190(1), 58-64

Smoes, M. L. 1980. Chemical waves in the oscillatory Zhabotinskii system. A transition from temporal and spatio-temporal organization. In: Dynamics of Synergetic Systems. (H. Haken, ed.). Springer Series in Synergetics, Vol. 6, Springer, Berlin, Heidelberg, 80-96

Smoljaninov, V. V. Mathematical Models of Biological Tissues. Moscow, Nauka. 367 p.

Steeb, W. H. 1977. Nonlinear systems of differential equations involving limit cycles and conservative Hamiltonian systems. J. Phys. A., Math. Gen. 10, L221

Stern, K. H. 1954. The Liesegang phenomenon. Chem. Rev. 54, 79-99

Stern, K. H. 1967. A Bibliography of Liesegang Rings. 2nd ed. U.S. GPO, Washington, D. C., 61 p.

Stewart, P. A. 1964. The organization of movement in slime mold plasmodia. In: Primitive Motile Systems in Cell Biology. Academic Press, New York, 69-76

Stewart, P. A., Stewart, B. I. 1961. Circular streaming patterns in a slime mold plasmodium. Nature 192, 1206-1207

Strogatz, S. H., Prueitt, M. L., Winfree, A. T. 1984. Exotic shapes in chemistry and biology. IEEE Comp. Graph. Appl. 272, 66-69

Suwa, N., Takahashi, T. 1971. Morphological and Morphometrical Analysis of Circulation in Hypertension and Ischemic Kidney. Urban & Schwarzenberg, München. 132 p.

Tanaka, T., Sun, S. T., Hirokawa, Y., Katayama, S., Kucera, J., Hirose, Y, Amiya, T. 1987. Mechanical instability of gels at the phase transition. Nature 325, 96-98

Teplov, V. A., Beylina, S. I., Evdokimov, M. V., Prieizzhev, A. V., Romanovskii, Yu. M. 1981 a. Autowave mechanisms of intracellular motility. In: Autowave Processes in Systems with Diffusion. Inst. Appl. Phys. USSR, Gorky, Acad. Sci. Press, 190-201 (Russian)

Teplov, V. A., Beylina, S. I., Matveeva, N. B., Layrand, D. B. Lednev, V. V. 1981 b. Adaptation process in chemotaxis of *Physarum polycephalum*. In: Progress in Protozoology. Abstracts. Warszawa, 363

Teplov, V. A., Budnitsky, A. A. 1973. Device for registration of mechanical activity of plasmodial strands. In: Biophysics of Living Cell. Pushchino, Biol. Res. Centre USSR Acad. Sci Press, Vol. 4, 165-170 (Russian)

Teplov, V. A., Matveeva, N. B., Zinchenko, V. P. 1973. Free calcium level oscillations in myxomycete plasmodium during protoplasm shuttle streaming. In: Biophysics of Living Cell. Pushchino, Biol. Res. Centre USSR Acad. Sci. Press, vol. 4, 110-115 (Russian)

Teplov, V. A., Romanovskii, Yu. M. 1986. Mechanochemical distributed autooscillations in cell motility. Autowave phenomena in contractile activity of Physarum polycephalum plasmodium. Pushchino, Biol. Res. Centre USSR, Acad. Sci. Press. Preprint. Inst. Biophys. USSR Acad. Sci. 48 p. (Russian)

Thompson d'Arcy, W. 1942. On Growth and Form. Cambridge Univ. Press, Cambridge. 1116 p.

Tickle, C., Alberts, B., Wolpert, I., Lee, J. 1982. Local application of retinoic acid to the limb bud mimics the action of the polarizing region. Nature 296, 564-566

Trebin, H. R. 1982. The topology of non-uniform media in condensed matter physics. Adv. Phys 31, 195-254

Troy, W., 1985. Mathematical analysis of the Oregonator model of the Belousov-Zhabotinskii reaction. In: Oscillations and Traveling Waves in Chemical Systems (R. J. Field, M. Burger, eds.) New York, Wiley, 145-170

Tuller, B., Kelso, J. A. S., Harris, K. S. 1982. Interarticulator phasing as an index of temporal regularity in speech. J. Exp. Psycholog. Hum. Percept. Perform. 8, 460-472

Turing, A. M. 1952. The chemical basis of morphogenesis. Phil. Trans. Roy. Soc., ser. B, 237, 37-72

Tyson, J. J. 1976. The Belousov-Zhabotinskii Reaction. In: Lecture Notes in Biomathematics. (A. Levin, ed.). Springer, Berlin, 128 p.

Tyson, J. J. 1982. Periodic phenomena in Physarum. In: Cell Biology of Physarum and Didymium. Academic Press, New York, Vol. 1, 61-110

Tyson, J. J. 1985. A quantitative account of oscillations, bistabiltty and traveling waves in the Belousov-Zhabotinskii reaction. In: Oscillations and Traveling Waves in Chemical Systems (R. J. Field, M. Burger, eds.). New York, Wiley, 93-144

Tyson, J. J., Fife, P. C. 1980. Target patterns in a realistic model of the Belousov-Zhabotinskii model. J. Chem. Phys. 73, 2224-2237

Ueda, T., Goetz von Olenhusen, K., Wohlfarth-Bottermann, K. E. 1978. Reaction of the contractile apparatus in Physarum to injected calcium, ATP, ADP, and 5'-AMP. Cytobiol. 18, 76-94

Ueda, T., Kobatake, Y. 1982. Chemotaxis in plasmodia of *Physarum polycephalum*. In: Cell Biology of Physarum and Didymium. Academic Press, New York, Vol. 1, 111-143

Ungvarai-Nagy, Zs., Müller, S. C. 1988. Wave propagation in non-equilibrium systems: New experimental results. In: Propagation in Non-Equilibrium Systems. Springer Proc. Physics. (J. E. Wesfreid, ed.). Springer, Berlin, Heidelberg. (in press)

Vacquier, V. D. 1981. Dynamic changes of the egg cortex. Develop. Biol. 84, N1, 1-26

van Schaftingen, E., Hers, H.-G. 1981. Phosphofructokinase-2, the enzyme that forms fructose 2,6-bisphosphate from fructose 6-phosphate and ATP. Biochem. Biophys. Res. Commun. 101, 1078-1084

Vasil'yev, V. A., Romanovskii, Yu. M., Yakhno, V. G. 1979. Autowave processes in distributed kinetic systems. Sov. Phys.-Usp. 22, 615-639

Vasilev, V. A., Romanovskii, Yu. M., Chernavskii, D. S., Yakhno, V. G. 1986. Autowave Processes in Kinetic Systems. Spatial and Temporal Self-Organization in Physics, Chemistry, Biology and Medicine. VEB Deutscher Verlag der Wissenschaften, Berlin. 248 p.

Venzl, G., Ross, J. 1982. Comments on pattern formation in precipitation processes. J. Chem. Phys. 77, 1308-1313

Vidal, C., Hanusse, P. 1986. Non-equilibrium behaviour in isothermal liquid chemical systems. Inter. Rev. Phys. Chem. 5, 1-55

Vidal, C., Noyau, A. 1980. Some differences between thermokinetic and chemical oscillating reactions. J. Am. Chem. Soc. 102, 666-671

Vidal, C., Pacault, A. (eds.). 1984. Non-Equilibrium Dynamics in Chemical Systems. Springer Series in Synergetics, Vol. 27. Springer, Berlin, Heidelberg. 255 pages. See also: Vols. 6, 12, 19, 28 in this series

Vohmann, H. J. 1977. Abbau und Speicherung von Reservekohlenhydraten bei *Saccharomyces carlsbergensis*. Thesis, Bonn

Volkenstein, M. V. 1978. General Biophysics. Nauka, Moscow, 590 p. (Russian)

Volkenstein, M. V. 1984. The essence of biological evolution. Sov. Phys.-Usp. 27, 515-546

Volovik, G. E., Lavrentovich, O. D. 1983. Topological dynamics of defects: boojums in nematic drops. ZhETF 85, 1997 (Sov. Phys. JETP 58, 1159-1166)

Wagner, C. 1950. Mathematical analysis of the formation of periodic precipitations. J. Colloid Sci. 5, 85-97

Walgraef, D. 1984. Pattern formation in chemical systems - The effect of convection. In: Non-Equilibrium Dynamics in Chemical Systems. Springer Series in Synergetics, Vol. 27. (C. Vidal, A. Pacault, eds.). Springer, Berlin, Heidelberg, 114-117

Wallace, A. H. 1968. Differential Topology. First Steps. Benjamin, New York, Amsterdam. 130 p.

Walter, R., Lamprecht, I. 1978. Modern theories concerning the growth equations. In: Thermodynamics of Biological Processes. (I. Lamprecht, A. I. Zotin, eds.). Walter de Gruyter, Berlin, New York, 143-162

Weaire, D. 1974. Some remarks on the arrangement of grains in a polycrystal. Metallogr. 7, 157-160

Weaire, D., Rivier, N. 1984. Soap, cells and statistics - random patterns in two dimensions. Contemp. Physics 25, 59-99

Weibel, E. R. 1963. Morphometry on the Human Lung. Academic Press, New York. 151 p.

Weibel, E. R. 1979. Stereological methods. Academic Press, London. 415 p.

Weibel, E. R. 1980. Design and structure of the human lung. In: Pulmonary Diseases and Disorders. (A. P. Fishman, ed.). McGraw-Hill, New York. 224-271

Weitz, D. M., Epstein, I. R. 1984. Spatial waves in the reaction of chlorite with iodide. J. Phys. Chem. 88, 5300-5304

Welsh, B., Gomatam, J., Burgess, A. 1983. Three-dimensional chemical waves in the Belousov-Zhabotinskii reaction. Nature 304, 611-614

West, B. J., Bhargava, V., Goldberger, A. L. 1986. Beyond the principle of similitude. J. Appl. Physiol. 60, 1089-1097

Westerhoff, H. V., Chen, Y.-D. 1984. How do enzyme activities control metabolite concentrations? Eur. J. Biochem. 142, 425-430

Wilson, K. 1975. The renormalization group: critical phenomena and the Kondo problem. Reviews of Modern Physics 47, 773-840

Wilson, T. A. 1967. Design of the bronchial tree. Nature 18, 668-669

Winfree, A. T. 1972. Spiral waves of chemical activity. Science 175, 634-636

Winfree, A. T. 1973. Scroll-shaped waves of chemical activity in three dimensions. Science 181, 937-939

Winfree, A. T. 1974. Rotating chemical reactions. Scient. Amer. 230(12), 82-95

Winfree, A. T. 1974. Rotating solutions to reaction-diffusion equations. SIAM AMS Proc. 8, 13-31

Winfree, A. T. 1978. Stably rotating patterns of reaction and diffusion. In: Theoretical Chemistry: Periodicities in Chemistry and Biology, Vol. 4 (H. Eyring, D. Henderson, eds.). New York, Academic Press. 1-51

Winfree, A. T. 1980. The Geometry of Biological Time. New York, Springer, 530 pp.

Winfree, A. T. 1985. Organizing centers for chemical waves in two and three dimensions. In: Oscillations and Traveling Waves in Chemical Systems (R. J. Field, M. Burger, eds.). Wiley, New York, 441-472

Winfree, A. T., Strogatz, S. H. 1983 a. Singular filaments organize chemical waves in three dimensions. I. Geometrically Simple Waves. Physica 8D, 35-49

Winfree, A. T., Strogatz, S. H. 1983 b. Singular filaments organize chemical waves in three dimensions. II. Twisted Waves. Physica 9D, 65-80

Winfree, A. T., Strogatz, S. H. 1983 c. Singular filaments organize chemical waves in three dimensions. III. Knotted Waves. Physica 9D, 335-345

Winfree, A. T., Strogatz, S. H. 1984. Singular filaments organize chemical waves in three dimensions. IV. Wave taxonomy. Physica 13D, 221-233

Wirges, H. P. 1980. Experimental study of self-sustained oscillations in a stirred tank reactor. Chem. Eng. Sci. 35, 2141-2146

Wohlfarth-Bottermann, K. E. 1964. Differentiations of the ground cytoplasm and their significance for the generation of the motive force of ameboid movement. In: Primitive Motile Systems in Cell Biology. Academic Press, New York, 79-109

Wohlfarth-Bottermann, K. E. 1979. Oscillatory contraction activity in Physarum. J. Exp. Biol. 81, 15-32

Wohlfarth-Bottermann, K. E., Fleischer, M. 1976. Cycling aggregation patterns of cytoplasmic F-actin coordinated with oscillating tension force generation. Cell Tiss. Res. 165, 327-344

Wolf, K. V., Stockem, W., Wohlfarth-Bottermann, K. E. 1981. Cytoplasmic actomyosin fibrils after preservation with high pressure freezing. Cell Tiss. Res. 217, 479-495

Wolfram, S. 1984. Universality and complexity in cellular automata. Physica 10D, 1-35

Wolpert, L. 1969. Positional information and the spatial pattern of cellular differentiation. J. Theor. Biol. 25, N1, 1-47

Wolpert, L. 1971. Positional information and pattern formation. curr. top. dev. biol. 6, 183-224

Wood, P. M., Ross, J. 1985. A quantitative study of chemical waves in the Belousov-Zhabotinskii reaction. J. Chem. Phys. 82, 1924-1936

Wooten, F., Winer, K., Weaire, D. 1985. Computer generation of structural models of amorphous Si and Ge. Phys. Rev. Lett. 54, 1392-1395

Yoshimoto, Y., Kamiya, N. 1978. Studies on contraction rhythm of the plasmodial strand. I. Synchronization of local rhythms. Protoplasma 95, 89-99

Yoshimoto, Y., Kamiya, N. 1984. ATP- and calcium-controlled contraction in a saponin model of *Physarum polycephalum*. Cell. Struct. Funct. 9, 135-141

Yoshimoto, Y., Matsumura, F., Kamiya, N. 1981. Simultaneous oscillations of Ca^{2+} efflux and tension generation in the permealized plasmodial strand of Physarum. Cell Motility 1, 433-443

Zaika, V. E. 1975. Balanced animal growth equations. In: Quantitative Aspects of Organisms Growth. (A. I. Zotin, V. A. Grudnitsky, G. A. Klevezal, eds.). Nauka, Moscow, 25-33 (Russian)

Zaikin, A. N., Kawczynski, A. L. 1977. Spatial effects in active chemical systems. 1. Model of Leading Center. J. Non-Equil. Thermodyn. 2, 39-48

Zaikin, A. N., Zhabotinskii, A. M. 1970. Concentration wave propagation in two-dimensional liquid-phase self-oscillating system. Nature 225, 535-537

Zeeman, E. C., Buneman, P. 1968. Tolerance spaces and the brain. In: Towards a Theoretical Biology: 1. Prolegomena. (C. H. Waddington, ed.). Edinburgh Univ. Press, Edinburgh, 140-151

Zeldovich, B. Ya., Saakyan, D. B., Sobelman, I. I. 1977. On the energy difference between left and right rotating molecules due to parity violation in weak interaction of electrons with nuclei. Pis'ma Zh. Teor. Eksp. Fiz. 25, 106-109 (Russian)

Zeldovich, Ya. B. 1949. On a theory of combustion of nonmixed gases. Zh. Tekn. Fiz. 19, 1199-1209 (Russian)

Zhabotinskii, A. M. 1964. Periodic liquid-phase oxidation Reactions. Dokl. Akad. Nauk. SSSR 157, 392

Zhabotinskii, A. M. 1964. Periodic processes of the oxidation of malonic acid in solution. Biofizika 9, 306-311 (Russian)

Zhabotinskii, A. M. 1967. Spatial behavior of an oscillating chemical reaction in a homogeneous system. In: Oscillating Processes in Biological and Chemical Systems (G. M. Frank, ed.). Moscow, Science Publ., p. 252

Zhabotinskii, A. M. 1971. Spreading of concentration waves in two-dimensional liquid phase of auto-oscillating systems. J. Fiz. Chim. Akad. Nauka SSSR 55, 265

Zhabotinskii, A. M., Zaikin, A. N. 1971. Spreading of concentration waves in two-dimensional liquid phases of auto-oscillating systems.In: Oscillating Processes. (E. E. Sel'kov, ed.). Pushchino-on-Oka, 279-283

Zierep, J., Örtel, H. (eds.). 1982. Convective Transport and Instability Phenomena. G. Braun, Karlsruhe, 577 p.

Zotin, A. I. 1972. Thermodynamic Aspects of Developmental Biology. Karger, Basel. 159 p.

Zotin, A. I. 1985. Thermodynamics and growth of organisms in ecosystems. In: Ecosystem Theory for Biological Oceanography. (R. E. Ilanowicz, T. Platt, eds.). Depart. Fish. Oceans, Ottawa, 27-37

Zotina, R. S. , Zotin, A. I. 1978. Differential equations of developmental biology. In: Thermodynamics of Biological Processes. (I. Lamprecht, A. I. Zotin, eds.). Walter de Gruyter, Berlin, New York, 121-133

Zykov, V. S. 1980. Kinematics of stationary circulation in excitable medium. Biofizika 25, 319-322 (Russian)

Subject Index